MTP International Review of Science

Mass Spectrometry

MTP International Review of Science

Publisher's Note

The MTP International Review of Science is an important new venture in scientific publishing, which we present in association with MTP Medical and Technical Publishing Co. Ltd. and University Park Press, Baltimore. The basic concept of the Review is to provide regular authoritative reviews of entire disciplines. We are starting with chemistry because the problems of literature survey are probably more acute in this subject than in any other. As a matter of policy, the authorship of the MTP Review of Chemistry is international and distinguished; the subject coverage is extensive, systematic and critical; and most important of all, new issues of the Review will be published every two years.

In the MTP Review of Chemistry (Series One), Inorganic, Physical and Organic Chemistry are comprehensively reviewed in 33 text volumes and 3 index volumes, details of which are shown opposite. In general, the reviews cover the period 1967 to 1971. In 1974, it is planned to issue the MTP Review of Chemistry (Series Two), consisting of a similar set of volumes covering the period 1971 to 1973. Series Three is planned for 1976, and so on.

The MTP Review of Chemistry has been conceived within a carefully organised editorial framework. The over-all plan was drawn up, and the volume editors were appointed, by three consultant editors. In turn, each volume editor planned the coverage of his field and appointed authors to write on subjects which were within the area of their own research experience. No geographical restriction was imposed. Hence, the 300 or so contributions to the MTP Review of Chemistry come from many countries of the world and provide an authoritative account of progress in chemistry.

To facilitate rapid production, individual volumes do not have an index. Instead, each chapter has been prefaced with a detailed list of contents, and an index to the 13 volumes of the MTP Review of Physical Chemistry (Series One) will appear, as a separate volume, after publication of the final volume. Similar arrangements will apply to the MTP Review of Organic Chemistry (Series One) and to subsequent series.

Butterworth & Co. (Publishers) Ltd.

Physical Chemistry
Series One
Consultant Editor
A. D. Buckingham
Department of Chemistry
University of Cambridge

Volume titles and Editors

Inorganic Chemistry
Series One
Consultant Editor
H. J. Emeléus, F.R.S.
Department of Chemistry
University of Cambridge

Volume titles and Editors

1 **MAIN GROUP ELEMENTS— HYDROGEN AND GROUPS I–IV**
Professor M. F. Lappert, *University of Sussex*

2 **MAIN GROUP ELEMENTS— GROUPS V AND VI**
Professor C. C. Addison, F.R.S. and Dr. D. B. Sowerby, *University of Nottingham*

3 **MAIN GROUP ELEMENTS— GROUP VII AND NOBLE GASES**
Professor Viktor Gutmann, *Technical University of Vienna*

4 **ORGANOMETALLIC DERIVATIVES OF THE MAIN GROUP ELEMENTS**
Dr. B. J. Aylett, *Westfield College, University of London*

5 **TRANSITION METALS—PART 1**
Professor D. W. A. Sharp, *University of Glasgow*

6 **TRANSITION METALS—PART 2**
Dr. M. J. Mays, *University of Cambridge*

7 **LANTHANIDES AND ACTINIDES**
Professor K. W. Bagnall, *University of Manchester*

8 **RADIOCHEMISTRY**
Dr. A. G. Maddock, *University of Cambridge*

9 **REACTION MECHANISMS IN INORGANIC CHEMISTRY**
Professor M. L. Tobe, *University College, University of London*

10 **SOLID STATE CHEMISTRY**
Dr. L. E. J. Roberts, *Atomic Energy Research Establishment, Harwell*

INDEX VOLUME

Organic Chemistry
Series One
Consultant Editor
D. H. Hey, F.R.S.
Department of Chemistry
King's College, University of London

Volume titles and Editors

1 **STRUCTURE DETERMINATION IN ORGANIC CHEMISTRY**
Professor W. D. Ollis, F.R.S., *University of Sheffield*

2 **ALIPHATIC COMPOUNDS**
Professor N. B. Chapman, *Hull University*

3 **AROMATIC COMPOUNDS**
Professor H. Zollinger, *Swiss Federal Institute of Technology*

4 **HETEROCYCLIC COMPOUNDS**
Dr. K. Schofield, *University of Exeter*

5 **ALICYCLIC COMPOUNDS**
Professor W. Parker, *University of Stirling*

6 **AMINO ACIDS, PEPTIDES AND RELATED COMPOUNDS**
Professor D. H. Hey, F.R.S. and Dr. D. I. John, *King's College, University of London*

7 **CARBOHYDRATES**
Professor G. O. Aspinall, *University of Trent, Ontario*

8 **STEROIDS**
Dr. W. D. Johns, *G. D. Searle & Co., Chicago*

9 **ALKALOIDS**
Professor K. F. Wiesner, F.R.S., *University of New Brunswick*

10 **FREE RADICAL REACTIONS**
Professor W. A. Waters, F.R.S., *University of Oxford*

INDEX VOLUME

Physical Chemistry
Series One

Consultant Editor
A. D. Buckingham

MTP International Review of Science

Volume 5

Mass Spectrometry

Edited by **A. Maccoll**
University of London

Butterworths · London
University Park Press · Baltimore

THE BUTTERWORTH GROUP

ENGLAND
Butterworth & Co (Publishers) Ltd
London: 88 Kingsway, WC2B 6AB

AUSTRALIA
Butterworth & Co (Australia) Ltd
Sydney: 586 Pacific Highway 2067
Melbourne: 343 Little Collins Street, 3000
Brisbane: 240 Queen Street, 4000

NEW ZEALAND
Butterworth & Co (New Zealand) Ltd
Wellington: 26–28 Waring Taylor Street, 1

SOUTH AFRICA
Butterworth & Co (South Africa) (Pty) Ltd
Durban: 152–154 Gale Street

ISBN 0 408 70266 4

UNIVERSITY PARK PRESS

U.S.A. and CANADA
University Park Press Inc
Chamber of Commerce Building
Baltimore, Maryland, 21202

Library of Congress Cataloging in Publication Data

Maccoll, Allan.
Mass spectrometry.

(Physical chemistry, series one, v. 5) (MTP international review of science)
Includes bibliographies.
1. Ions. 2. Mass spectrometry. I. Title.
QD453.2.P58 Vol. 5 [QD561] 541′.3′08s [545′.35]
72–2330
ISBN 0–8391–1019–7

Filmset by Photoprint Plates Ltd., Rayleigh, Essex
Printed in England by Redwood Press Ltd., Trowbridge, Wilts
and bound by R. J. Acford Ltd., Chichester, Sussex

Consultant Editor's Note

The MTP International Review of Science is designed to provide a comprehensive, critical and continuing survey of progress in research. The difficult problem of keeping up with advances on a reasonably broad front makes the idea of the Review especially appealing, and I was grateful to be given the opportunity of helping to plan it.

This particular 13-volume section is concerned with Physical Chemistry, Chemical Crystallography and Analytical Chemistry. The subdivision of Physical Chemistry adopted is not completely conventional, but it has been designed to reflect current research trends and it is hoped that it will appeal to the reader. Each volume has been edited by a distinguished chemist and has been written by a team of authoritative scientists. Each author has assessed and interpreted research progress in a specialised topic in terms of his own experience. I believe that their efforts have produced very useful and timely accounts of progress in these branches of chemistry, and that the volumes will make a valuable contribution towards the solution of our problem of keeping abreast of progress in research.

It is my pleasure to thank all those who have collaborated in making this venture possible – the volume editors, the chapter authors and the publishers.

Cambridge

A. D. Buckingham

Preface

In its broadest sense, mass spectrometry is concerned with the production of a beam of ions and the analysis of the beams so produced in terms of mass-to-charge ratio. The beam comprises molecular ions, and also fragment ions which are derived from the molecular ion by unimolecular decomposition. The totality of ions produced under specified conditions of ionisation from a given molecular species, together with their relative abundances, constitutes the mass spectrum of the substance. Such mass spectra provide the basis for a qualitative (structural) analysis of the substance or a quantitative analysis of mixtures of substances. These applications are largely outside the scope of the present volume; rather its concern will be the investigation of the chemical physics of gas-phase ions.

As in the chemistry of neutral species, two problems arise, namely, the energetics of ion production and the lifetimes of ions with respect to the various channels of decomposition. Usually only low pressures are involved and so the theory of unimolecular reactions plays a very important role in the understanding and prediction of mass spectra. Under certain conditions, bimolecular ion–molecule processes can be favoured, and this greatly enlarges the range of ionic processes that can be studied and also leads to a novel method of production of mass spectra.

Although of wide availability, the commercial mass spectrometer is a compromise instrument, designed primarily for the production of reproducible mass spectra. It is due in the main to the ingenuity of several generations of mass spectrometrists, that it can be so widely adapted to the investigation of the chemistry of ionic species. The alternative approach, of building specialised equipment for the study of gas-phase ions, is costly and time-consuming, but in certain instances is unavoidable. Thus in a review of the field, it is necessary at times to go beyond what has conventionally come to be known as mass spectrometry.

In planning a volume of this type, there will almost inevitably be a less than complete coverage of the field. In the last decade or so, mass spectrometry has passed through a period of unprecedented growth. Old techniques and concepts have been modified, new techniques and concepts developed. It is sincerely hoped that this volume catches the spirit of the pioneers in the field, and will enable those who come after to assess the present position, and chart new forays into the unknown.

London A. Maccoll

Contents

1
Theory of Mass Spectra

A. L. WAHRHAFTIG
University of Utah, Salt Lake City

1.1 INTRODUCTION

This review attempts to cover only a fraction of the subject implied by its title, being limited to the consideration of the fragmentation of polyatomic ions for which many competing reactions are possible. The ionisation process itself will receive meagre mention (see Chapters 2 and 3) and no discussion; its aftermath is the subject of this article. Theoretical studies pertinent primarily to the mass spectra of diatomic and small polyatomic molecules also will not be included for two reasons. All major aspects of the fragmentation of such simple systems appear to be interpretable in terms of potential curves (or surfaces), known at least in part from optical spectroscopy and not known for larger ions, and there seems to be little hope that the rigorous treatment that is possible with simple ions can be extended soon to much larger ions. 'Chemical Ionisation' (see Chapter 5), that is ionisation in the presence of a large excess of a reagent gas in a source where ion–molecule reactions are maximised rather than minimised, also will not be considered in this review.

Many papers not specifically about the theory of mass spectra are closely related to this subject. A selection, necessarily arbitrary, has been made of articles on rate theory in general, ion and activated-complex configurations, potential energy surfaces, fragmentation pathways and related material that seemed most pertinent.

It will be obvious that the quasi-equilibrium theory of mass spectra has been taken as the point of reference for most discussions. The reviewer admits his bias, but does hope it has not unduly affected his judgement.

1.2 UNIMOLECULAR DISSOCIATIONS

Before considering theories of mass spectra, let us note the difference between the dissociations of ions in the mass spectrometer and unimolecular dissociations in the gas phase at one atmosphere as normally observed. In the typical mass spectrometer source, the pressure is so low that an ion once formed in general is either detected or lost by collision with a wall without interaction with any other molecule or ion. The dissociation of an ion in this case is clearly a unimolecular process, but it is quite different from the thermal unimolecular reaction which is unimolecular with first-order kinetic behaviour only as long as the gas pressure is sufficiently high to make energy transfer for both activation and deactivation fast relative to dissociation. The essential difference is that normally chemical reactions occur *at a temperature*; in the mass spectrometer each ionised molecule is an *isolated system.* The quasi-equilibrium assumed in conventional absolute rate theory applied to unimolecular dissociations holds if collisions between molecules are frequent relative to the rate of dissociation of molecules with enough energy to dissociate. The distribution of energy over the assembly of molecules, and over the internal degrees of freedom of each molecule, may then reasonably be assumed to correspond closely to that of thermal equilibrium, as the removal of energetic molecules by reaction will be only a small perturbation on the system. In the mass spectrometer each ion is formed with some specific amount of internal energy (and angular momentum); these are *conserved independently* for each ion in all its subsequent dissociations. There are no external perturbations changing its energy or aiding the re-distribution of energy among the internal degree of freedom of the ion.

The loss of an electron by a molecule due to electron bombardment or photon absorption is rapid compared to the rate of any subsequent dissociation. The ion formed will in general not be in its ground state. Hence, the starting point for a theory of mass spectra must be a collection of ions with a distribution of energy and hence also an initial distribution over vibrational levels, and quite possibly also with a distribution over electronic, or more exactly, over vibronic, states. An observed mass spectrum then corresponds to the ionic products obtained after a time determined by the geometry and operating conditions of the mass spectrometer; each ion dissociates as an isolated system.

1.2.1 Quasi-equilibrium theory

The first theoretical discussion of the fragmentation process in a polyatomic molecule after ionisation and of the resultant mass spectrum of the molecule appeared in 1952[1]. This theory, now generally known as the quasi-equilibrium theory of mass spectra (QET), is based upon the following assumptions.

(a) The time required for dissociation of the initial molecular ion is relatively long compared to the time of interaction leading to its formation.

(b) The rate of dissociation of the ion is slow relative to the rate of re-distribution of energy of the internal degrees of freedom, both electronic and vibrational, of the ion.

(c) Each dissociation process may be described as a motion along a reaction coordinate separable from all other internal coordinates through a critical 'activated-complex' configuration.

The first two assumptions permit one to discuss the dissociation of the ion in terms of its internal energy but without regard to how that energy was acquired. As noted above, the first assumption should generally hold in the cases of electron and photon ionisation; it may hold for simple charge exchange, but is not necessarily valid for field ionisation (see Chapter 4).

The second and third assumptions are analogous to, and in the spirit of, the assumptions made in the absolute reaction rate theory for chemical reactions[2]. The rate of energy flow within an ion is now a specific property of that ion, since there are no external perturbations. One can easily conceive of specific excitation in one portion of a very large molecule such that assumption (b) would not hold. Thus, the range of its validity is a matter of interest.

The third assumption is common to the reaction rate theories of Eyring *et al.*[2] and of Rice, Ramsperger and Kassel as re-formulated by Marcus and Rice (RRKM theory)[3,4].

The quasi-equilibrium theory of mass spectra has been the subject of several reviews[5–9] as well as the basis for numerous calculations[8,9], so only a very brief outline will be given here. On the basis of the assumptions stated above, if an isolated system has internal energy E, its rate of dissociation via a reaction path i with activation energy ε_i is

$$k_i(E) = \frac{\sigma}{h}\frac{W_i^{\ddagger}(E-\varepsilon_i)}{\rho(E)} \tag{1.1}$$

where $W_i^{\ddagger}(E-\varepsilon_i)$ is the number of states of the activated complex with energy less than or equal to $E-\varepsilon_i$, $\rho(E)$ is the density of states for the system with energy E, σ is the number of equivalent paths for the reaction, and h is Planck's constant. An equivalent equation was obtained by Marcus and Rice[3] using a different approach in a discussion of thermal reactions. In principle, rates for all possible paths for a parent molecule ion with energy E can thus be calculated, and so the amount of each primary product formed during the source residence time obtained. The excess energy above ε_i will be distributed between product ion and neutral fragment plus relative translational and rotational energies so that the same equation may be applied, with consideration of the fluctuations expected in the energy division, to secondary reactions. Further reaction may be treated in similar fashion.

The calculated 'breakdown graphs' for fragmentation as a function of internal energy may then be compared directly with first-derivative curves of photon ionisation efficiency curves, with the second-derivative curves of electron ionisation efficiency curves or with charge-exchange mass spectra[10]. For comparison with an observed mass spectrum, the calculated fragmentation is integrated over energy using a suitable energy distribution function.

Careful consideration of the nature of the approximations and assumptions of the QET was given by Rosenstock and Krauss in their reviews[5,6]. The questions noted by them are still far from answered. More recent discussions are those by Rosenstock[9] on mass spectral theory in general, and by Vestal[8] on the calculations of breakdown graphs and mass spectra based upon this method. Some questions concerning the QET can be answered. It does not claim to permit the interpretation of the formation of all ions produced in a mass spectrometer; rather, it was noted[1] that there would certainly be some ionisation, in particular double ionisation, that would not give dissociation interpretable by a statistical theory. The serious errors resulting from the application of classical statistics to the calculation of the density of states of a collection of oscillators when the energies and oscillator frequencies are those appropriate to a dissociating polyatomic molecule-ion have been discussed adequately in previous articles and reviews.

1.2.1.1 Energy level density expressions

Several methods for calculating number and density of energy levels for systems of harmonic oscillators or harmonic oscillators plus free rotors have been developed[11–16, 112, 113]. Seven of the formulae developed have been compared with exact counting of states by Forst and Prášil[17], who applied them to two different model systems. These were taken to be representative of small and large molecules, with five and 15 internal vibrational degrees of freedom, respectively, plus 2–5 or 12 free internal rotors. Their 'large' system corresponds only to a molecule such as C_2H_6; it would be of interest to see a similar comparison for a molecule such as heptane or ethyl n-butylamine with the order of 60 internal degrees of vibrational freedom. The accuracy and ease of application to computer calculations of several of the methods are such that the factor limiting the reliability of the rate calculations is our rather complete lack of quantitative knowledge concerning the vibrational frequencies of ions and our even smaller exact knowledge of activated-complex configurations for large molecules. To date, one can only say that the vibrational frequencies and moments of inertia for internal rotations have been 'reasonable', in the sense that the parameters assumed were internally consistent, consistent with known molecular frequencies, and within bounds set by general knowledge of force constants and bond lengths. In addition, anharmonicities and the effect of finite vibrational amplitudes are ignored in the usual model.

While methods have been devised to include the effect of anharmonicity in multiple excitations of a set of independent oscillators[15, 18–20], neither the mathematical techniques nor the data are available for a model with all the interactions between normal modes that must occur when amplitudes of motion approach those required to give dissociation. In their studies of the effects of anharmonicity and the exclusion of disallowed states from the state density function, Forst and Prášil[20] found that errors resulting from the neglect of these two effects not only were of opposite sign but in the cases they examined were of similar magnitude and so tended to cancel.

1.2.2 Non-quasi-equilibrium theory

A quantum-mechanical theory for unimolecular reactions which does not involve the assumptions of absolute rate theory but uses the theory of resonance scattering applied to a system of overlapping resonances was developed by Mies and Krauss[21]. One very interesting result was that under certain circumstances there would be large deviations from the simple exponential decay normally assumed. This quantum-mechanical approach has been applied to a harmonic oscillator model by Knewstub[22], developing in greater detail the further assumptions necessary for QET to be valid. The evidence, together with some consequences, for a 'loose' reaction complex with two free relative rotations of the separating particles is stated in the above article. A detailed discussion of the interrelation of the Gioumousis and Stevenson theory of ion–molecule reactions and the nature of the reaction coordinate in the vicinity of the activated complex for ion dissociations is presented in a related paper[23].

Later studies by Mies[24], while concerned primarily with thermal reactions, include a discussion of the basic nature of an 'activated complex' and a 'reaction coordinate'. Another highly mathematical discussion of the motion of a large model system along a reaction coordinate is that of Fischer, Hofacker and Seiler[25].

An elaboration of his earlier work, including the relationship of his approach to that of Mies and Krauss, has been given by Rice[26].

Two other papers[27, 28] have been published which use the language of quantum mechanics and resonances to discuss mass spectra. This reviewer is unable to understand the connection between the basic theory and the claimed application.

An entirely different approach to unimolecular reaction rate theory is that of Gelbart, Rice and Freed[29], who use the language of stochastic processes and consider the events leading to a dissociation a Markov chain. Their description of a reacting system is not one that can be simply correlated with the physical parameters of a real molecule but it does include explicitly the idea of competing dissociations and does not make any assumptions regarding energy randomisation. They found that localised excitation of the type that might well occur in the ionisation of a molecule could result both in an overall dissociation rate not describable by a simple rate constant and in the occurrence of competing reactions to extents greatly at variance with the prediction of any quasi-equilibrium theory. This paper includes an excellent discussion of the problems existing in a calculation of competition between reaction pathways if detailed account is taken of the energy flow in the dissociating system.

A discussion of paraffin hydrocarbon fragmentation by Finney and Hall[30] is based upon the incorporation into a Slater type unimolecular reaction rate theory of the idea that dissociation required both the elongation of a bond and the separation of the mass centres of the resulting particles by critical amounts. The model calculations are of interest, but their applicability to a normal paraffin is doubtful since it is not a linear chain of carbon atoms but a chain with tetrahedral bond angles. The resulting vibrational normal modes associated with bond stretching are all essentially the same frequency, completely different from those for the linear model. Also the problem noted in Slater type critical state theories is not present in the transition state theories.

Many articles have appeared on correlation of energy levels in the fragmentation of small ions. In general, these have discussed systems of 2–5 atoms which are not being covered in this review. Lindholm has found many new Rydberg series in the ultraviolet spectra of a number of molecules and correlated these with the results of photoelectron spectroscopy and molecular orbital calculations. A recent series of papers include studies of pyridine, furan, thiophene, pyrrole, and cyclopentadiene[31, 32]. In these papers he also assumes that mass spectral data, in particular charge-exchange fragmentation, may be used by correlating an ionisation primarily from a particular bond with the production of the fragment ion corresponding to the breaking of that bond. This correlation does not exist in general[1] and no reason is given why it should exist in the above compounds. The change in calculated electron density on vertical ionisation is not necessarily related to the relative bond strengths in the ion, which are determined by the changes occurring in electronic structure as one bond is stretched and the others readjusted for minimum system energy.

This latter problem has been considered in detail by Lorquet *et al.*[33, 34], who have calculated a partial potential surface for ethylamine ion with the C—C and C—N bond lengths as variables. In the ground state of this ion, their calculations indicate the positive charge is largely localised on the nitrogen atom, *c.* 70% in the lone pair, 15% in the C—N bond, and only 2% in the C—C bond. This is reflected in the shape of the potential surface for *very small* changes in bond length, for which the C—N bond force constant is much less than the C—C force constant. For displacements larger than *c.* 0.1 Å, however, the energy starts to rise with increase in C—N bond length far more rapidly than for increase in C—C bond length, so that the calculated bond dissociation energies are 0.9 and 4.7 eV, respectively. The qualitative picture of the process has been understood for many years. The first (rough) approximation to the electronic structure of the ethylamine ion is one with the molecule minus a lone pair electron and with the bonds essentially unchanged from their normal strengths. When sufficient internal vibrational energy exists to give dissociation, then the energies of the possible products must be examined since activation energies are generally very small, if not zero, for an ion–molecule reaction, the reverse of a dissociation. Since $NH_2{=}CH_2^+$, isoelectronic with C_2H_4, is an extremely stable ion, the potential surface must reflect the fact that as the C—C bond is stretched, the electron distribution changes with the C—N bond tending toward its eventual double bond in $CH_2NH_2^+$.

There is no reason to believe this sort of behaviour is unusual. Quite the contrary, the large number of rearrangements observed by the organic mass spectrometrists where the 'driving force' is the formation of a particularly stable ion or neutral species as one of the products indicates the lack of relationship between initial charge density in the ground state equilibrium configuration of the parent molecule-ion and its eventual ion fragmentation.

While admittedly very approximate, calculations of the intersection of the ground state and the lower envelope of the excited states of the ethylamine ion taken as functions of the C—C and one N—H bond distances[34] indicate that a radiationless transition to the ground state can occur here even though the first excited state lies *c.* 1.35 eV above the ground state. However, as the transition is likely only for energies well above that for C—C bond breaking in the excited state, that disocciation would be favoured.

Detailed calculations of the changes in electronic structure occurring in the reaction

$$H_2C{=}O^+H \rightarrow HC{\equiv}O^+ + H_2$$

as the distance between the H_2 and the C—O bond is increased and the new bonds are formed in the four-centre reaction complex have been made by Smyth and Shannon[35]. Changes in atomic net charges, overlap populations between reacting atoms and orbital energies and charge distributions with reaction distance are given. Changes in net charge are *far* from monotonic in going from reactant to products; rather the charge on the oxygen atom decreases to about half its initial value, then increases to a still greater value, while the charge on the carbon atom approximately triples and then drops back to a final value lower than in the reactant ion. Analogous calculations of partial potential surfaces using valence-bond theory for the loss of H and of H_2 from methane (neutral) following photo-excitation have been reported by Karplus and Bersohn[36].

Calculations, indicating that *all* configurations of the tetramethylene diradical, $(CH_2)_4$, which would be formed by a single bond break in cyclobutane, are unstable with respect to dissociation into two ethylene molecules, are stated to have startling implications concerning non-concerted reactions[37]. In view of the long-range forces existing between ion and neutral fragments in an ionic dissociation, it is unlikely that an opened cyclobutane ion is similarly unstable, otherwise this calculation would indicate many activated-complex structures assumed in QET calculations might also be erroneous.

Papers demonstrating a linear relationship between amount of some fragment ion and some calculated property such as bond order or some empirical parameter for a series of homologous compounds or a compound with a set of substituents have appeared in too great a quantity to list individually. In general, these are cases where the calculated property or the parameter is simply related to the activation energy for the dissociation, and where, as noted in the section in this review on organic mass spectrometry, the dissociation scheme for the molecule is such that there is a simple correlation between activation energy and relative fragmentation in the group of compounds studied.

A discussion of mass spectra resulting from simple bond scission in terms primarily of fast fission determined by charge distribution in the molecular ion has been given by Hirota and co-workers[38, 39]. His argument that the nature of the potential energy surface can be ignored in a fast fission[40] is, as noted by Lorquet[40], not justified. Also, it is difficult to understand his argument that the QET should hold for the spectra observed at low electron energy but that his theory holds for 80 eV spectra, when his theory is based solely upon the electron distribution in the ion ground state while the effect of increasing electron energy is to form ions, at least initially, in excited electronic states. It is stated that his theory should apply only to highly excited electronic states[41], but in another article[47] the same theory with an arbitrary modification as to the probability that each fragment will retain the charge is applied to *low* voltage alkane mass spectra. Additional references may be found in a set of three communications on the charge distributions in some alkanes and their mass spectra[42–44].

The fragmentation pathways of CrO_2F_2 and CrO_2Cl_2 ions have been discussed in terms of dissociation dependent upon the particular vibronic level excited, with dissociation rapid compared to energy equilibration[45]. This is, of course, the reverse of the QET assumption. It is quite possible that these molecules are too small and symmetrical for the QET assumption to be applicable. However, this reviewer has reservations concerning the theory presented. Major emphasis is placed upon the similarity in shape of all molecule and fragment ion ionisation efficiency curves for 2 V above threshold, but this is to be expected for such curves obtained using a conventional electron bombardment ion source with a large thermal energy spread[7]. Also, their 'deconvolution' procedure was applied at 0.25 V steps with resultant relatively poor energy resolution, as noted by the authors. If the application of the method would yield agreement with ionisation efficiency curves obtained with monoenergetic electrons, the work would be far more convincing. The fragmentations of $W(CO)_6$, $Mo(CO)_6$ and $Cr(CO)_6$ have also been treated by an elaboration of the same method[46].

1.2.3 Related rate theory

A complete coverage of all research related to theories of mass spectra would require reference to almost all work on reaction rates in general, and especially reference to many of the theoretical discussions of gas-phase reaction rates resulting from chemical activation which do not correspond to a thermal equilibrium. While, in general, such studies have been assumed to lie outside the scope of this review, it is obvious that any assumptions regarding activated complexes used in the QET, or a similar theory, should be consistent with the findings of studies on radical recombinations and on the unimolecular dissociation of molecules chemically activated, such as by addition of an H atom to an α-olefin[48], by insertion of a methylene radical in a molecule[49], or by one of several photochemical techniques[50]. Also the questions of energy randomisation and of the effect of rotation apply equally to applications of the Eyring absolute reaction rate theory, RRKM theory, and the QET theory of mass spectra[51–53].

Of special interest is the unambiguous observation of non-randomisation of internal energy in a radical decomposition observed by Rynbrandt and Rabinovitch[54, 55]. The molecule of structure (1) is formed with vibrational

```
CF2—CF—CF—CF2
  \  /    \  /
  CH2      CD2
```

(1)

excitation initially localised in either of the cyclopropyl rings by addition of CH_2 or CD_2 to the appropriate substrate and then, if not stabilised by collision, suffers loss of CF_2 to give a stable product. The mass spectrometric analysis of the products obtained as a function of pressure of the reaction mixture permitted the determination of the extent to which the dissociation favoured the ring initially excited. Analysis of the data gave an effective rate for intramolecular energy relaxation of about $1.1 \times 10^{12}\ s^{-1}$. As noted by the authors, this is sufficiently large to indicate that the RRKM (and hence QET) postulate of energy randomisation has practical validity.

The comments by Tschuikow-Roux[56] on the lack of significance of the Kassel parameter usually called *s*, the number of effective oscillators, in various applications of thermal unimolecular rate theory also apply to its use in QET calculations of mass spectra.

1.3 RATE CONSTANTS

A new method for the direct measurement of the rate constant for the dissociation of ions formed by charge exchange over a range of k greater than 10^3 has been described by Andlauer and Ottinger[57]. The data for the reaction: $C_6H_5CN^+ \rightarrow C_6H_4^+ + HCN$ lie on a smooth curve (log k *v.* excitation energy of the parent ion) quite similar to those obtained by QET calculations. The dissociations of benzene ion to yield $C_6H_5^+$ and $C_4H_4^+$ were also studied. It was noted in the early days of the QET that molecules exhibiting fluorescence, such as benzene, might well give molecule–ions with groups of non-interacting attractive electronic states[58]. Indirect evidence for the existence of such states in benzene and the resultant complications in the applications (or lack of applicability) of the QET have been reviewed[6]. However, the calculations of Vestal[8] indicated that it was not necessary to invoke isolated electronic states in order to obtain reasonable agreement with the benzene mass spectrum including metastable ion peaks. The direct measurements of Andlauer and Ottinger prove conclusively that the two dissociations studied are *not* competitive and that the calculations by Vestal were oversimplified.

The data for production of $C_4H_4^+$ ion is certainly interpretable in terms of the QET. However, that for $C_6H_5^+$ production implies an extremely small (or zero) transition rate from states yielding $C_6H_5^+$ to those yielding $C_4H_4^+$, and also an unusually small rate of increase in dissociation rate to $C_6H_5^+$ with increasing internal energy. It is difficult to see how this rate dependence can be rationalised on the basis of any model usually postulated. Possible interpretations are that the rate-determining step is a radiationless transition, slow and depending only slightly on excess energy, but still fast compared to any radiationless transition to the state yielding $C_4H_4^+$ ion, or that there exists a manifold of states for which the radiative half-life is short and which have as their final state one dissociating to yield $C_6H_5^+$.

1.4 METASTABLE IONS

The need to develop critical experiments to test the fundamental hypothesis of the QET, that the dissociation of an ion depends only on its energy content and not on the mode of acquisition of that energy, was noted by Rosenstock and Krauss[59]. One such experiment is the comparison of the metastable decompositions (see Chapter 8) for an ion derived from a variety of sources. In a given mass spectrometer experiment, ions observed as metastable are those with a limited range of dissociation rates; the energy equilibration hypothesis then implies that their energies lie within a narrow band, independent of the overall energy distribution for the ion. Hence, while the overall fragmentation of the ion may vary greatly with source, its metastable transitions should be essentially invariant. Rosenstock *et al.*[60] found that the ratio of the competing metastable losses of C_2H_4 and C_3H_6 from hexyl ions from ten different molecules was constant to 3 %. It should be noted that variation in time scale can greatly affect the ratio of competing metastable ions[61].

Shannon and McLafferty[68] noted that the kinetic energy release in a metastable decomposition of the $C_2H_5O^+$ ion was in most cases independent of the origin of the ion and took this as evidence that the $C_2H_5O^+$ ions in general had the same structure, with the exceptions most likely resulting from a different non-equilibrating structure. This principle, the basis for several more recent articles by McLafferty and co-workers and others, is the subject of a thorough analysis in terms of the QET by Yeo and Williams[87].

They have calculated the ratio of competing metastable peak intensities as a function of the internal energy of the decomposing ion and also the fraction of a given metastable peak obtained by decomposition of ions of specified internal energy *v.* that energy. Assumed in the calculations was a time scale typical for a particular mass spectrometer. Rates were calculated using a simplified, but adequate, rate expression involving a frequency factor which was taken as a parameter in the calculations. It has long been realised that the dissociation rates observed by measurements of metastable-ion intensities were determined by the geometry and operating conditions of the mass spectrometer used[1]. It then follows directly from QET that the specific internal energies of ions dissociating to yield observed metastable peaks will also depend on instrument geometry and operating conditions, as calculated in detail by Yeo and Williams. The question is: what variations in relative metastable intensities are to be expected under one specific set of operating conditions in a particular instrument due to variation in the internal energy distribution of the reactant ion? Here, the curves of Figure 3 of the above authors[87] are very helpful in obtaining an understanding of the various possibilities. One problem in the discussion in the paper is the meaning of the phrase, 'significantly different metastable ratios'. The conclusion of the paper contains a reasonable answer to the problem.

The observation of two-step metastable decompositions, $A^+ \rightarrow B^+ \rightarrow C^+$, with the first and second reactions occurring in the first and second field-free regions of a double-focusing mass spectrometer, respectively (hence on about the same time scale), seemed initially to be inconsistent with the QET predictions. Since only ions A^+ with energy in a relatively narrow band just slightly above the minimum required would dissociate to yield B^+ in the first region, it seemed unlikely that the B^+ ions formed would have sufficient energy to dissociate further. Experimental studies of the metastable peak for the consecutive reactions, $C_7H_7^+$ (from toluene) $\rightarrow C_5H_5^+ \rightarrow C_3H_3^+$ and for the individual-reaction metastable ions in both field-free regions, indicated these were true unimolecular dissociations and not just collision induced. QET calculations of the relative magnitudes of the several metastable peaks[62] were in reasonable agreement with the experimental results.

The results of a detailed study of the electron-impact mass spectrum of methane and all the deuteriomethanes, including the 70 eV spectrum, the normal metastable peaks and the collision-induced metastable peaks have been compared with a QET calculations by Hills, Vestal and Futrell[63]. Reasonable agreement is obtained for the mass spectra and for the relative amounts of the several metastable ions, but the simple calculation gives absolute rates too high for metastable ions to be observed in significant amount for any of the methanes, contrary to experiment. One possibility, that one is observing tunnelling through a barrier which is most probably due to rotation, was proposed by Rosenstock[9]; Hills *et al.* propose that the rate near threshold might be much lower than calculated due to a small transmission coefficient, a quantity usually assumed to be unity in the QET, which they compute for a square well. Another possibility is that the effect of the many equivalent potential minima for CH_4^+ ion upon its density of states function has been grossly underestimated. Evidence for this last point of view is to be found in the detailed *ab initio* study of the geometry of the CH_4^+ ion[64], in which it is found that the most stable configuration was that of a regular tetrahedron compressed along one of the S_4 axes, hence of symmetry D_{2d} with three equivalent potential minima. Only slightly higher in energy (0.12 eV) is another configuration of minimum energy of C_{3v} symmetry, corresponding to a lengthening of one C—H bond and a slight flattening of the remaining CH_3 group; this requires the existence of an additional four equivalent potential minima. Other computed energies of interest relative to the most stable configuration are: square planar (D_{4h}) configuration, 1.14 eV; tetrahedral (T_d) configuration, 1.44 eV; $CH_3^+ + H$, 0.46 eV. The finding that the energy for the tetrahedral structure is above that for the dissociation to $CH_3^+ + H$ is consistent with the mass spectral and photo-ionisation data. Very similar values for the energies of the above structures were obtained by

Dixon[111] in another *ab initio* calculation of the Jahn–Teller distortions expected in CH_4^+ ion.

The metastable decompositions of $C_2H_6^+$ and $CH_3CD_3^+$ have been examined at high sensitivity and the results discussed in terms of isotope effect and the extent of scrambling of H and D before the decomposition[65]. It is noted that the usual assumption for an alkane – no scrambling in parent ion, complete scrambling in fragment ions before a secondary reaction – is a reasonable first approximation. However, in agreement with prior work on deuterated butanes[66], some scrambling does occur in CH_3CD_3 before the C—C bond breaks. In some fragment ions, particularly C_2H_4 analogues, scrambling, while appreciable, is not complete. It is noted that the results are qualitatively in agreement with the QET, but that in some finer details the breakdown might be governed by a non-statistical mechanism. However, the one difficulty mentioned, the existence of consecutive metastable ions, has been shown to be consistent with the QET in another case[62].

Metastable transitions of trimethyl phosphate after electron ionisation and of buta-1,3-diene after photon ionisation have been studied over the time ranges of 2–9 and 1–8 μs, respectively[67]. QET calculations were made using a number of different assumptions concerning frequencies in normal and activated states and the structure of the activated state itself. Excellent agreement with experiment was obtained after normalising the calculated value to the experimental result at 2 μs. The results were remarkably insensitive to variation of the frequency patterns.

The QET calculations of the mass spectrum of propane, including the metastable peaks and the effect of deuterium substitution, were among the first given in support of that theory. A new study of the mass spectra of C_3H_8, $CH_3CD_2CH_3$, $CD_3CH_2CD_3$ and C_3D_8 by Vestal and Futrell[61] is similar to the original, but makes use of the much better and more varied data presently available and of additional data obtained by them, and employs a reasonably accurate method for calculating state densities and sums. While these new calculations still required many assumptions, the arbitrariness has been greatly reduced, the amount of data fitted greatly increased, and the satisfactory fit by the QET calculations would appear to be excellent confirmation of the QET model.

The results of a careful study of kinetic energy release in a number of metastable transitions in paraffin hydrocarbons are in agreement with the postulate of energy randomisation[69]. Other aspects of the decomposition process are discussed in this paper, including the fact that the kinetic energy distribution, and hence metastable-ion peak shape, can vary and is far too complex to be characterised solely by a mean or maximum kinetic energy. Thus the constancy of peak shape for each metastable dissociation reaction is even stronger support for the postulate. The similarity in peak shapes for the $C_4H_9^+ \rightarrow C_3H_5^+ + CH_4$ metastable decomposition from n-butane, isobutane, neopentane and n-hexane strongly supports the idea that the $C_4H_9^+$ ions from at least two of these molecules have rearranged so that the butyl ions from all four have a common structure (or mixture of structures) before subsequent decomposition.

The kinetic energy distribution for the metastable transition, $CH_2OH^+ \rightarrow CHO^+ + H_2$, in methanol has been carefully studied and an attempt made to interpret it in terms of the potential energy surface previously mentioned[35]. The kinetic energy distribution can be rationalised in part by assuming the metastable reaction for this system occurs by tunnelling through the barrier on the potential surface since for such a small system ions with a minimum of energy in excess of barrier height would decompose too rapidly to be observed as metastable ions. However, difficulties arise when an attempt is made to include the data available on deuterated methanols.

The variation of the kinetic energy released in the metastable transition, $C_4H_4N_2^+ \rightarrow C_3H_3N^+ + HCN$ in pyrazine, with change in mean lifetime of the dissociating ions has been determined and discussed in terms of QET by Beynon *et al.*[70].

The observation that s-triazole has a metastable decomposition involving loss of N_2, and that there is a release of 1.41 eV in translational energy in the process, is discussed by Shannon[71]. He suggests, on the basis of a correlation diagram for the molecular orbitals calculated for s-triazole, an assumed ring intermediate, and products, $C_2H_3N^+ + N_2$, that the reaction is slow in spite of being from a repulsive state (in order to have 1.41 eV in kinetic energy) because of a symmetry-forbidden crossing of levels. However, the position of the molecular orbitals for C_2H_3N are questionable in that they imply an ionisation potential for this fragment appreciably larger than that for N_2. Since the bond strength for N_2 is exceptionally large compared to other bonds, it is more likely that the large energy release is due to the fact that in this particular reaction the activation energy for the reverse reaction is unusually large for an ion–molecule reaction.

1.4.1 Collision-induced mestastable ion decompositions

In a study of the collision-induced decomposition of aromatic ions in the second field-free region of a double-focusing mass spectrometer, Jennings found that the resulting reactions were similar to those occurring after electron excitation[72]. He concluded that the excitation on collision was largely electronic, since only collisions involving negligible momentum transfer would give ion trajectories permitting their detection as mass peaks. Energy randomisation would then give fragmentation as in the source, except for differences in time scale for fragmentation and in the distribution over energy.

Beynon *et al.* have noted that collision-induced dissociations provide a very useful means of determining the routes by which ions normally decompose in the source[73]. Since information on reaction paths is essential for the application of the QET, this method should greatly assist the theoretical interpretation in many cases. However, the extremely great sensitivity of the mass spectrometer can lead one astray unless relative abundance data, at least semi-quantitative in nature, is examined[63]. It seems that practically all possible rearrangement–fragmentation paths occur to some extent; to attempt to explain those occurring to the extent of 1 in 10^3 or less of total ionisation is hopeless considering our present lack of detailed knowledge of the many excited states of high multiplicity which might easily yield anomalous rearrangements.

1.5 ORGANIC MASS SPECTROMETRY

A large fraction of the papers relating to mass spectrometry since the publication of the pioneering work of Beynon[74] have been in the area generally called 'organic mass spectrometry'. A discussion of the relevance to mass spectral theory of any appreciable fraction of the enormous numbers of papers on the relationship of the structure of organic molecules to their mass spectra is impossible in a review of reasonable length. A few arbitrarily selected topics only will be presented.

The earliest workers in the field, primarily interested in hydrocarbon analysis, saw few regularities and many strange products in mass spectral fragmentations and generally felt that the language of ordinary chemical reactions. Since, as previously discussed, the situation in a mass spectrometer dous body of knowledge exists, interpretable on the assumption that most, if not essentially all, of ion dissociations can be rationalised by arguments very similar to, and frequently the same as, the ones applied to ordinary organic reactions. Since, as previously discussed, the situation in a mass-spectrometer ion source is the exact opposite of thermal equilibrium, a satisfactory theory of mass spectra should serve as a basis for discussions of both the similarities and the differences of mass spectrometric and thermal unimolecular dissociations.

The QET seems in general to pass this test. The usual expressions for unimolecular decomposition for a system in thermal equilibrium may be obtained by integrating the QET expression (equation (1.1)) over all energies, E, assuming a Boltzman distribution for the energy probability function[75]. Thus, *if* (a) there are no complications from secondary reactions of the ions first formed by parent-ion dissociation, and (b) there are no effects from irregularities, maxima and minima, in the parent-ion energy distribution function, one may expect the following analogies. The relative amounts of two products formed by analogous reactions in a mass spectrometer will depend upon the difference in their activation energies, $\varepsilon_i - \varepsilon_j$, similar to the dependence of the relative amounts of two products in a thermal reaction upon the difference, $\Delta H_i^\ddagger - \Delta H_j^\ddagger$. If only one reaction is important, the relative amounts of parent ion and fragment will change with ε for the fragmentation in fashion analogous to the effect of $\Delta H^\ddagger$ on the rate of a thermal reaction. The nature of the activated complex, 'tight' or 'loose', with more rings or with more free rotations, will affect the values of $W^\ddagger$ and so of k and amount of product in fashion entirely analogous to the effect of the same changes upon $\Delta S^\ddagger$ and so upon product formation in thermal reactions.

In many other cases, the special nature of the dissociation in the mass spectrometer must be considered. Often the effect of a change in a functional group is to lower the (lowest) ionisation potential of the molecule without significantly changing the inner ionisation potentials. Then, the average ion energy will be shifted to a higher value. While activation energies will be similarly shifted, those for fragmentations yielding an ion containing the functional group will be increased to a lesser extent so a net increase in dissociation is to be expected. Most cases require a more detailed consideration of the change of the energy distribution function. Several aspects of the general problem have recently been discussed by McLafferty[76].

The relationship of the QET to the simpler mechanistic interpretations of mass spectral decompositions, with particular regard to substituent effects, has been summarised by McLafferty *et al.*[77]. It is noted that the well-known general effects of change in energy distribution and changes in nature of activated complex and activation energy in the QET can be combined with the mechanistic predictions as to the nature of these changes with substituent to obtain agreement with experiment in cases where the simple mechanistic ideas alone were insufficient. New and excellent evidence for the need to consider the actual energy distribution curve for the parent ion, $P(E)$, is the correlation of the change in metastable-ion intensity with temperature with $P(E)$. The compounds studied, *m*- or *p*-substituted 1,2-diphenylethanes, have a deep minimum in the $P(E)$ *v.* E curve *c.* 2 V above the ionisation potential; the onset potential for the lowest energy dissociation varies from below to above the potential at the minimum. For those compounds where $P(E)$ has a large negative slope at the metastable-ion appearance potential, there is a large (five to twelvefold increase for $\Delta T = 200$ °C) positive temperature coefficient for the metastable-peak intensity. Where $P(E)$ has a positive slope so that an increase in initial thermal energy decreases $P(E)$ at the metastable-ion appearance potential, a negative temperature coefficient is observed.

The limitations of the concept of 'charge localisation' in the interpretation of mass spectra are well presented by Bentley, Johnstone and Mellon[108].

Calculations of the rates of the two major competing primary dissociation reactions for methyl *o*-toluate, methyl salicylate, *p*-chloroaniline, benzene and diethyl ketone using a simple form of the QET have been reported by Yeo and Williams[78]. They find the agreement with experimental mass spectra obtained with 12–20 V electrons reasonable considering the approximate nature of their calculations, provided they assumed that apparently simple bond scissions in some cases actually proceed by a more complicated mechanism.

The qualitative effect of 'tight' and 'loose' complexes in the QET is used by Tou in a discussion of the mass spectra of a dimethylthiocarbamate and a phosphorothioate[79]. The dissociations of deuterated $C_3H_6O^+$ ions have been interpreted as indicating a lack of energy randomisation in a reaction intermediate[80]. While this is a possible explanation, the observed results could also be due to a reaction mechanism more complicated than those proposed.

1.6 EFFECT OF TEMPERATURE

A detailed study of the effect of thermal energy on fragment-ion ionisation efficiency curves by Chupka[81] presents in detail the basis for, and approximations involved in, the usual assumption of additivity of thermal energy and energy transferred by the bombarding photon or electron in applications of the QET. In particular, it is shown that the thermal energy shifts observed in photo-ionisation studies of a number of alkanes are in agreement with the assumption of additivity in determining the ion energy distribution required in the QET[82].

1.7 KINETIC ENERGY OF FRAGMENTS

The kinetic energy distribution of the $C_{12}H_5$ fragment ions from anthracene, phenanthrene and diphenylacetylene has been studied by Rowland[83]. Similarity in kinetic energy distributions and energy requirements for formation of the activated complex indicates that all three have the same dissociation path. By assuming a tight activated complex so that vibrational energy in the activated complex may be transferred to kinetic energy in the products, with the reverse activation energy obtained from appearance potentials and thermochemical data, and with reasonable (although arbitrary) assumptions regarding state density functions, a calculated kinetic energy distribution is obtained which is in excellent agreement with that deduced from the experimental data.

Calculations of the translational energy resulting from some ionic fragmentations using direct counting of states and two alternative methods in the QET and comparison with experimental results have been given by Spotz, Seitz and Franklin[84]. They find that better agreement with experiment is obtained if it is assumed that the energy in three vibrational degrees of freedom is eventually transformed into relative translation of the products. Franklin and co-workers[85] also have discussed the need to consider the kinetic energy distribution of fragment-ion energies in calculations of ionic and radical heats of formation.

An analysis of the kinetic energy of fragmentation of ions which is interpreted as indicating a lack of energy randomisation has been given[86]. The evidence for this conclusion would seem to be rather weak, considering the scatter in the data and the assumptions in the treatment.

1.8 REARRANGEMENT REACTIONS

Rearrangement reactions as well as dissociation reactions need be considered in the interpretation of mass spectral data[1]; the effect of competition between rearrangement and dissociation was well covered in the review by Vestal[8]. In an application of the QET to the dissociation of cyclobutane ions by Lifshitz and Tiernan[88], in which excellent agreement with experiment was obtained, it is noted that comparisons of the fragmentation of several C_4H_8 isomers following charge-exchange ionisation also are indicative of competition between fragmentation and rearrangement. The differences in the mass spectra of six C_4H_8 compounds are interpreted by Meisels, Park and Giessner[89] as arising primarily from the differences in heats of formation of the molecule-ions, with the energy required for randomisation of structures appreciably less than the energy for all major dissociation pathways.

The value of 12.43 kcal mol^{-1} obtained by an *ab initio* calculation of the activation energy for hydrogen migration in the $C_2H_5^+$ ion[90] is far less than the energy for any dissociation of this ion and also less than the dissociation energy for most processes for larger fragment ions. This is consistent with the usual finding of rapid H–D interchange on deuterated fragment ions.

A thorough *ab initio* study of the structures and relative stabilities of various forms of the $C_3H_7^+$ ion indicates that the most stable form, as expected, is the 2-propyl[91]. Most interesting is the fact that two other conformations, 1-propyl and corner-protonated cyclopropane, with the methyl group in each case in either of the two possible orientations with C_s symmetry, have energies in the very narrow range 16.9–17.4 kcal mol^{-1} greater than that for 2-propyl. In addition, the energy for interconversion among all these forms, for methyl rotation or, more important, a 2 → 1 methyl shift on 1-propyl, is only about 0.5 kcal mol^{-1}. The energy required for a 3 → 1 hydrogen shift via an edge-protonated cyclopropane is appreciably higher, approximately 10 kcal mol^{-1}, hence 27 kcal mol^{-1} relative to the energy of 2-propyl, but still small compared to the energy for further fragmentation of this ion. Rapid rearrangement by these two processes would completely scramble both carbons and hydrogens. The one additional activation energy of great interest, which hopefully will appear in the more complete paper promised, is the one for the reaction, 2-propyl → 1-propyl.

A study of the reactions of $C_3H_7^+$ ions from various sources with furan to yield the addition ion, $C_7H_{11}O^+$, in the ion-cyclotron resonance mass spectrometer (see Chapter 6) leads Gross to conclude that at least two, and perhaps three, different forms of the propyl ion exist[92]. He argues convincingly against the possibility that his data is interpretable solely on the basis of differences in internal energy in a common $C_3H_7^+$ ion, but most of it is in agreement with the strong dependence of abundance on internal energy for an ion formed by simple addition, as noted in Section 1.11 of this review. The difficulty in the simple internal energy explanation would be resolved if there were a small activation energy for the 16 kcal mol^{-1} exothermic reaction, 1-propyl$^+$ → 2-propyl$^+$, and both these ions, if of sufficiently low internal energy, formed addition ions with furan stable on the time scale of the experiment.

The equilibrium configurations and energies for all possible C_1 and C_2 hydrocarbon molecules, radicals and singly-charged cations have been calculated via an *ab initio* molecular orbital method, with generally very good agreement with experimental results where available[93]. While it seems unlikely that anyone will attempt the calculation of the complete, 27 dimensional potential surface for the propane ion soon, it is evident that the calculation of the critical regions of the surface required for what might be called an *ab initio* QET calculation is now within the realm of possibility.

Studies of molecules with ^{13}C labelling have indicated the existence of such extensive rearrangement of the carbon skeleton of benzothiophene[94] and benzene[95] before fragmentation that the carbons are essentially completely scrambled. Since the energy to fragment is certainly greater than the energy necessary for ring opening, it is probable that the reactions of ring opening, then re-closing to other positions (necessarily to form a five-membered ring to account for the benzene ion data) is rapid relative to the fragmentation of the opened ring. A similar mechanism would appear to be indicated for the carbon atom randomisation observed before subsequent dissociation of $C_7H_7^+$ ions, as the data of Siegel[96] on the mass spectrum of 2,6-[$^{13}C_2$]-toluene indicates that the randomisation is greater than can be obtained by an initial random insertion of the α-carbon into the ring.

The observation by Safe[110] of complete H–D randomisation of an ethynyl deuterium in phenylacetylene and several ring substituted phenylacetylenes is noted as being consistent with rapid equilibration of cyclic and acyclic species. Insufficient data is available to attempt an explanation of the changes in relative rates of rearrangement and fragmentation responsible for the reduction of H–D scrambling to 80% and 70%, respectively, prior to some secondary dissociations of *o*- and *m*-methoxyphenylacetylenes.

Further evidence for independent scrambling of carbon and hydrogen atoms in the benzene ion before fragmentation to produce $C_4H_4^+$ or $C_3H_3^+$, with some evidence that the carbon scrambling reactions which preserve C—H bonds occur to a greater extent than H scrambling reactions is presented in an analysis of the fragmentation of 1,2-[$^{13}C_2$]-3,4,5,6-[$^2H_4$]-benzene[97]. This note also contains a large number of references to related studies.

The scrambling of hydrogen and deuterium as reflected in the dissociations of the molecule-ions of toluene and cycloheptatriene has been studied by Howe and McLafferty[98]. The relative loss of H or D from the parent ion of α-[$^2H_3$]-toluene, ring-[$^2H_5$]-toluene and 2,4,6-[$^2H_3$]-toluene, determined for (a) the normal metastable ions, (b) the collision-induced metastable ion, (c) the normal mass spectrum, and (d) loss before secondary metastable decomposition by C_2H_2 loss, is interpreted as resulting from the dissociation of parent ions of different average energy as indicated by QET. The resultant data, and similar data for two deuterated cycloheptatrienes, is interpreted in terms of an isotope effect, i = (loss of H)/(loss of D), and a parameter, α, the fraction of ions undergoing H–D scrambling (apparently assumed to be either complete or zero) before dissociation. The changes in the values of i and α are qualitatively consistent with the QET if it is assumed that the rearrangement reaction has an activation energy smaller than that for H loss; in particular, for toluene i decreases from 2.8 to 1.4, and α decreases from 1.00 (complete scrambling) to 0.72, with change from the lowest energy to the highest energy group of ions described above.

This H–D scrambling on the parent ion cannot arise from the ring opening–re-closing mechanism described above, but must be from rearrangements such as those indicated by Howe and McLafferty. However, they note that there is additional rapid scrambling on the $C_7H_7^+$ ion before its secondary fragmentation.

1.9 NEGATIVE IONS

While many experimental papers have been published on negative ion formation and dissociation, theoretical discussions were limited to attempts to derive values for the energies associated with the reactions, without regard for the kinetics. The first complete study of the kinetics of negative ion breakdown is that of Lifshitz *et al.*[99], who discuss the fragmentation of sulphur hexafluoride, perfluoro-1,2-dimethylcyclobutane, perfluoromethylcyclohexane and perfluoro-1,1-dimethylcyclopentane. The mass spectra, including metastable peaks, were obtained over a range of electron energy with an RPD source and at several temperatures and the breakdown curves calculated. The widths of metastable peaks as a function of temperature and the changes in fragmentation with electron energy were in agreement with QET calculations.

The negative-ion mass spectra of *m*- and *p*-dinitrobenzene have been reported by Brown and Weber[100]. Their analysis of the fragmentation patterns of the two compounds, including the energies released in the metastable transitions, appears to be consistent with QET predictions.

The studies of Bowie and co-workers[109] appear to confirm the idea that the principles governing negative-ion fragmentation are similar to those for positive ions.

1.10 FIELD IONISATION

Generally, discussions of mass spectra produced by field ionisation have treated matters specific to the field-ionisation process itself. The field-ionisation mass spectrometer is unique in that data are interpretable in terms of average rate constants for dissociation *v.* time since formation of the ion dissociating can be obtained over a range from $< 10^{-10}$ s to $> 10^{-6}$ s[101, 114]. The data obtained for several molecules indicates the existence of a continuous spectrum of rate constants, consistent with the QET. The probability distribution for internal energy of parent ions produced by field ionisation also can be derived from the data; it is found that while the mean internal energy of the ions formed by 70 eV electrons in normal hexane is over 6 eV, the mean energy after field ionisation is 0.48 eV.

The drastic reduction in relative intensities of ions produced by rearrangement reactions relative to those produced by simple bond break in the comparison of field-ionisation mass spectra to low-energy electron impact mass spectra is explained by Beckey[102] as resulting from the very short time available for dissociations that are to appear at the normal mass positions in a field-ionisation mass spectrum. Curves for reaction rate *v.* internal energy for typical rearrangement and simple bond-fission dissociations as given by the QET are used to show how the different time scales for the instruments affect the results obtained.

1.11 ION–MOLECULE REACTIONS

The QET has been applied by Butrill[103] to the calculation of the reaction product distribution from ion–molecule reactions (see Chapter 5) in ethylene and in ethylene–acetylene mixtures. The reactions studied are:

$$C_2H_4^+ + C_2H_4 \rightarrow [C_4H_8^+] \begin{array}{l} \nearrow C_3H_5^+ + CH_3 \\ \searrow C_4H_7^+ + H \end{array}$$

and

$$\left.\begin{array}{c} C_2H_4^+ + C_2H_2 \\ \text{or} \\ C_2H_2^+ + C_2H_4 \end{array}\right\} \rightarrow [C_4H_6^+] \begin{array}{l} \nearrow C_3H_3^+ + CH_3 \\ \searrow C_4H_5^+ + H \end{array}$$

and the analogous reactions in perdeuterated compounds. The reactant ions were formed by low-energy (11 V) electrons and so had relatively little internal energy. Thus, the internal energy in the activated complex was primarily that of the exothermicity of its formation with only small contributions from internal energy and relative translational energy of reactants and is a known quantity. The agreement with experiment obtained in all cases was excellent. The agreement could be fortuitous for one set of reactions, but equally good agreement was obtained for the perdeutero compounds where the fragmentation ratios were significantly different. Since all parameters used in the rate calculations for the perdeutero molecules were fixed by the values taken for the prior calculations and the standard effects on frequency of change in mass, this is a fairly rigorous test of the QET theory. Also, since it would seem likely that the collisions of ($C_2H_2^+ + C_2H_4$) and ($C_2H_4^+ + C_2H_2$) would give an activated $C_4H_6^+$ with quite different initial distributions of internal energy, the applicability of the theory implies that energy randomisation is rapid relative to dissociation in the reaction complex.

The fact that ion-cyclotron resonance (ICR) mass spectrometry permits the study of ion–molecule reactions at a very low relative translational kinetic energy has been used to elucidate the exact fragmentation pathways in one of the reaction sequences for the fragmentation of ethers at normal electron energies[104]. The interest was in the original location of the hydrogens in the water lost in the sequence,

$$((CH_3)_2CH)_2O^+ \rightarrow CH_3CH{=}O^+{-}CH(CH_3)_2 \rightarrow C_5H_9^+ + H_2O$$

With electron ionisation, the competing secondary reaction to $C_3H_6 +$ CH_3CHOH^+ is favoured even at quite low electron energies. In addition, any excess energy in the reacting ions will tend to increase the relative rate of H–D rearrangement and so complicate the interpretation of data on deuterated compounds. An ICR study of the reaction sequence

$$(CH_3)_2CHOH^+ \rightarrow CH_3CH\overset{+}{O}H + CH_3$$

$$+ (CH_3)_2CHOH$$

$$\rightarrow CH_3CH\overset{+}{O}CH(CH_3)_2 + H_2O$$

$$\rightarrow C_5H_9^+ + H_2O$$

in several deuterated propanols permitted the fragmentation paths of the low energy $C_5H_{11}O^+$ ion to be followed. The indicated major pathway is the loss of the proton on the oxygen in the ion plus the hydroxyl from the alcohol in the ion–molecule reaction, and protons from two different methyl groups, at least one of which is part of the 2-propyl group, in the final loss of water. In addition, of the order of 15% of the fragmentation is either via paths involving other protons or subsequent to a rearrangement.

In an ICR study of the reaction with NH_3 of $C_3H_6^+$ ions formed at low electron voltages from cyclopropane and from propene it was found that the cyclic-ion NH_3 complex dissociates to yield CH_4N^+ and CH_5N^+, but that these ions are not obtained from the propene ion under the same conditions[105]. The authors note the reaction might be diagnostic as to internal energy content rather than structure, but give a reasonable if not conclusive argument in support of their belief that it is in fact structure sensitive.

The complex formed by an ion–molecule collision necessarily contains the energy to dissociate, to reactants in all cases, and to other products if of similar or lower energy. As a result, the only ionic species generally observed as products have been fragments of the collision complex but not the collision complex itself. In a study of the reactions of the cyclohexene ion, Ausloos and co-workers[106] found that ionisation of cyclo-C_6H_{10}(I_z = 8.9 eV) by 10.0 eV photons at pressures of 5–40 mtorr in a mass spectrometer ion source gave yields of 80–90% dimer, $C_{12}H_{20}^+$, with $C_6H_8^+$ the only other major product ion. With 11.6–11.8 eV photons, the $C_{12}H_{20}^+$ yield is reduced to *c.* 30%, with dissociation to $C_{11}H_{17}^+$ giving the major product (*c.* 50% of total ionisation, 10–20 mtorr). It would appear that the $C_{12}H_{20}^+$ ion is large enough to have as internal energy approximately one eV more than that required for dissociation without decomposing to an appreciable extent on the time scale of the mass spectrometer used, but that very little additional energy is required to make the dissociation via CH_3 loss rapid.

A detailed ICR study of the styrene ion–molecule reaction indicates that here also a dimer ion can be formed which is stable, in this case on the longer time scale of the ICR mass spectrometer, about a millisecond[107]. The relative abundance of dimer increases drastically as the energy of the ionising electrons approaches the threshold value. Also cyclo-octatetraene, a higher energy isomer of styrene, gave no stable dimer. Studies of deuterated styrenes and of possible dimer molecules indicated that most of the observed ion–molecule products must come from a very specific structure most likely 1-phenyl tetralin, which presumably is also the structure of the $C_{16}H_{16}^+$ ions detected as such.

$$\text{-CH=CH}_2{}^+ + \text{-CH=CH}_2 \longrightarrow +$$

Qualitatively, the stability of the above ions would appear to be consistent with the QET, but calculations need to be made. An alternative would be that in these systems the collision yields a dimer ion in a specific excited electronic state stable to dissociation and with radiative half-life short compared to internal conversion of energy to vibration.

References

1. Rosenstock, H. B., Wallenstein, M. B., Wahrhaftig, A. L. and Eyring, H. (1952). *Proc. Nat. Acad. Sci.*, **38,** 667
2. Glasstone, S., Laidler, K. J. and Eyring, H. (1941). *Theory of Rate Processes,* 184. (New York: McGraw-Hill)
3. Marcus, R. A. and Rice, O. K. (1951). *J. Phys. and Colloid Chem.*, **55,** 894
4. Marcus, R. A. (1952). *J. Chem. Phys.*, **20,** 359
5. Rosenstock, H. M. and Krauss, M. (1963). *Mass Spectrometry of Organic Ions.* (F. McLafferty, editor). (New York: Academic Press)
6. Rosenstock, H. M. and Krauss, M. (1963). *Advances in Mass Spectrometry,* Vol. 2, p. 251. (Oxford: Pergamon Press)
7. Wahrhaftig, A. L. (1964). *NATO Advanced Study Institute on Mass Spectrometry,* 137. (R. I. Reed, editor). (New York: Academic Press)
8. Vestal, M. L. (1968). *Fundamental Processes in Radiation Chemistry,* Chapt. 2. (P. Ausloos, editor). (New York: Interscience)
9. Rosenstock, H. M. (1968). *Advances in Mass Spectrometry,* Vol. 4, 523. (London: The Institute of Petroleum)

10. Morrison, J. D. (1957). *J. Appl. Phys.*, **28,** 1409
11. Vestal, M., Wahrhaftig, A. L. and Johnston, W. H. (1962). *J. Chem. Phys.*, **37,** 1276
12. Haarhoff, P. C. (1963). *Molec. Phys.*, **7,** 101
13. Thiele, E. (1963). *J. Chem. Phys.*, **39,** 3258
14. Tou, J. C. and Wahrhaftig, A. L. (1968). *J. Phys. Chem.*, **72,** 3034
15. Hoare, M. R. and Ruijgrok, T. H. W. (1970). *J. Chem. Phys.*, **52,** 113
16. Hoare, M. R. (1970). *J. Chem. Phys.*, **52,** 5695; (1971). ibid., **55,** 3058
17. Forst, W. and Prášil, Z. (1969). *J. Chem. Phys.*, **51,** 3006
18. Schlag, E. W., Sandsmark, R. A. and Valance, W. G. (1964). *J. Chem. Phys.*, **40,** 1461
19. Wilde, K. A. (1964). *J. Chem. Phys.*, **41,** 448
20. Forst, W. and Prášil, Z. (1970). *J. Chem. Phys.*, **53,** 3065
21. Mies, F. H. and Krauss, M. (1966). *J. Chem. Phys.*, **45,** 4455
22. Knewstub, P. F. (1971). *Int. J. Mass Spectrom. Ion Phys.*, **6,** 229
23. Knewstub, P. F. (1971). *Int. J. Mass Spectrom. Ion Phys.*, **6,** 217
24. Mies, F. H. (1969). *J. Chem. Phys.*, **51,** 787, 798
25. Fischer, S. F., Hofacker, G. L. and Seiler, R. (1969). *J. Chem. Phys.*, **51,** 3951
26. Rice, O. K. (1971). *J. Chem. Phys.*, **55,** 439
27. Alexandru, Gr. (1970). *Int. J. Mass Spectrom. Ion Phys.*, **4,** 1
28. Alexandru, Gr. (1971). *Int. J. Mass Spectrom. Ion Phys.*, **6,** 125
29. Gelbart, W. M., Rice, S. A. and Freed, K. F. (1970). *J. Chem. Phys.*, **52,** 5718
30. Finney, R. D. and Hall, G. G. (1970). *Int. J. Mass Spectrom. Ion Phys.*, **4,** 489
31. Jonsson, B.-Ö., Lindholm, E. and Skerbele, A. (1969). *Int. J. Mass Spectrom. Ion Phys.*, **3,** 385
32. Derrick, P. J., Åsbrink, A., Edqvist, O., Jonsson, B.-Ö. and Lindholm, E. (1971). *Int. J. Mass Spectrom. Ion Phys.*, **6,** 161, 177, 191, 203
33. Lorquet, J. C., Lorquet, A. J. and Leclerc, J. C. (1968). *Advances in Mass Spectrometry*, Vol. 4, 569. (London: Institute of Petroleum)
34. Leclerc, J. C. and Lorquet, J. C. (1967). *J. Phys. Chem.*, **71,** 787
35. Smyth, K. C. and Shannon, T. W. (1969). *J. Chem. Phys.*, **51,** 4633
36. Karplus, S. and Bersohn, R. (1969). *J. Chem. Phys.*, **51,** 2040
37. Hoffmann, R., Swaminathan, S., Odell, B. G. and Gleiter, R. (1970). *J. Amer. Chem. Soc.*, **92,** 7091
38. Hirota, K., Fujita, I., Yamamoto, M. and Niwa, Y. (1970). *Proceedings International Conference on Mass Spectroscopy, Kyoto,* 1203. (K. Ogata and T. Hayakawa, editors). (Baltimore: University Park Press)
39. Hirota, K., Fujita, I., Yamamoto, M. and Niwa, Y. (1970). *J. Phys. Chem.*, **74,** 410
40. Discussion of Reference 33
41. Discussion of Reference 38
42. Santoro, V. and Spadaccini, G. (1969). *J. Phys. Chem.*, **73,** 462
43. Lorquet, J. C. (1969). *J. Phys. Chem.*, **73,** 463
44. Hirota, K., Niwa, Y. and Yamamoto, M. (1969). *J. Phys. Chem.*, **73,** 464
45. Flesch, G. D. and Svec, H. J. (1971). *J. Chem. Phys.*, **54,** 2681
46. Flesch, G. D. and Svec, H. J. (1971). *J. Chem. Phys.*, **55,** 4310
47. Saito, M., Fujita, I. and Hirota, K. (1970). *J. Phys. Chem.*, **74,** 3147
48. Larson, C. W. and Rabinovitch, B. S. (1970). *J. Chem. Phys.*, **52,** 5181
49. Johnson, R. L., Hase, W. L. and Simons, J. W. (1970). *J. Chem. Phys.*, **52,** 3911
50. Kirk, A. W. and Tschuikow-Roux, E. (1969). *J. Chem. Phys.*, **51,** 2247
51. Hase, W. L. and Simons, J. W. (1971). *J. Chem. Phys.*, **54,** 1277
52. Oref, I., Schuetzle, D. and Rabinovitch, B. S. (1971). *J. Chem. Phys.*, **54,** 575
53. Spicer, L. D. and Rabinovitch, B. S. (1970). 'Elementary Gas Reactions' in *Annual Review of Physical Chemistry,* **21,** 349
54. Rynbrandt, J. D. and Rabinovitch, B. S. (1971). *J. Chem. Phys.*, **54,** 2275
55. Rynbrandt, J. D. and Rabinovitch, B. S. (1971). *J. Phys. Chem.*, **75,** 2164
56. Tschuikow-Roux, E. (1969). *J. Phys. Chem.*, **73,** 3891
57. Andlauer, B. and Ottinger, C. (1971). *J. Chem. Phys.*, **55,** 1471
58. Krauss, M., Wahrhaftig, A. L. and Eyring, H. (1955). *Annual Review of Nuclear Science,* **5,** 241
59. Reference 5, page 51
60. Rosenstock, H. M., Dibeler, V. H. and Harllee, F. N. (1964). *J. Chem. Phys.*, **40,** 591
61. Vestal, M. and Futrell, J. H. (1970). *J. Chem. Phys.*, **52,** 978

62. Hills, L. P., Futrell, J. H. and Wahrhaftig, A. L. (1969). *J. Chem. Phys.*, **51**, 5255
63. Hills, L. P., Vestal, M. L. and Futrell, J. H. (1971). *J. Chem. Phys.*, **54**, 3834
64. Arents, J. and Allen, L. C. (1970). *J. Chem. Phys.*, **53**, 73
65. Löhle, U. and Ottinger, C. (1969). *J. Chem. Phys.*, **51**, 3097
66. McFadden, W. H. and Wahrhaftig, A. L. (1956). *J. Amer. Chem. Soc.*, **78**, 1572
67. Knewstubb, P. F. and Reid, N. W. (1970). *Int. J. Mass Spectrom. Ion Phys.*, **5**, 361
68. Shannon, T. W. and McLafferty, F. W. (1966). *J. Amer. Chem. Soc.*, **88**, 5021
69. Khodadadi, G., Botter, R. and Rosenstock, H. M. (1969). *Int. J. Mass Spectrom. Ion Phys.*, **3**, 397
70. Beynon, J. H., Hopkinson, J. A. and Lester, G. R. (1968). *Int. J. Mass Spectrom. Ion Phys.*, **1**, 343
71. Shannon, T. W. (1970). *Int. J. Mass Spectrom. Ion Phys.*, **3**, App. 12
72. Jennings, K. R. (1968). *Int. J. Mass Spectrom. Ion Phys.*, **1**, 227
73. Beynon, J. H., Caprioli, R. M. and Ast, T. (1971). *Int. J. Mass Spectrom. Ion Phys.*, **7**, 92
74. Beynon, J. H. (1960). *Mass Spectrometry and Its Application to Organic Chemistry.* (Amsterdam: Elsevier)
75. Magee, J. L. (1952). *Proc. Nat. Acad. Sci. USA*, **38**, 764
76. McLafferty, F. W. (1970). *Proceedings International Conference on Mass Spectroscopy, Kyoto*, 70. (K. Ogata and T. Hayakawa, editors). (Baltimore: University Park Press)
77. McLafferty, F. W., Wachs, T., Lifshitz, C., Innorta, G. and Irving, P. (1970). *J. Amer. Chem. Soc.*, **92**, 6867
78. Yeo, A. N. H. and Williams, D. H. (1970). *J. Amer. Chem. Soc.*, **92**, 3984
79. Tou, J. C. (1971). *J. Phys. Chem.*, **75**, 1903
80. McAdoo, D. J., McLafferty, F. W. and Smith, J. S. (1970). *J. Amer. Chem. Soc.*, **92**, 6343
81. Chupka, W. A. (1971). *J. Chem. Phys.*, **54**, 1936
82. Tou, J. C., Hills, L. P. and Wahrhaftig, A. L. (1966). *J. Chem. Phys.*, **45**, 2129
83. Rowland, C. G. (1971). *Int. J. Mass Spectrom. Ion Phys.*, **7**, 79
84. Spotz, E. L., Seitz, W. A. and Franklin, J. L. (1969). *J. Chem. Phys.*, **51**, 5142
85. Franklin, J. L. and Haney, M. A. (1970). *Proceedings International Conference on Mass Spectroscopy, Kyoto*, 909. (K. Ogata and T. Hayakawa, editors). (Baltimore: University Park Press)
86. LeRoy, R. L. (1970). *J. Chem. Phys.*, **53**, 846
87. Yeo, A. N. H. and Williams, D. H. (1971). *J. Amer. Chem. Soc.*, **93**, 395
88. Lifshitz, C. and Tiernan, T. O. (1971). *J. Chem. Phys.*, **55**, 3555
89. Meisels, G. G., Park, J. Y. and Giessner, B. G. (1970). *J. Amer. Chem. Soc.*, **92**, 254
90. Pfeiffer, G. V. and Jewett, J. G. (1970). *J. Amer. Chem. Soc.*, **90**, 2143
91. Radom, L., Pople, J. A., Buss, V. and Schleyer, P. v. R. (1971). *J. Amer. Chem. Soc.*, **93**, 1813
92. Gross, M. L. (1971). *J. Amer. Chem. Soc.*, **93**, 253
93. Lathan, W. A., Hehre, W. J. and Pople, J. A. (1971). *J. Amer. Chem. Soc.*, **93**, 808
94. Cooks, R. G. and Bernasek, S. L. (1970). *J. Amer. Chem. Soc.*, **92**, 2129
95. Horman, I., Yeo, A. N. H. and Williams, D. H. (1970). *J. Amer. Chem. Soc.*, **92**, 2131
96. Siegel, A. S. (1970). *J. Amer. Chem. Soc.*, **92**, 5277
97. Perry, W. O., Beynon, J. H., Baitinger, W. E., Amy, J. W., Caprioli, R. M., Renaud, R. N., Leitch, L. C. and Meyerson, S. (1970). *J. Amer. Chem. Soc.*, **92**, 7236
98. Howe, I. and McLafferty, F. W. (1971). *J. Amer. Chem. Soc.*, **93**, 99
99. Lifshitz, C., Peers, A. M., Grajower, R. and Weiss, M. (1970). *J. Chem. Phys.*, **53**, 4605
100. Brown, C. L. and Weber, W. P. (1970). *J. Amer. Chem. Soc.*, **92**, 5775
101. Tenschert, G. and Beckey, H. D. (1971). *Int. J. Mass Spectrom. Ion Phys.*, **7**, 97
102. Beckey, H. D. (1970). *Proceedings International Conference on Mass Spectroscopy, Kyoto*, 1154. (K. Ogata and T. Hayakawa, editors). (Baltimore: University Park Press)
103. Butrill, S. E., Jr. (1971). *J. Chem. Phys.*, **52**, 6174
104. Lehman, T. A., Elwood, T. A., Bursey, J. T., Bursey, M. M. and Beauchamp, J. L. (1971). *J. Amer. Chem. Soc.*, **93**, 2108
105. Gross, M. L. and McLafferty, F. W. (1971). *J. Amer. Chem. Soc.*, **93**, 1267
106. Lesclaux, R., Searles, S., Sieck, L. W. and Ausloos, P. (1970). *J. Chem. Phys.*, **53**, 3336
107. Wilkins, C. L. and Gross, M. L. (1971). *J. Amer. Chem. Soc.*, **93**, 895
108. Bentley, T. W., Johnstone, R. A. W. and Mellon, F. A. (1971). *J. Chem. Soc. B*, 1800
109. Ho, A. C., Bowie, J. H. and Fry, A. (1971). *J. Chem. Soc. B*, 530
110. Safe, S. (1971). *J. Chem. Soc. B*, 962

111. Dixon, R. N. (1971). *Molec. Phys.,* **20,** 113
112. Hoare, M. R. and Pal, P. (1971). *Molec. Phys.,* **20,** 695
113. Lin, S. H. (1971). *Molec. Phys.,* **20,** 953
114. Derrick, P. J. and Robertson, A. J. B. (1971). *Proc. Roy. Soc. London,* **A324,** 491

2
Ionisation and Appearance Potentials

J. D. MORRISON*
La Trobe University, Bundoora, Victoria

*Presently on sabbatical leave at the Department of Chemistry, University of Utah, Salt Lake City, Utah, U.S.A.

2.1 INTRODUCTION

In the application of the mass spectrometer to the study of molecular energetics, several distinct phases of development can be discerned.

The first work was devoted almost exclusively to the measurement of critical potentials and the potentials at which the various ions in mass spectra first appeared. These appearance potentials could be coupled with spectroscopic or thermochemical data or other appearance potentials to give values for bond dissociation energies. These values were sometimes reasonable; at other times discrepancies arose and the method fell into some discredit.

In the second phase, the methods used to interpret the data were examined, and it was shown that with greater knowledge of the processes occurring very precise information could be obtained about the energy states of molecular ions.

With the advent of photoionisation, and more recently of photoelectron spectroscopy methods capable of giving the same information but with even greater precision in energy, there has been less interest in the general use of the mass spectrometer in this type of study. Electron impact coupled with mass analysis still possesses unique advantages when it is desired to study the energetics of dissociation, in that it is an easy matter to produce a beam of electrons of any energy and there is positive identification of products.

At the same time, the much wider availability of mass spectrometers and the very great ease with which it is possible to measure a rough appearance potential has encouraged quite a few to use them for this purpose. It is very disappointing to notice that the same errors in interpretation are often being made at the present which caused electron impact measurements to have a bad name in the past. A measurement of a threshold for ionisation, as such, means little unless there is a thorough understanding of the processes involved.

It is proposed, therefore, to devote some space to the interpretation of the data, before dealing with the latest results.

2.1.1 Processes of ionisation induced by electron impact

When a beam of electrons is directed through a gas, some electrons in the beam may interact with the molecules present and be scattered. At low electron velocities this scattering is elastic, but at high velocities energy may be transferred from the incident particles to the gas molecules. When the energy transferred is great enough, ionisation or ionisation-fragmentation of the molecules may occur.

Charged species may result in the following ways:

(i) a Direct positive ionisation:

$$M + e \rightarrow M^+ + 2e$$

$$M + e \rightarrow M^{n+} + (n+1)e$$

(i) b Direct positive ionisation with fragmentation:

$$M + e \rightarrow M^+ + 2e \rightarrow F_1^+ + F_2 + 2e$$

(ii) a Excitation, also sometimes ionisation, followed by further autoionisation:

$$M + e \rightarrow M^{**} + e \rightarrow M^+ + 2e$$

$$M + e \rightarrow M^{+**} + 2e \rightarrow M^{++} + 3e$$

(ii) b Excitation, followed by ion pair production

$$M + e \rightarrow M^* + e \rightarrow F^+ + F^- + e$$

(iii) Electron capture, usually followed by fragmentation

$$M + e \rightarrow M^- \rightarrow F_1^- + F_2$$

Each of these processes has been observed, and since it is able to identify the charged products the mass spectrometer is ideally suited to their study.

2.2 IONISATION MEASUREMENTS

2.2.1 Ionisation cross-sections and ionisation efficiencies

For a beam of energetic particles travelling through a gas of uniform density, a total scattering cross-section σ_t may be defined as:

$$\sigma_t = \frac{1}{nl} \ln \frac{I_{e1}}{I_{e2}} \qquad (2.1)$$

where n = number of molecules of gas cm^{-3}, I_{e1}, I_{e2} are the number of particles $cm^{-2} s^{-1}$ in the beam before and after it has travelled for a distance of l cm. The number of particles scattered from the beam is $I_{e1} - I_{e2}\ s^{-1}$. The total scattering cross-section may be split into the sum of the cross-sections for elastic scattering, σ_{el}, scattering leading to excitation, σ_{ex}, and scattering leading to ionisation, σ_i.

$$\sigma_t = \sigma_{el} + \sigma_{ex} + \sigma_i$$

The mass spectrometer is able to give information regarding σ_i, and in some cases also regarding σ_{ex}.

It is not practicable in the mass spectrometer to measure σ_i directly, since the gas density in the source is not uniform, nor can the path length effective in ionisation be readily determined. The measurement most readily carried out with the mass spectrometer is the relative ionisation efficiency, defined as the ion current produced by a given beam of ionising particles with a given incident energy divided by that beam current. Provided the gas density is low, so that the fraction of the incident ionising beam which is scattered is small, the relative ionisation efficiency, I, is closely proportional to the ionisation cross-section. Considering only two-body collisions, where I_i is proportional to the number of scattered electrons $(I_{e1} - I_{e2})$, the ionisation efficiency at a given energy is

$$I = \frac{I_i}{I_{e1}} \propto \frac{I_{e1} - I_{e2}}{I_{e2}} \cdot \frac{\sigma_i}{\sigma_t}$$

Provided I is less than 1 %

$$\frac{I_{e2}}{I_{e1}} = \exp(-nl\sigma_t) \doteqdot 1 - nl\sigma_t$$

$$\therefore \quad \sigma_t \propto 1 - \frac{I_{e2}}{I_{e1}} \doteqdot (I_{e1} - I_{e2})/I_{e2}$$

$$\therefore \quad I \propto \sigma_i$$

This approximation would still be reasonable, even if I were to be great as 10 % in practice I rarely exceeds 0.1–1.0 %.

2.2.2 Measurement and calculation of ionisation cross-sections

In spite of the experimental difficulties, there has been renewed interest both in the calculation and the measurement of the maximum value of σ_i for each element. Otvos and Stevenson[1] in 1956 proposed that the maximum total cross-section, Q, for single positive ionisation of an atom by a electron would be given by a weighted sum of contributions from the outer electron shells, the weights being the mean square radii of each shell.

$$Q = \sum_i N_i q_i = \Sigma A N_i (r_i^2)$$

Where q_i is the cross-section for ionising one electron in the ith shell, N_i is the number of electrons in the shell, r_i^2 is the mean-square radius, and A an empirical constant. This calculation has been modified and refined by Mann[2], by Lin and Stafford[3], and by Tiwari, Rai and Rustgi[4]. A good summary is given by Moiseivitch and Smith[5]. Detailed experimental measurements of ionisation cross-sections have been made recently by several workers[6–9] and summarised by Kieffer[10].

For molecules, the simple additivity rule of Otvos and Stevenson gives poor agreement with experiment and better values can be obtained by relating the ionisation cross-section to the atomic polarisabilities[11].

The cross-sections for dissociative electron capture in a series of molecules have been studied by Christophorou and Stockdale[12]. The absolute cross-sections as such affect the measurement of ionisation threshold potentials only in so far as if they are low, it may be difficult to obtain good ionisation curves. A knowledge of the detailed form of the relative cross-sections for different substances is of great importance in the interpretation of the data to yield threshold values.

2.2.3 Total and mass-analysed ionisation efficiency curves

The primary data in mass spectrometric studies are the ionisation efficiency curves, which relate the relative ionisation efficiency to the energy of the incident ionising beam. The mass analyser permits the total ionisation efficiency curve for a given molecule to be separated into contributions from the parent ion and the various other ions which can be formed by fragmentation processes. However, while the primary process of energy transfer from beam to molecules, causing ionisation, is usually extremely rapid, and the total ionisation efficiency has definite significance, the processes of fragmentation can occur over times comparable with the time required for the ions to drift out of the source and pass through the mass analyser. Accordingly, the distribution of the ions observed between parent and fragments may differ depending on the conditions under which the source is operated. An ion which, after receiving an initial acceleration, then dissociates to a second species before being accelerated further, and mass analysed, will appear as a so-called 'metastable' ion, or may not reach the ion detector at all ('missing metastable')[13].

2.2.4 Instrumental factors

2.2.4.1 Electron energy scale

For measurements of ionisation efficiencies to be meaningful, it is essential that the energy of the ionising beam is well defined.

The nominal energy of the ionising beam has superimposed on it contributions due to contact potential differences between cathode and the ion box, space charge and field penetration (see Figure 2.1). The first two vary from one sample gas to another and with the previous history of the source, so that the nominal energy scale can be treated only as a relative one. It is customary to calibrate it by always including a rare gas together with the sample to be studied, and recording both ionisation efficiencies. The appearance potential of the sample ions can thus be related to that for production of the rare gas ions.

Calibration of the energy scale poses a particular problem in the study of negative ions. A very satisfactory solution is the double mass spectrometer

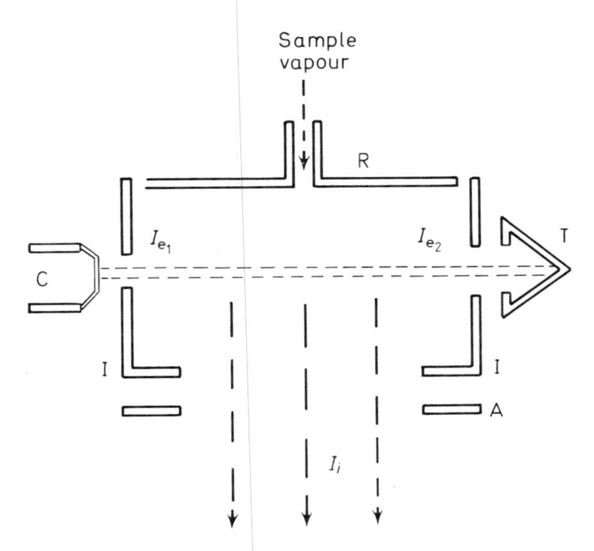

Figure 2.1 Diagram of simple ionisation source used in mass spectrometry
C = cathode; I = ion box; R = ion repeller electrode; T = electron trap; and A = ion accelerating electrode

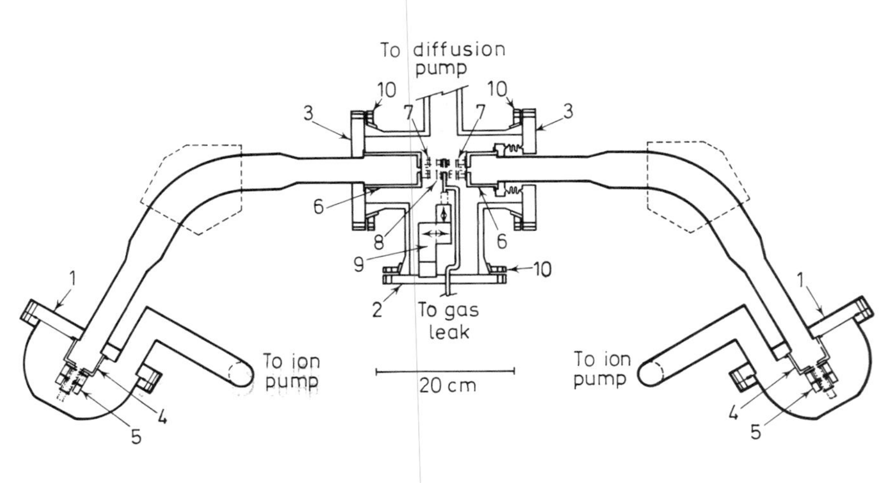

Figure 2.2 Mass spectrometer designed for simultaneous collection of positive and negative ions; 1, collector flange; 2, ion box flange; 3, collimation flange; 4, collector cylinder; 5, collector assembly; 6, collimation cylinder; 7, ion collimation assembly; 8, ion box – ion extraction assembly; 9, ion box – ion extraction assembly mount; 10, flange pressure rings
(From Svec and Flesch[14], by courtesy of Elsevier)

built by Svec and Flesch[14] (see Figure 2.2), which enables the simultaneous recording of both positive and negative ions. The energy scale obtained for negative ions is claimed to be accurate to within 0.01 eV. When measurements are made using a source of the type shown in Figure 2.1, the ionisation efficiencies for all positively charged parent and fragment ions rise from zero in the energy region usually ~9–30 eV, reach a maximum value somewhere in the range 30–200 eV, then fall back to zero very gradually over some thousands of eV. Negatively charged ions may have curves of this type, or may consist of a peak or series of peaks each extending over a range of 0.5–4 eV. In no case is the threshold sharply defined, even for monatomic gases. This is primarily due to the spread in energies of the electrons emitted from a hot cathode. This spread is quasi-Maxwellian in form[15]

$$n(U) \propto U \exp(-U/kT)$$

where $n(U)$ is the number of electrons possessing energy U, so that a beam of electrons accelerated from a cathode at 2500 K to a potential of 10 V will comprise electrons with energies ranging from 10 up to at least 12 eV. Superimposed on this thermal spread will be contributions due to the fact that the potential within the ionising region is not uniform. This may arise from the penetration of the fields due to the ion-accelerating or the electron-trap electrodes, the cross-field due to the repeller electrode, potential variations in the electron beam due to space charge, variable charge distribution on the surface of the ion box or even just the effect of the beam passing through an aperture. If a magnetic field is used to align the electron beam, additional variations in energy may arise. In a carefully designed system, the sum of these effects will give an energy spread with a half width of at least 0.4–0.5 eV.

In addition to the electron energy spread, there is a Doppler effect due to the thermal motion of the molecules of the target gas. At room temperatures this contributes an additional energy spread of the order of 0.02 eV half-width for He or H_2. This effect is much less for all other gases, so that it is not normally significant compared with the thermal spread due to electron emission.

Because of the total energy spread of the ionising beam, even for the process $He + e \rightarrow He^+ + 2e$, which has a sharply defined energy required for ionisation of 24.5 eV, the threshold of the ionisation efficiency curve is not clearly defined. As usually measured, the efficiency for $He \rightarrow He^+$ rises gradually from zero at ~22–23 eV, and curves upwards for some 1–2 eV before becoming an approximately linear increase.

2.2.4.2 Requirements of apparatus

It is desirable in this work to use a mass analyser giving ion peaks with flat tops, so that they will remain focused over the duration of a series of measurements. The ion source must be designed and operated so that the ionisation region is as nearly field-free as possible. Pulse techniques have been employed, most recently by Williams and Hamill[16], to remove the effect of the repeller cross-field and give more efficient ion extraction. The current in the electron beam should have a density of not more than 1 mA cm^{-2} to minimise space charge. The pressure of the sample gas in the source should not be greater than 2×10^{-6} mmHg to minimise processes involving ion–molecule collisions and this should be verified by checking that the ion current is linearly proportional to the ionising current.

Cleanliness of the source is of critical importance, especially when it is desired to study the form of the ionisation efficiency at higher energies. Surface films may show up as a hysteresis effect if a given $I(E)$ curve is scanned first by lowering the ionising energy and then by raising it[17] or may give rise to spurious structure in the curve[18]. These effects can be avoided only by baking the source prior to each run and frequent physical abrasion of the walls of the ion source with carborundum paper.

2.2.4.3 Reduction of energy spread of ionising beam

(a) *Reduction of cathode temperature*—The energy spread of the ionising beam can be reduced by lowering the cathode temperature; thoriated iridium[19], lanthanum boride[20] and carbon filaments have been used to achieve this. Dispenser cathodes, and indirectly heated oxide cathodes[21, 22], have given good results, reducing the half width of the energy spread to ~0.15 eV. These have the disadvantage of rapidly causing serious contamination of the source and have not been popular for this reason. It is possible that electrons emitted by a photoelectric process, or by a solid state tunnel cathode[22a], might have a smaller energy spread, but they have not been applied in energetic studies in mass spectrometry.

(b) *Retarding potential difference technique*—The retarding potential difference (RPD) method devised by Fox *et al.*[23] has been used by many as a method of virtually reducing the effective energy spread. By applying a variable retarding potential to an electrode in the path of the electron beam, varying amounts of the electron energy distribution are able to pass into the source, causing a change in the ionisation effeciency. The amount of this change gives the ionisation efficiency produced by a narrow band of electron energies. Various ingenious methods have been used to carry this out in an automatic fashion[24, 25]. The RPD method is capable of giving effective energy spreads of ~0.03 eV half width, but can in some cases cause serious distortion of the shape of the $I(E)$ curves at energies above the threshold[26, 27]. A detailed discussion of the errors possible has been given recently by Gordon *et al.*[28].

(c) *Attempts to avoid aperture effects*—The simple ionisation source shown in Figure 2.1 has many drawbacks. It is very desirable to have a collimated electron beam but the use of a magnetic field to achieve this can lead to periodic focusing of the electrons in the beam and may cause spurious discontinuities in the ionisation produced at certain energies[29–31]. The use of apertures to collimate the beam can also lead to errors, while space charge can cause marked differences between the energy of electrons in the centre of the beam and those at its edges. The use of apertures was minimised in the RPD ion source used by Cloutier and Schiff[32] and gave what are still some of the best data possible by this method.

(d) *Electron velocity selectors*—The most direct approach is to use an electron velocity filter. Little use has been made of the magnetic type, since the instruments of Lawrence[33] and Nottingham[34]. The cylindrical condenser, first used by Hughes and Rojansky[35] has been combined with a mass spectrometer ion source and, with progressive refinement, is capable of giving directly an electron beam with an energy spread at half width of 0.04 eV at a current of 10^{-8} A [36–44]. Attempts to increase this current much above this level invariably causes the electron energy spread to increase also, due to space charge. The hemispherical condenser is theoretically more efficient and the design due to Simpson[45, 46] is capable of producing a beam homogeneous in energy to 0.005 eV. The difficulties of combining this

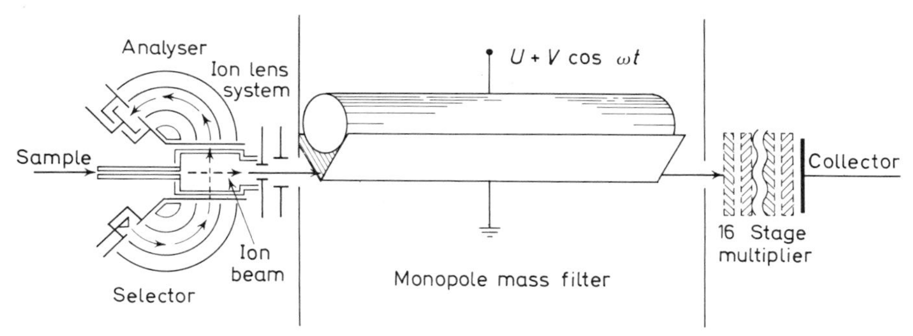

Figure 2.3 Schematic diagram of mass spectrometer using electron energy selector (From Brion and Thomas[43], by courtesy of Elsevier)

selector with a mass spectrometer cause the resulting energy spread to be perhaps only marginally better than with the cylindrical condenser[47, 48]. A trochoidal crossed magnetic and electrostatic field selector has been constructed by Stamatovic and Schulz[49]. This instrument gives an effective half-width in the electron energy of 0.07 eV. The parallel plate condenser should be capable of acting as an energy selector but the results have been disappointing[50, 51].

Because the electron currents obtainable are so small, the use of ion detectors with electron multipliers and of counting methods become necessary. All of these electrostatic filters are affected adversely by the magnetic fields used in a conventional mass analyser[52]. The latest instruments make use of a quadrupole[48] or monopole[43] mass filter for ion analysis (see Figure 2.3).

(e) *Analytical methods*—The observed ionisation efficiency at nominal electron energy V, $i(V)$, is the result of convoluting the energy spread $m(U)$ with a 'true' ionisation efficiency $I(E)$,

$$\text{i.e.} \quad i(V) = m(U) * I(E)$$

Provided the energy distribution $m(U)$ has sharp features and therefore mathematically possesses high-frequency components, its effect can be removed by a deconvolution process, using Fourier transforms[53, 54]

$$I(E) = T^{-1}\left[\frac{T\{i(V)\}}{T\{m(U)\}}\right]$$

It is not even necessary to know $m(U)$ exactly, although the better it is known the better the results. It is essential, for successful application of this method, that incompatible scatter in the data be removed[55].

In practice, scatter in the experimental values of $i(V)$ seriously limits the extent to which the energy spread can be reduced. Recently, by making use of a computer to scan $i(V)$ curves many times and sum the results, data with a sufficiently low scatter have been obtained to make the process worth while[56].

The iterative method of deconvolution due to van Cittert has also been used very successfully by Thynne and McNeil[57, 58] to study the ionisation efficiency curves for electron capture processes.

An empirical, and very simple, method of processing ionisation efficiency data has been claimed to give significant reduction in the energy spread. This method, termed the EDD method[59], involves multiplying the ionisation efficiency curve by an arbitrary factor, usually ~ 0.63, displacing it by a tenth of an eV or so on the energy axis and subtracting it from the original curve. The resulting curve is undoubtedly markedly sharpened[60–62]. The method seems to have no theoretical justification but appears to work.

It is in some ways a mathematical analogue of the RPD method, and its success could raise the question whether the RPD method would not work for the same reason.

2.3 INTERPRETATION OF IONISATION EFFICIENCY DATA

2.3.1 Detailed shape of ionisation efficiency curves

As usually measured, the efficiency curves for production of the parent ions of atoms and molecules, both singly and multiply charged, and of fragment ions from molecules, are superficially remarkably similar to each other, and it was long believed that the only useful value which could be obtained was the threshold or appearance potential.

The appearance potential should measure the minimum energy required for the process of formation of each ion. For parent ions it should give the lowest ionisation potential of the molecule, while for fragments it should measure the energy of the chemical change required to disrupt the molecule and form neutral fragments plus the energy to ionise at least one of them. However, while the appearance potentials for some fragment ions could be used to deduce reasonable values for bond dissociation energies, e.g. by writing equations of the type:

$$A(\mathrm{CH}_3^+) = I(\mathrm{CH}_3) + D(\mathrm{CH}_3\text{—H})$$

or to calculate heats of formation of ions[63], in other cases the disagreement was marked. This was not due entirely to the electron energy spread, since even when using closely monenergetic beams, thresholds, particularly for multicharged, and for fragment ions, were still diffuse. To clarify this, it is necessary to look more closely at the primary process of energy transfer from the ionising beam to the molecule.

2.3.1.1 *Ionisation at threshold—collision factor*

While for many processes of single positive ionisation by electron impact the cross-section appears to increase linearly with excess energy above the threshold, there are sufficient exceptions to suggest that more complicated behaviour often applies.

Theoretically the problem is that of the approach of an electron to a neutral molecule followed, after interaction, by the separation of a positive ion and two electrons. After separation the molecule-ion must be in a stationary state and therefore the amount of energy transferred to it is sharply defined. Because of the great difference in mass, momentum transfer from electron to molecule in the collision is a second-order effect. Since ionisation by electron impact is possible at all energies above the initial amount required for the process, excess energy must be carried off in the form of internal molecular energy, which must be quantised, or as translational energy. Because of the difference in mass between molecule and electrons, the latter must carry off almost all the non-quantised translational energy. A molecule-ion formed in an excited state may revert to its lowest state by fluorescence, or undergo subsequent dissociation, but both these processes occur at least 10^{-12} s later.

Wigner considered the ionisation process as taking place in three distinct steps—approach, the formation of a collision complex and dissociation of this complex. He postulated that the probability of a given process of ionisation is determined mainly by the conditions necessary for successful dissociation and these will depend on the number of ways in which the excess energy can be dissipated[64]. The number of degrees of freedom for removal of the excess energy is one less than the number of electrons leaving the complex; this would therefore lead to linear dependence on the excess energy for direct single ionisation and an nth power dependence for n-fold ionisation. A simple phase-space argument due to Wannier[65] illustrates this. To the degree of precision possible where electron beams with energy spreads of from 0.1 to 0.5 eV were the only ones available, this was shown to be the case for helium, where only one process of ionisation is possible in the energy region 24.5–50 eV. It was also consistent with the interpretation of the curves of the other rare gases, of molecules and of processes of multiple ionisation.

Other processes of ionisation exhibit differing behaviour but, in each case (negative ion formation by electron capture, excitation followed by ion pair formation or autoionisation), the number of degrees of freedom in the primary rate-determining process is consistent with the power law followed above threshold.

(a) *Direct single ionisation*—The same reasoning can be applied to processes of excitation and ionisation induced by photon impact and again proves correct in its essentials. In the case of a single ionisation by photon impact, the ionised electrons carry off all the excess energy and the detection of bands of such electrons with differing energies is the basis of photoelectron spectroscopy. One would expect that the cross-section for an inelastic collision should be not dependent solely on the allocation of the excess energy. A photon wave packet traverses the molecule at the speed of light while an electron wave travels more slowly at a speed dependent on its energy. The differing force fields operating in the approach of an electron and a photon to a molecule, and in the retreat of one or more electrons from either a neutral or charged species, should show up as differences in detail. In view of this, it is truly surprising that the differences found are so small. The first differential of the electron impact ionisation efficiency curve is remarkably close to that of photon impact; this has been discussed by Collin and Delwiche[66]. The only really marked discrepancies arise where a particular transition is allowed for electrons, but forbidden by selection rules for photons. It has been suggested by Lloyd[67] that in the ionisation of certain transition metal complexes, electron impact excites only the second and higher states of the ion, whereas photon impact reaches all the states. All of the electron impact work on these has been carried out with unfiltered beams so this can not be taken as proven.

Wannier, in a detailed attempt to include electron correlation in the retreat zone, predicted in 1955 that the cross-section for direct single ionisation should behave as the 1.127th form of the excess energy[68]. More recent calculations by Temkin[69] and Omidvar[70] also suggest a power law greater than unity.

Much effort has gone into attempts to establish whether this is the case or not. McGowan and Clarke[71] claim that the ionisation of atomic hydrogen near the threshold obeys a 1.127 power law. Brion and Thomas[43,72], using their electron selector, claim that their data for $He \rightarrow He^+$ indicates a slightly higher than 1.127 power law in the first 5 eV above threshold, and that this data can be fitted very accurately by a function based on a classical model due to Gryzinski[73] (see Figure 2.4). Marchand *et al.*[44], in a very detailed

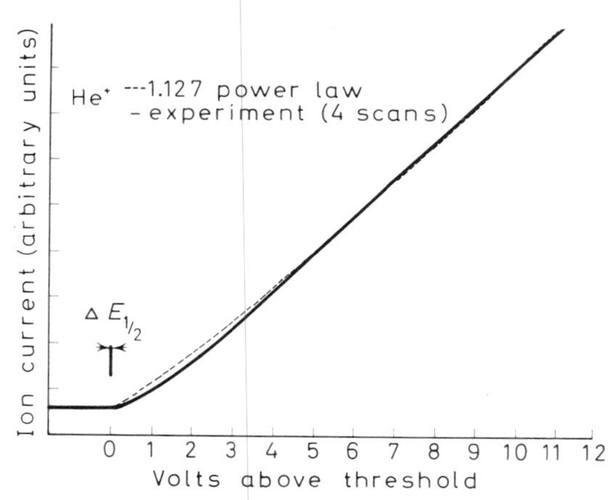

Figure 2.4 Ionisation efficiency curve for helium recorded using electron energy selector. A curve with a 1.127 power law is superimposed on the data (From Brion and Thomas[43], by courtesy of Elsevier)

and careful study using a molecular beam to reduce the Doppler effect, found that their data for He^+ could best be fitted with a 1.17 power law.

In spite of this evidence, it is the opinion of the reviewer that these higher power laws are not yet established. Some curvature will occur at threshold for electron impact due to the Doppler effect. At room temperature for H or He this will cause an effective increase in energy spread of from 0.02–0.03 eV, although it will be much less for any other molecule. More significantly, if it is the assumption that the true ionisation efficiency is a linear function of the excess energy, the second differential of the curve should give the energy distribution in the ionising beam[74,75]. If this is done to data obtained from

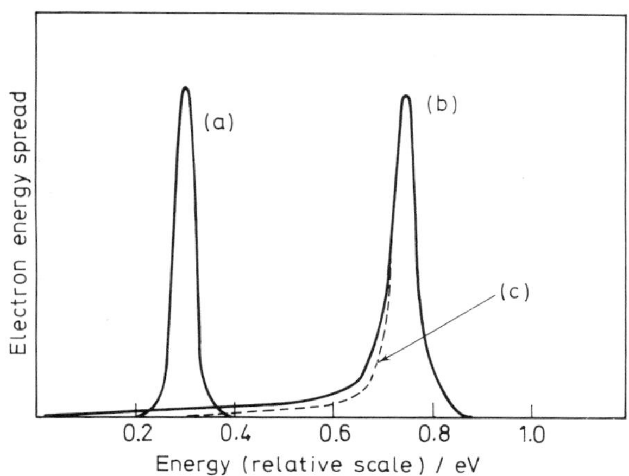

Figure 2.5 (a) Symmetrical electron energy distribution with half width of 0.05 eV.
(b) 2nd differential of the convolution of distribution (a) with the curve $y = x^{1.127}$, reversed on the energy scale.
(c) 2nd differential of data of Brion and Thomas[43] for He^+, smoothed and reversed on energy scale
(From Giessner and Miesels[76], by courtesy of Elsevier)

a source using an electron selector, the curve obtained consists of a sharp peak and a tail extending over a range of some 3–5 eV to lower energies[76] (see Figure 2.5). Depending on the degree of collimation employed in the ion source and how recently the source has been baked the extent of this tail can be varied. The possibility cannot be disregarded that some electrons in the beam have lost energy either in slightly inelastic collisions or by passage through apertures, and that while the half width is still 0.04 eV the total spread extends over some 3–4 eV.

(b) *Multiple ionisation*—The most recent work by Brion and Thomas on He[24,43] confirms a square-law behaviour found previously, although Marmet and his colleagues[77] suggest that a slightly higher power than two may apply.

No work has been done on higher ionisation using monoenergetic electrons, earlier studies suggest that the *n*th power law holds up to at least fourfold ionisation and possible even higher.

(c) *Excitation leading to autoionisation*—In the case of excited states possessing relatively long lifetimes before either autoionising or undergoing fluorescence the cross-section for the subsequent autoionisation seems to be fairly well represented by a step function of the excess energy. The shorter the lifetime, the closer the threshold behaviour should tend to that of direct single ionisation. As Fano has shown[78], configuration interaction between the discrete excited state and the continuum states of the ion can modify the cross-sections for both over a range of energies in a way which makes it not possible to distinguish sharply between them.

In some cases, the presence of an autoionising state can even cause a decrease in the gradient of the curve for the underlying direct ionisation. Nevertheless, the presence of such anomalous behaviour in the cross-section can be recognised and identified as autoionisation[79]. The autoionisation of numerous optically forbidden states has been observed in the rare gases[17, 80, 81]. Features are present in the efficiency of production of doubly and higher-charged

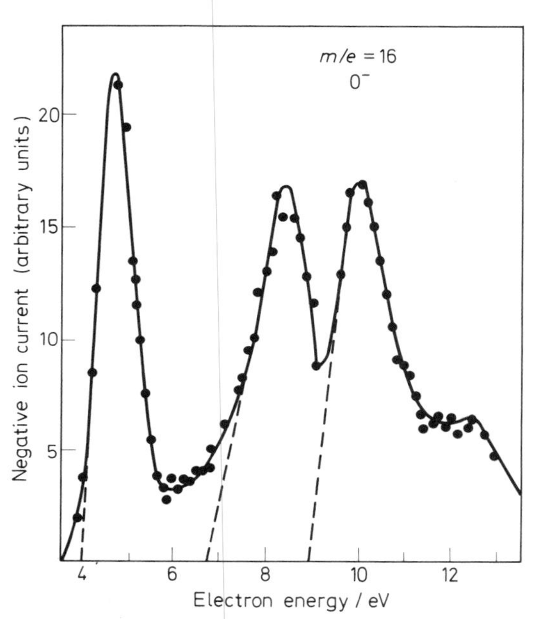

Figure 2.6 Formation of ions of $m/e = 16$ from ketene by dissociation attachment. At least four states are present in the energy range shown
(From Collin and Locht[168], by courtesy of Elsevier)

be tentatively assigned to Auger processes of the type:

Process	*Law*
$A+e \rightarrow A^{+*}+2e \rightarrow A^{2+}+3e$	Linear
$A+e \rightarrow A^{**}+e \rightarrow A^{2+}+3e$	Step

(d) *Excitation followed by ion pair dissociation*—In this case, theory predicts a square root dependence on the excess energy[83]. No recent work has been carried out with monoenergetic beams and it is difficult to separate the threshold law from the Franck–Condon factor (q.v.). What evidence there is might suggest a rounded step beginning at the threshold and becoming constant about 1 eV higher.

(e) *Electron capture*—In the case of the process $SF_6 + e \rightarrow SF_6^-$, the cross-section is found to be an extremely sharp peak at zero energy. The sharpness of this peak and its high cross-section have made SF_6 an extremely useful detector for zero energy electrons. This has formed the basis of the SF_6 scavenger method for studying electronic excitation by electron impact[84–89]. This molecular negative ion state is of particular interest in that it is unstable and autoionises with a lifetime of ~25 μs[90].

In other cases of resonance capture coupled with dissociation, Franck–Condon considerations apply and while the data are consistent with a threshold law consisting of a sharp peak at a critical energy, it cannot be proved that this is so (see Figure 2.6).

2.3.1.2 *Breakdown of threshold behaviour*

In those cases where the possible states of the ion are well separated, notably in He and the other rare gases, it is observed that the threshold behaviour is followed for *c.* 20 eV for He and progressively less as the atom becomes heavier; it becomes ~4 eV for Xe. Beyond this region, the cross-section deviates from linearity, very gradually in the case of rare gases, more abruptly for some molecules. Its approximate form at higher energies can be fitted empirically by the curve $I(E) = E^{-1} \log E/I_c$ (reference 91) where I_c is the ionisation potential.

In the theory of threshold behaviour described, the deviation can be assumed to be due to breakdown of the mechanism of sharing the energy excess. Recent experiments show that for incident electron energies of *c.* 60 eV, there is a greater probability that the two electrons emitted from the complex will have widely differing energies than that they will both have the same energy, which supports this view[92].

For double ionisation, the cross-section at higher energies takes the form $\sigma \propto E^{-1}$ and this difference has also been used to identify Auger processes[93] where the primary step is single ionisation.

2.3.2 Additivity of ionisation cross-sections

It is usually assumed that when there are two or more states of an ion possible, the probabilities for ionisation to each are independent of each other and the resulting ionisation efficiency is the simple sum of these probabilities. There are some indications that this is not always strictly so and that, in the dissociation of the collision complex to form the ion in various states, competition or interaction between dissociation channels may result[94]. It is almost always the case that all processes above the lowest appear to have much lower probabilities, which can be explained if the overall cross-section for formation of the collision complex is an essentially smooth function of the incident energy and new channels can open up only at the expense of already existing ones with thresholds at lower energy. If this is indeed so it will seriously affect the measured relative transition probabilities as measured both by electron and photon impact. Unfortunately, the experimental evidence is not good enough to confirm this.

One undoubted effect of this kind is that where the presence of an autoionising state interacts markedly with the underlying continuum of a lower ionised state[95–97]. Such processes have been discussed by Fano[78] in terms of configuration interaction.

2.3.3 Identification of processes by threshold behaviour

Since the number of electrons leaving the collision complex is the predominant factor in determining the threshold behaviour, the detailed form of the ionisation efficiency curve at threshold can be used to identify the process. Processes of electron capture and ion pair production may be distinguished in the curve for production of a negative ion, and similarly

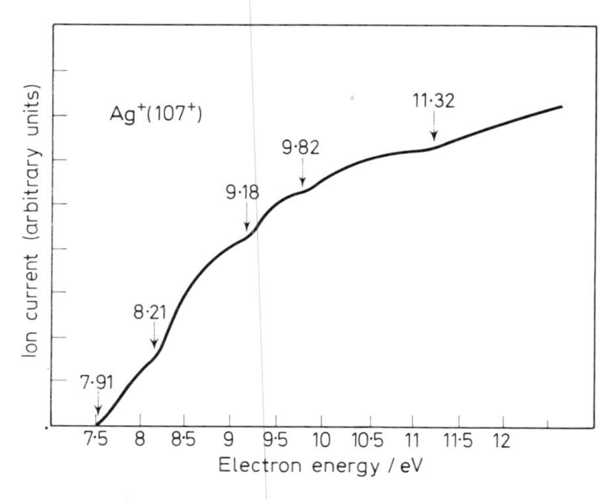

Figure 2.7 Ionisation efficiency for Ag showing structure due to autoionisation (From Cabaud *et al.*[187], by courtesy of Elsevier)

autoionisation, and ion pair production may be distinguished from processes of direct single and double ionisation. From the shape of the curves for production of Sr^{2+} and Ca^{2+} [82], it appears likely that some of the double ionisation originates in a primary process of excitation at *c.* 25–26 eV. followed by double autoionisation, rather than in two consecutive single ionisations (see Figure 2.7).

2.3.4 Half lives of autoionising states

In the case of single autoionisation, the half life of the excited state formed initially can vary in different atoms and molecules from 10^{-6} to 10^{-15} s (reference 98) i.e. it becomes indistinguishable from direct single ionisation. These lifetimes, and those for possible competing processes, have been studied recently by Berry[99] and also by Lifshitz[100]. The detailed form of the $I(E)$

curves for Xe^+ obtained by photon impact[97] shows clearly the difference in lifetimes, and hence in the shape of the curve, for different types of transition, and also the lengthening which occurs in the mean lifetimes of a series of similar states as they are farther from the threshold of the lower ionised state. The same effect is present but less obvious in the curve obtained for Xe^+ by electron impact[101, 102].

The autoionisation lifetimes for negative ions can be measured by the effect on the peak shape in the ion cyclotron resonance spectrometer[103].

Autoionisation is an important contributor to the total single ionisation in the case of many molecules. With poor energy resolution, it obscures many of the higher threshold values and makes interpretation very difficult. A comparison of the ionisation efficiency curves with the corresponding photoelectron energy spectra is often most useful, since the latter method does not normally detect autoionisation[104].

2.3.5 Franck–Condon factors affecting the ionisation efficiency curve shape

The primary process of energy transfer from an electron beam to a molecule is extremely rapid ($\sim 10^{-16}$ s) so that only electronic readjustments are possible. Such a transition is said to be vertical. This may be an unstable configuration and the nuclei will then vibrate about a new equilibrium position in the changed potential field of the electrons, causing further readjustment of the latter. For diatomic molecules, the relative electronic transition probabilities can be very readily calculated from the overlaps of the eigenfunctions of the ground vibrational level of the neutral molecule and those of the vibrational levels of each state of the ion. Detailed calculations of this kind have been carried out for many simple molecules[105–109]. A review of these has been given by Rosenstock and Botter[110].

For small molecules, the majority of the molecules are initially in their lowest vibrational state. Depending on whether the potential curve for the upper state lies directly above that of the ground state, or is displaced to greater internuclear spacing, the ionisation transition probability will take the forms shown in Figure 2.8.

Where the process is that of single ionisation by electron impact, the sum of the separate vibronic ionisation probabilities leads to $I(E)$ curves as shown.

Particularly in the case of larger molecules, significant numbers of molecules are in higher vibrational levels, producing an exponential foot to the curve, analogous to the hot bands in spectroscopy.

It is most important to distinguish between the adiabatic, $e_0v_0 \rightarrow e_1v_0$ ionisation energy, and the vertical ionisation energy, which is the energy difference between e_0v_0 and the point at which the value for r_e, the equilibrium internuclear spacing in the neutral molecule, cuts the potential energy surface for the ionised state. The vertical value as such may not coincide with any vibronic level of the ion. It does have theoretical significance because it is the energy required to remove one electron and readjust the orbital electron cloud without changing the positions of any of the nuclei. As such, Koopmann's theorem equates it to $-E_n$, the orbital energy of the most loosely bound electron in the molecule[111, 111a].

Much of the confusion regarding the accuracy of threshold values from electron impact work has arisen because the value which is being measured is not defined. If there were no energy spread in the electron beam the threshold of the ionisation efficiency would set an upper limit to the adiabatic first ionisation energy. It may not be possible to observe the actual $''v_0 \rightarrow {'v_0}$ transition.

Some of the discrepancies in the past have arisen because spectroscopic values themselves have not always been clearly specified. Where a complete

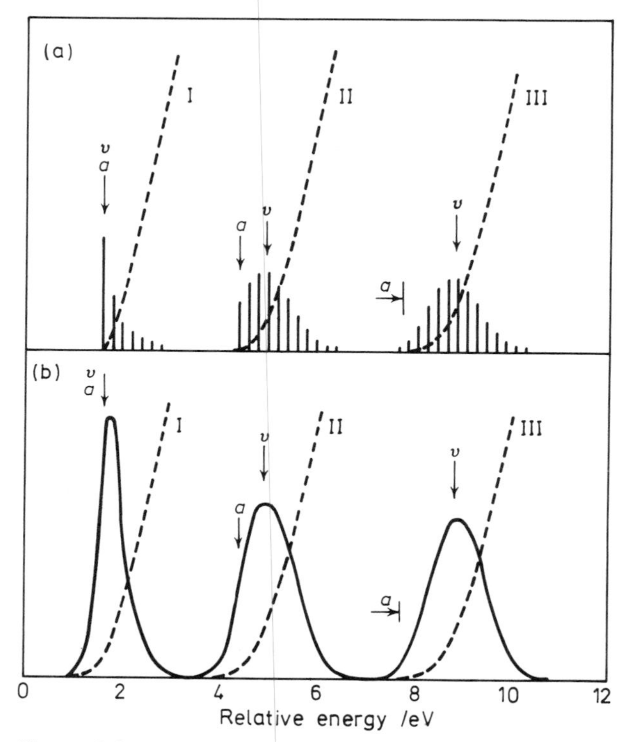

Figure 2.8 (a) Electronic transition probability curves to be expected when the potential energy curve for the ionised state has its minimum: (I) directly above the minimum for the curve for the ground state; (II) and (III) displaced progressively to larger values of r_e. Superimposed on each is the expected ionisation efficiency using monoenergetic electron impact. The positions of the vertical (v) and adiabatic (a) ionisation potentials I.P. are indicated.
(b) The effects of convoluting curves (I)-(III) with a Maxwellian energy distribution for a cathode at 2300 K.

vibronic assignment has been carried out a Rydberg extrapolation of the $e_n v_0$ terms will give the adiabatic value . In such a case, the adiabatic and vertical values are close together or coincide. Where this has not been done, the ionisation potential obtained by Rydberg extrapolation of the band maxima will tend to give a vertical value.

At ion source temperatures of 300–500 K the neutral molecules will populate several vibrational levels and a distribution of rotational levels within each of these. With the energy resolution at present attainable for electron impact coupled with mass analysis, rotational fine structure is not resolved. Transitions from excited vibrational levels of the neutral molecule can introduce an uncertainty in estimates of the upper limit to the adiabatic ionisation energy of $\sim$0.02 eV.

The mass spectrometer is essentially a rate-measuring device, and because the rate of dissociation of a polyatomic molecule depends on the excess energy above the dissociation limit additional curvature arises in the ionisation efficiencies of such fragmentations.

In its most recent form, the statistical theory of mass spectra (q.v.) is able to account for the shape of the ionisation efficiency curves obtained for fragment ions reasonably well.

In the case particularly of molecules containing π orbitals the ionisation efficiency curves are often complicated by other structure. There are numerous excited neutral states of the molecule lying above the lowest state of the ion and many of these can autoionise. This autoionisation can often be recognised as such by the differing threshold behaviour. It can happen that the autoionising levels are close together and their contributions add together to give an apparent linear increase. In both cases, the autoionisation may be much more probable than, and obscure completely, the underlying direct ionisation. In the simple molecules CH_4, NH_3 and H_2O, it has been found that strong autoionisation contributes to the ionisation efficiencies of both the parent and fragment ion species at energies in the neighbourhood of 20 eV.

2.3.6 Measurement of appearance potentials

The first attempts to obtain threshold potentials from ionisation efficiency curves were concerned solely with finding an empirical method which would give reproducible values. These methods ought to be only of historical interest. However, since all of them are still actively being employed it is worthwhile to review briefly the errors which can arise by their use. The merit of a method can be judged by whether the values it gives are in good agreement with those from other sources, e.g. spectroscopic onset of ionisation continua or extrapolation of Rydberg series, photoionisation or photoelectron measurements, or indirectly by consistency with thermochemical data.

2.3.6.1 Vanishing current methods

In the vanishing current (V.C.) methods the energy at which the ionisation efficiency first rises above zero, or more correctly, above the noise level of detection is taken as the appearance potential; sometimes the slope of the linear part of the curves is normalised to a constant value, sometimes not[112]. This method gives reasonable values for the differences between the lowest ionisation potentials of the rare gases, but is much less satisfactory in deriving the ionisation energies of molecules. Not all curves have the same shape, for the reasons outlined in Sections 2.3.1 and 2.3.5 above and in such cases the point at which the $i(V)$ curve emerges from the background scatter can depend markedly on the signal-to-noise ratio of the data.

For molecules, depending on whether the electronic transition probability is of types I and II, or III (Figure 2.8), the threshold should give either the adiabatic ionisation energy or set an upper limit to it. Even if this method is not very good it is better than some of those which follow.

2.3.6.2 Linear extrapolation methods

The linear extrapolation method (L.E.)[113] takes advantage of the fact that in very many cases the $i(V)$ curves are linear or approximately so from about 1 to at least 10 eV above the threshold. An extrapolation to the energy axis gives values for the rare gases which are accurate to 0.02 eV. In a variant of this, the extrapolated difference method[114, 115], the ionisation efficiency curves for the calibrating gas and for the sample are recorded at the same time and are normalised to make the linear portion of the curves parallel. A plot is then made of the energy difference between the curves for each value of the ion current and is extrapolated to zero ion current. The resulting difference is taken as the true difference in energy between the appearance potentials of the calibrating and the sample ions. Again, systematic discrepancies arise when the ionisation potentials for molecules determined in this way are compared with spectroscopic values. In the case of curves of type I, quite a good value for the vertical I.P. is obtained, but for types II and III the value may be too high by up to 0.6 eV.

2.3.6.3 Logarithmic methods

The critical slope method[116] was the first attempt to account explicitly for the effect of the electron energy spread. A plot of the logarithm of the ionisation efficiency *v.* energy is linear for the first 1–2 eV, with slope proportional to the temperature of the cathode. In this method, it was assumed by Honig that $I(E)$ for single ionisation is a square-law curve and that therefore the correct threshold should occur at the point where the gradient of the straight line in the semilog plot falls to $\frac{2}{3}$ of its initial value. This method is easy to apply, reproducible and good for rare gases, but again much less satisfactory for molecules. Simpler variants of the method have been used[117, 118]. The point at which each semilog curve ceases to be linear can give equally good values in many cases[119]. As with the L.E. method, curves of types II and III can give measured values which are too high. The nature of the errors has been discussed in detail by Barfield and Wahrhaftig[119a].

2.3.6.4 Derivative methods

If the assumption is made not that the $I(E)$ curve for a given ion is a linear function of E, but that $I(E)$ may be made up as the sum of a series of ionisation probability curves, each having a separate threshold energy E_n and each a linear function of the excess energy above E_n,

$$\text{i.e.} \quad I(E) = P_1(E-E_1)+P_2(E-E_2)+\ldots.$$

then it may be shown that the second differential of $I(E)$, $\Delta^2 I(E)/\Delta E^2$ consists of a series of peaks at the threshold values E_1, E_2, etc., with heights determined by the probabilities P_n. The effect of the electron energy spread is to smear these peaks out into distributions which are the same as the energy distribution, but reversed on the energy scale. Even if $I(E)$ is not exactly linear, provided there is a discontinuity in gradient at each E_c, the second differential consists of a set of peaks on a slowly varying base line[120].

Provided the separate probability curves are not too close together and provided one assumes that $P(E)$ for a single process is linear, the position of the maximum of the lowest peak gives the ionisation threshold.

For molecules, the second differential curve gives the electronic transition probability curves for each state of the ion. Even though the vibrational structure may not be resolved, the position of the peak maximum gives the vertical I.P. The upper limit to the adiabatic I.P. lies at approximately one half of the width of the energy distribution above the apparent threshold on the low-energy side of the peak.

Until recently, it has been difficult to obtain data with low enough scatter to use this method successfully. In such cases, the positions of inflections in the first derivative curve can be taken as the vertical values[121–124]. In the case of resonance capture, the ionisation efficiency gives the electronic transition probability directly and can be interpreted in the same way as the second differential curves for direct single ionisation.

Extensive tabulations of ionisation and appearance potentials and derived information have been published[63, 125].

2.3.7 Fragment ion kinetic energies

When transition occurs to a potential curve well above its dissociation limit, only fragment ions are observed, and they may possess appreciable amounts of kinetic energy. This is particularly so in the case of a process where a doubly charged molecule ion dissociates to two singly charged fragments. The mass spectrometer has the advantage of not only being able to identify the particular dissociation but also to obtain kinetic energies, but it is disappointing how few accurate measurements have been made. This has probably been due to the fact that many magnetic mass analysers discriminate against such ions. This criticism does not apply to either the quadrupole or monopole instruments, and they could be used very effectively in this type of study. The amount of kinetic energy carried off by the charged product in a dissociation can be measured by analysing the ion peak shape[126, 127] by the application of a retarding potential in front of the ion collector; this was most recently done by Newton and Sciamanna[128]. A more precise method is to use an electrostatic sector[129, 130].

It is seldom that fragment ions from polyatomic molecules are observed with appreciable amounts of kinetic energy and it would seem that radiationless transitions and predissociations are preferred processes.

Such processes are often much slower than direct fragmentation and their occurrence is confirmed by the existence of numerous metastable ions. There has been a considerable resurgence of interest in the energies of such ions, in particular the metastable ions arising in dissocation processes with lifetimes longer than 10^{-8} s. By modifying the voltages on the electric sector in a double-focusing mass spectrometer it has proved possible to determine the kinetic energy distribution in such ions to a precision of around $\sim$0.02 eV[131–135].

The main uncertainty in the calculation of the total kinetic energy release in the dissociation of a polyatomic molecule is the identification of the neutral fragments formed at the same time. Haney and Franklin[136, 137] have postulated an empirical relation between the excess energy and translational energy of the products of dissociation.

2.3.8 Calculation of dissociation energies

For diatomic molecules a knowledge of the vertical ionisation potential of the state giving rise to a fragment, $A(A^+)$, and of the kinetic energy, $E_k(A^+)$, of the fragment ion formed at this energy, will permit a determination of the energy of the dissociation limit for the state. This then gives a direct measurement of the dissociation energy of the neutral molecule in its ground state, D_0(A—B), since

$$A(A^+) - E_k(A^+)[1 + M_A/M_B] = I(A^+) + D_0(A-B) + E(B)$$

M_A and M_B are the masses of fragments A and B, respectively; $I(A^+)$, the ionisation energy of atom A can be obtained from spectroscopic tables[138]. E(B) is the energy of excitation of atom B, if non-zero.

In the case of polyatomic molecules, such a simple relationship does not exist. The products of dissociation have many degrees of internal freedom and part, if not all, of the excess energy above the dissociation limit can persist as vibrational excitation of the products. Because of the large number of degrees of internal freedom possible, large molecules may possess as much as 1–2 eV of internal energy, even at room temperature. A molecule such as n-decane also possesses very many electronic states of the ion closely spaced over an energy range of some 4–6 eV above the lowest ionisation potential.

The term bond energy must be used very carefully in the context of a polyatomic molecule. Is it to include reorganisation energy of the fragments, or not? As pointed out by Field and Franklin[139] and others, ionic heats of formation are of more significance and may be readily derived from appearance potential data.

2.4 RECENT RESULTS

2.4.1 Ionisation potentials of free radicals

Much of the earlier work on the ionisation potentials of free radicals, carried out with conventional ion sources, and using logarithmic methods of interpretation, gave values which have since been found to be high by from 0.1–0.8 eV. More recent studies, using the RPD method[140, 141] and photo-ionisation[142] and photoelectron spectroscopy, indicate that the ionisation efficiency curves may possess considerable structure in the immediate neighbourhood of the thresholds. Unfortunately, apart from the fact that the thresholds were lower there was not very good correspondence between the curves obtained by electron impact and those using photons. By use of a mass spectrometer with an electron selector, Lossing and his colleagues have recently re-examined[143–146] a number of these radicals. In each case, the form of the ionisation efficiency curve is quite consistent with the shape of the curve obtained by photon impact with an allowance for the differing threshold laws. The ionisation energies found are in excellent agreement with the photon impact (PI) values. These are listed in Table 2.1.

Table 2.1 Adiabatic ionisation potentials of free radicals measured by electron impact

Radical	*Electron Analyser*	*RPD*	*PI*
CH_3	9.84[143]	9.87[141]	9.82[147, 142]
C_2H_5	$\leqslant$8.38[143]	8.34[141]	$<$8.4[147]
n-C_3H_7	$\leqslant$8.10[143]	8.13[141]	$<$8.1[147]
s-C_3H_7	$\leqslant$7.55[143]	7.57[141]	$<$7.5[147]
n-C_4H_9	$\leqslant$8.01[143]	8.01[141]	
i-C_4H_9	$\leqslant$8.01[143]		
s-C_4H_9	$\leqslant$7.41[143]		
t-C_4H_9	$\leqslant$6.93[143]		
Vinyl	8.95[145]		
Allyl	$\leqslant$8.07[145]		
Benzyl	$\leqslant$7.27[145]		

In this work, the radicals are formed in the mass spectrometer by a pyrolysis apparatus. A most interesting method has been developed by Beck *et al.*[148–151], which involves two electron beams, one to form neutral fragments, and one to ionise them. This method has been used to study a whole series of radicals[152, 153]. The extrapolated difference method was used to interpret the results, so that the values found for the ionisation potentials can be taken only as upper limits and may be higher than the adiabatic values by up to 0.5 eV.

2.4.2 Negative ion studies

A series of studies[154–160] has been undertaken by MacNeil and Thynne of processes giving rise to negative ions in halogenated molecules. The molecules studied include CF_4[154], CHF_3[155], SiF_4[154], C_2F_6[155], CF_3CH_3[155], C_2H_4[157], COF_2[158], CF_3OF[158]. Ion formation was studied over an energy range to 70 eV using a time-of-flight instrument. Deconvolution was used to reduce the electron energy spread. The data is discussed in terms of dissociative resonance and ion pair processes and estimates are made of the C—F bond strength.

A very detailed discussion of the shape of the ionisation efficiency curves for dissociative attachment in diatomic molecules in terms of potential energy curves is given by Locht and Momigny[161,162]. Earlier work is reviewed. In all the cases of resonance capture, the curves are consistent with direct reflections of the ground-state expectation function in the upper-state potential curve. The authors establish criteria for obtaining the appearance potentials of negatively charged fragment ions and consider in detail the processes observed in CO, H_2, NO and O_2.

The formation of C^- and O^- from CO has also been studied by Stamatovic and Schulz[49], using an electron selector, and compared with earlier work. It is suggested that fine structure in the curve for O^- indicates a possible radiationless transition at *c.* 10 eV.

Van Brunt and Kieffer[163] have studied the angular distribution of the O^- ions found by dissociative attachment in O_2, and found a strong dependence on energy.

Negative ion formation has been examined in several polyatomic molecules, SO_2 [164], NO_2 [164], XeF_6 [166], XeF_4 [165], NH_3 [167], ketene[168] and dimethyl sulphoxide[209]; a wealth of states were detected.

Many of the discrepancies in the study of negative ions can be attributed to the fact that not a few of the negative ion states have short lifetimes and either dissociate or autoionise back to the neutral molecule. SF_6 is the classical example of this. Depending on the lifetime of the state, the dwell time in the mass spectrometer, and the means for ion detection, these parent ions may be recorded with varying abundance.

Such ions have been studied in N_2O and CO_2 [169], in cyclic fluorocarbon molecules[179] and in perfluorocarbons[171], where the resonances are detected using the SF_6 scavenger method. Negative ion breakdown curves have been constructed for SF_6 and a number of fluorocarbons[172]. The dissociation rate constants are calculated as a function of internal energy of the parent ion and used to estimate kinetic shifts and thermochemical appearance potentials.

Dougherty[173] reports the observation of a doubly charged negative ion. No measurement was made of the ionisation efficiency. It is probable that this ion was not formed directly by resonance capture by a negative ion but arose by a secondary ion–ion reaction in the source.

2.4.3 Energetics of inorganic and metal-organic molecules

It is only relatively recently that many studies have been made of inorganic molecules, probably because of the experimental difficulties involved in Knudsen cell sources. It is difficult to obtain good ionisation efficiency curves and most of the work has been concerned with obtaining threshold values by V.C. or L.E. methods.

The substances examined include the alkaline earths[82], the Group I [174] and Group IIA metal chlorides[175], Ga_2S_3, Ga_2Se_3 and Ga_2Te_3 [176], CrO_2Cl_2 and CrO_2F_2 [177], the carbonyls of Ni, Fe, Cr, Mo, W and V [178,180], mixed chlorides of Pb and alkali metals[181], borane carbonyl and diborane[182], pentaborane and five derivatives[183], transition metal complexes[184], alkyl borates[101] and organic silicon compounds[185,186]. A very good RPD study has been made of Ag metal[187] and strong autoionisation is evident in the curve. The ionisation efficiencies of the parent and fragment ions for the metal carbonyls are of unusual shape. This may be due to the fact that as the atomic number is higher, the range over which the simple threshold behaviour holds is less, it may be unresolved autoionisation, or it may be interaction effects of the kind described in Section 2.3.2.

A study of the di-metal decacarbonyls of Te, Mn, Re, ReMn [196, 197] indicates that metal–metal interactions modify the close relation observed between the ionisation potential of the free metal atom and that of the single metal carbonyls.

Chambers, Glockling and their co-workers have made an extensive series of studies of metal alkyls: Be [198], Ge [199–201], Sn [202, 203], Pb [203] and Al [204]. The extrapolated voltage difference method of interpreting appearance potentials is used and bond dissociation energies are calculated.

2.4.4 Other results

A detailed study of ionisation–dissociation and autoionisation in CH_4 [66] and C_2H_2 [188] and comparison with photon impact work suggested that optically forbidden transitions are significant. This finding for acetylene has been supported by Collin, Winters and Engerholm[189], also by Brion[190]. On the other hand Lossing[191], using an energy-resolved beam, found excellent agreement between electron and photon impact.

Other molecules studied using an energy selector are H_2, D_2, CH_4, CD_4 [192], CO, N_2 [77], N_2, O_2, SF_6 [43]. The appearance potentials of cyclic ethers have been used to derive heats of formation of numerous fragment ions[193].

The ionisation of cyclohexa-1-, and cyclohexa-1,4-diene, and of hexa-1,3,5-triene has been examined, and detailed fragmentation schemes constructed on the basis of the energetics[205].

An attempt has been made[194] to correlate the levels of a molecular ion AD^+(AD) with the levels observed for the same ion when it can arise from a molecule ABCD. It is claimed that in this way the energy levels of the neutral fragment BC can be measured.

The ionisation efficiency curves are measured using the RPD method, and are claimed to be made up of linear segments. The agreement between the electron impact curve even for H_2O^+ with photon impact data is not good and points to the need for care in accepting the structure in ionisation efficiency curves recorded with this method. Attempts have been made recently to analyse the kinetics of breakdown of the molecular ion[195] in a simple manner. Unfortunately, the data were recorded with an unfiltered beam, making it very difficult to be certain that much unresolved structure is not being ignored.

In simple molecules, H_2O, H_2S, C_2H_2, HCHO, Ficquet-Fayard and Guyon[206–208] have demonstrated that correlation diagrams may be used to predict the presence of structure due to pre-dissociation in ionisation efficiency curves.

2.5 FUTURE DEVELOPMENTS

Insofar as the measurements of valence state ionisation potentials are concerned, photoelectron spectroscopy gives a much more direct measurement of the electronic transition probability and electron impact methods are unlikely to compete in this area.

The ability of the mass analyser to resolve the total ionisation efficiency into its separate contributions from the parent and various fragment ions is invaluable in identifying the various dissociation limits for a molecule. More precise methods of determining fragment ion kinetic energies will allow the energies of dissociation to be calculated with more certainty.

Electron impact remains the only general method for the detailed examination of single ionisation at energies above 25 eV, of autoionisation, multiple ionisation and electron capture processes.

Detailed comparisons of electron impact ionisation efficiency curves with those from photon impact, and with photoelectron spectra, should shed much more light on the mechanism of the ionisation process and lead to interesting information on optically forbidden transitions.

The study of ionisation processes in inorganic substances has hardly begun and there is a vast amount of information regarding bond energies and structure to be obtained.

There is a need for better and more efficient ion sources, using more homogeneous beams of electrons. It is possible that by the application of more sophisticated data-processing techniques much more information can be obtained from present data.

References

1. Otvos, J. W. and Stevenson, D. P. (1950). *J. Amer. Chem. Soc.*, **78,** 546. (See also Fano, U. (1946). *Phys. Rev.*, **70,** 44)
2. Mann, J. B. (1967). *J. Chem. Phys.*, **46,** 1646
3. Lin, S. and Stafford, F. E. (1968). *J. Chem. Phys.*, **48,** 3885
4. Tiwari, P., Rai, D. K. and Rustgi, M. L. (1969). *J. Chem. Phys.*, **50,** 3040
5. Moiseivitch, B. L. and Smith, S. J. (1968). *Rev. Mod. Phys.*, **40,** 238
6. Fiquet-Fayard, F., Chiari, J., Muller, F. and Ziesel, J. (1968). *J. Chem. Phys.*, **48,** 478
7. Nygaard, K. J. (1968). *J. Chem. Phys.*, **49,** 1995
8. Gingerich, K. A. and Blue, G. D. (1970). *J. Chem. Phys.*, **53,** 4713
9. Rovner, L. H. and Norman, J. M. (1970). *J. Chem. Phys.*, **52,** 2946
10. Kieffer, L. J. (Jan. 1969). *Compilation of Low Energy Electron Collision Cross-section Data,* Part 1 JILA, Information Centre Rept. No. 6, Boulder, Colorado, U.S.A.
11. Lampe, F. W., Franklin, J. L. and Field, F. H. (1957). *J. Amer. Chem. Soc.*, **79,** 6129
12. Christophorou, L. G. and Stockdale, J. A. D. (1968). *J. Chem. Phys.*, **48,** 1956
13. Steiner, B., Geise, C. F. and Inghram, M. G. (1961). *J. Chem. Phys.*, **34,** 189
14. Svec, H. J. and Flesch, G. D. (1968). *Int. J. Mass Spectrom. Ion Phys.*, **1,** 41
15. Zworykin, V. K., Morton, G. A., Ramberg, E. G., Hillier, J. and Vance, A. W. (1945). *Electron Optics and the Electron Microscope,* 207. (New York: Wiley and Sons)
16. Williams, J. M. and Hamill, W. M. (1968). *J. Chem. Phys.*, **49,** 4467
17. Morrison, J. D. and Traeger, J. C. (1971). *Int. J. Mass Spectrom. Ion Phys.*, **7,** 391
18. Marmet, P. and Morrison, J. D. (1962). *J. Chem. Phys.*, **36,** 1238
19. Morrison, J. D. and Nicholson, A. J. C. (1959). *J. Chem. Phys.*, **31,** 1320
20. Buckingham, D. (1965). *Brit. J. Appl. Phys.*, **16,** 1821
21. Smith, L. G. (1937). *Phys. Rev.*, **51,** 263
22. Dorman, F. H. and Morrison, J. D. (1961). *J. Chem. Phys.*, **35,** 575

22a. Sprouse, J. F., Jackson, K. M., Raju, T. A., Testerman, M. K. (1971). *Rev. Sci. Instr.*, **42,** 114

23. Fox, R. E., Hickam, W. M., Grove, D. J. and Kjeldaas, T., Jr. (1955). *Rev. Sci. Instr.*, **26,** 1101
24. Brongesma, H. H. and Oosterhoff, L. J. (1967). *Chem. Phys. Lett.*, **1,** 169
25. Burns, J. F. (1961). *Nature (London)*, **192,** 651
26. Marmet, P. (1964). *Can. J. Phys.*, **42,** 2120
27. Grajower, R. and Lifshitz, C. (1968). *Israel J. Chem.*, **6,** 847
28. Gordon, S. M., Haarhoff, P. C. and Krige, G. J. (1969). *Int. J. Mass. Spectrom. Ion Phys.*, **3,** 13
29. Simpson, J. A. (1961). *Rev. Sci. Instrum.*, **32,** 1283
30. Henderson, E., Riley, P. N. K. and Sedgwick, R. D. (1966). *J. Sci. Instr.*, **43,** 726
31. Anderson, N., Eggleton, P. P. and Keesing, R. G. W. (1967). *Rev. Sci. Instr.*, **38,** 924
32. Cloutier, G. G. and Schiff, H. I. (1959). *J. Chem. Phys.*, **31,** 793
33. Lawrence, E. O. (1926). *Phys. Rev.*, **28,** 947
34. Nottingham, W. B. (1939). *Phys. Rev.*, **55,** 203
35. Hughes, A. L. and Rojansky, V. (1929). *Phys. Rev.*, **34,** 284
36. Clarke, E. M. (1954). *Can. J. Phys.*, **32,** 764
37. Marmet, P. and Kerwin, L. (1960). *Can. J. Phys.*, **38,** 787
38. Marmet, P. and Morrison, J. D. (1962). *J. Chem. Phys.*, **36,** 1238
39. Schulz, G. J. (1962). *Phys. Rev.*, **125,** 229
40. McGowan, J. W. and Fineman, M. A. (1965). *Proc. of 4th Int. Conf. on Physics of Electronic and Atomic Collisions Quebec*, p. 429. (New York: Science Book Crafters, Inc.)
41. Brion, C. E., Frost, D. C. and McDowell, C. A. (1966). *J. Chem. Phys.*, **44,** 1034
42. McGowan, J. W., Fineman, M. A., Clarke, E. M. and Hanson, H. P. (1968). *Phys. Rev.*, **167,** 52
43. Brion, C. E. and Thomas, G. E. (1968). *Int. J. Mass Spectrom. Ion Phys.*, **1,** 25
44. Marchand, P., Paquet, C. and Marmet, P. (1969). *Phys. Rev.*, **180,** 123
45. Simpson, J. A. and Kuyatt, C. E. (1963). *Rev. Sci. Instr.*, **34,** 265
46. Simpson, J. A. (1964). *Rev. Sci. Instr.*, **35,** 1698
47. Skeberle, A. M. and Lassettre, E. N. (1964). *J. Chem. Phys.*, **40,** 1271
48. Maeda, K., Semeluk, G. P. and Lossing, F. P. (1968). *Int. J. Mass Spectrom. Ion Phys.*, **1,** 395
49. Stamatovic, A. and Schulz, G. J. (1970). *J. Chem. Phys.*, **53,** 2663
50. Foner, S. N. and Nall, B. H. (1961). *Phys. Rev.*, **122,** 512
51. Hutchison, D. A. (1963). *Advan. Mass Spectrom.*, **2,** 527
52. Marmet, P., Morrison, J. D. and Swingler, D. L. (1962). *Rev. Sci. Intr.*, **33,** 239
53. Morrison, J. D. (1963). *J. Chem. Phys.*, **39,** 200
54. Dromey, R. G. and Morrison, J. D. (1971). *Int. J. Mass Spectrom. Ion Phys.*, **6,** 253
55. Dromey, R. G. and Morrison, J. D. (1970). *Int. J. Mass Spectrom. Ion Phys.*, **4,** 475
56. Dromey, R. G., Morrison, J. D. and Traeger, J. C. (1971). *Int. J. Mass Spectrom. Ion Phys.*, **6,** 57
57. MacNeil, K. A. G. and Thynne, J. C. J. (1969). *Int. J. Mass Spectrom. Ion Phys.*, **3,** 35
58. MacNeil, K. A. G. and Thynne, J. C. J. (1970). *Int. J. Mass Spectrom. Ion Phys.*, **4,** 434
59. Winters, R. E., Collins, J. H. and Courchene, W. L. (1966). *J. Chem. Phys.*, **45,** 1931
60. Collins, J. H., Winters, R. E. and Engerholm, G. G. (1968). *J. Chem. Phys.*, **49,** 2469
61. Fisher, I. P. and Henderson, E. (1967). *Trans. Faraday Soc.*, **63,** 1342
62. Gallegos, E. J. and Klaver, R. F. (1967). *J. Sci. Instr.*, **44,** 427
63. Franklin, J. L., Dillard, J. G., Rosenstock, H. M., Herron, J. T., Draxl, K. and Field, F. H. (1969). *Ionization Potentials, Appearance Potentials and Heats of Formation of Positive Ions*, National Standard Reference Data System, NSRDS–NBS 26. (Washington: U.S. Govt. Printing Office)
64. Wigner, E. P. (1948). *Phys. Rev.*, **73,** 1002
65. Wannier, G. H. (1955). *Phys. Rev.*, **100,** 1180
66. Collin, J. E. and Delwiche, J. (1967). *Can. J. Chem.* **45,** 1875
67. Lloyd, D. R. (1970). *Int. J. Mass Spectrom. Ion Phys.*, **4,** 500
68. Wannier, G. H. (1953). *Phys. Rev.*, **90,** 817
69. Temkin, A. (1966). *Phys. Rev. Lett.*, **16,** 835
70. Omidvar, K. (1967). *Phys. Rev. Lett.*, **18,** 153
71. McGowan, J. W. and Clarke, E. M. (1968). *Phys. Rev.*, **167,** 43
72. Brion, C. E. and Thomas, G. E. (1969). *Int. J. Mass Spectrom. Ion Phys.*, **2,** 414
73. Bauer, E. and Bartky, C. D. (1965). *J. Chem. Phys.*, **43,** 2466

74. Meisels, G. G., Park, J. Y. and Geissner, B. G. (1970). *J. Amer. Chem. Soc.*, **92,** 254
75. Morrison, J. D. (1953). *J. Chem. Phys.*, **21,** 1767
76. Giessner, B. G. and Meisels, G. G. (1970). *Int. J. Mass Spectrom. Ion Phys.*, **4,** 84
77. Marmet, P., Marchand, P. and Pacquet, C. (1970). *Recent Developments in Mass Spectroscopy, Proceedings of Int. Conf. on Mass Spectrometry,* 765, Kyoto, Japan, 1969. (Baltimore: University Park Press)
78. Fano, U. (1961). *Phys. Rev.*, **124,** 1866
79. Lossing, F. P., Emmel, R. H., Giessner, B. G. and Meisels, G. G. (1971). *J. Chem. Phys.*, **54,** 5431
80. Simpson, J. Arol., Chamberlain, G. E. and Mielczarek, S. R. (1965). *Phys. Rev.*, **139,** A1039
81. Grissom, J. T., Garrett, W. R. and Compton, R. N. (1969). *Phys. Rev. Lett.*, **23,** 1011
82. Okudaira, S., Nagata, T., Okuno, Y., Kaneko, Y. and Kanomato, I. (1970). *Recent Developments in Mass Spectroscopy, Proceedings of Int. Conf. on Mass Spectrometry,* Kyoto, Japan, 1969, 769. (Baltimore: University Park Press)
83. Geltman, S. (1956). *Phys. Rev.*, **102,** 171
84. Curran, R. K. (1963). *J. Chem. Phys.*, **38,** 780
85. Compton, R. N., Juebner, R. H., Reinhardt, P. W. and Christophorou, L. G. (1968). *J. Chem. Phys.*, **48,** 901
86. Brion, C. E. and Eaton, C. R. (1968). *Int. J. Mass Spectrom Ion Phys.*, **1,** 102
87. Franskin, M. J. H. and Collin, J. E. (1970). *Int. J. Mass Spectrom. Ion Phys.*, **4,** 451
88. Franskin, M. J. H. and Collin, J. E. (1970). *Int. J. Mass Spectrom. Ion Phys.*, **5,** 163
89. Franskin, M. J. H. and Collin, J. E. (1970). *Int. J. Mass Spectrom. Ion Phys.*, **5,** 255
90. Compton, R. N., Christophorou, L. G., Hurst, G. S. and Reinhardt, P. W. (1966). *J. Chem. Phys.*, **45,** 4634
91. Schram, B. L., van der Weil, M. J., de Heer, F. J. and Moustapha, H. R. (1966). *J. Chem. Phys.*, **44,** 49
92. Grissom, J. T., Compton, R. N. and Garrett, W. R. (1970). *Electron Impact Excitation and Ionisation of the Rare Gases* (Thesis), ORNL–TM–2618, March
93. Fiquet-Fayard, F., Chiari, J., Muller, F. and Ziegel, J. (1968). *J. Chem. Phys.*, **48,** 478
94. Morrison, J. D. and Traeger, J. C. (1970). *J. Chem. Phys.*, **53,** 4053
95. Garton, W. R. S. (1962). *J. Quant. Spectrosc. Rad. Transfer*, **2,** 335
96. Huffman, R. E., Tanaka, Y. and Larrabee, J. C. (1963). *J. Chem. Phys.*, **39,** 902
97. Nicholson, A. J. C. (1963). *J. Chem. Phys.*, **39,** 954
98. McGowan, J. W., Fineman, M. A., Clarke, E. M. and Hanson, H. P. (1968). *Phys. Rev.*, **167,** 52
99. Berry, R. S. (1966). *J. Chem. Phys.*, **45,** 1228
100. Lifshitz, C. (1969). *Israel J. Chem.*, **7,** 261
101. Murphy, C. B., Jr. and Enrione, R. E. (1970). *Int. J. Mass Spectrom. Ion Phys.*, **5,** 157
102. Morrison, J. D. (1964). *J. Chem. Phys.*, **40,** 2488
103. Henis, J. M. S. and Marie, C. A. (1970). *J. Chem. Phys.*, **53,** 2999
104. Brehm, B., Fuchs, V. and Kebarle, P. (1971). *Int. J. Mass Spectrom. Ion Phys.*, **6,** 279
105. Villarejo, D. (1968). *J. Chem. Phys.*, **49,** 2523
106. McCulloh, K. E. (1968). *J. Chem. Phys.*, **48,** 2090
107. McCulloh, K. E. and Rosenstock, H. M. (1968). *J. Chem. Phys.*, **48,** 2084
108. Rosenstock, H. M. (1971). *Int. J. Mass Spectrom. Ion Phys.*, **5,** 33
109. Botter, R. and Rosenstock, H. M. (1969). *J. Res. NBS,* **73A,** 313
110. Rosenstock, H. M. and Botter, R. (1970). *Recent Developments in Mass Spectroscopy, Proceedings of Int. Conf. on Mass Spectrometry,* Kyoto, Japan, 1969, 797. (Baltimore: University Park Press)
111. Koopmans, T. (1933). *Physica,* **1,** 104
111a. Newton, J. M. D. (1968). *J. Chem. Phys.*, **48,** 2825
112. Smyth, H. D. (1922). *Proc. Roy. Soc. (London)*, **102,** 283
113. Smith, P. T. (1930). *Phys. Rev.*, **36,** 1293
114. Warren, J. W. (1950). *Nature (London)*, **165,** 810
115. Warren, J. W. and McDowell, C. A. (1951). *Discuss. Faraday Soc.*, **10,** 53
116. Honig, R. E. (1948). *J. Chem. Phys.*, **16,** 105
117. Lossing, F. P., Tickner, A. W. and Bryce, W. A. (1951). *J. Chem. Phys.*, **19,** 1254
118. Kiser, R. W. and Gallegos, E. J. (1962). *J. Phys. Chem.*, **66,** 947
119. Morrison, J. D. and Nicholson, A. J. C. (1952). *J. Chem. Phys.*, **20,** 102

119a. Barfield, A. F. and Wahrhaftig, A. L. (1964). *J. Chem. Phys.*, **41,** 2947
120. Dorman, F. H. and Morrison, J. D. (1961). *J. Chem. Phys.*, **34,** 578
121. Dorman, F. H. (1964). *J. Chem. Phys.*, **41,** 2857
122. Dorman, F. H. (1965). *J. Chem. Phys.*, **42,** 65
123. Dorman, F. H. (1965). *J. Chem. Phys.*, **43,** 3507
124. Dorman, F. H. (1966). *J. Chem. Phys.*, **44,** 35
125. Vedeneyev, V. I., Gurvich, L. V., Kondratyev, N. V., Medvedev, V. A. and Frankevich, Ye. L. (1966). *Bond Energies, Ionisation Potentials and Electron Affinities.* (London: Arnold)
126. Fox, R. E. and Langer, A. (1950). *J. Chem. Phys.*, **18,** 460
127. Reichert, C., Fraas, R. E. and Kiser, R. W. (1970). *Int. J. Mass Spectrom. Ion Phys.*, **5,** 457
128. Newton, A. S. and Sciamanna, A. F. (1970). *J. Chem. Phys.*, **53,** 132
129. Tsuchiya, M., Preston, F. J., Eguchi, M. and Svec, H. J. (1970). *Recent Developments in Mass Spectroscopy, Proceedings of Int. Conf. on Mass Spectrometry,* Kyoto, Japan, 1969, 837. (Baltimore: University Park Press)
130. Stanton, H. E. and Monahan, J. E. (1964). *J. Chem. Phys.*, **41,** 3694
131. Rowland, C. G., Eland, J. H. D. and Danby, C. J. (1969). *Int. J. Mass Spectrom. Ion Phys.*, **2,** 457
132. Smith, K. C. and Shannon, T. W. (1969). *J. Chem. Phys.*, **51,** 4633
133. Rowland, C. G. (1971). *Int. J. Mass Spectrom. Ion Phys.*, **6,** 155
134. Rowland, C. G. (1971). *Int. J. Mass Spectrom. Ion Phys.*, **7,** 79
135. Beynon, J. H., Hopkinson, J. A. and Lester, G. R. (1969). *Int. J. Mass Spectrom. Ion Phys.*, **2,** 291
136. Haney, M. A. and Franklin, J. L. (1968). *J. Chem. Phys.*, **48,** 4093
137. Haney, M. A. and Franklin, J. L. (1969). *J. Chem. Phys.*, **50,** 2028
138. Moore, C. E. (1949). U.S. N.B.S. Circular 467
139. Field, F. H. and Franklin, J. L. (1957). *Electron Impact Phenomena,* Chapt. 4. (New York: Academic Press)
140. Melton, C. E. and Hamill, W. H. (1964). *J. Chem. Phys.*, **41,** 3464
141. Williams, J. M. and Hamill, W. H. (1968). *J. Chem. Phys.*, **49,** 4467
142. Chupka, W. A. and Lifshitz, C. (1968). *J. Chem. Phys.*, **48,** 1109
143. Lossing, F. P. and Semeluk, G. P. (1970). *Can. J. Chem.*, **48,** 955
144. McAllister, T. and Lossing, F. P. (1969). *J. Phys. Chem.*, **73,** 2996
145. Lossing, F. P. (1971). *Can. J. Chem.*, **49,** 357
146. Lossing, F. P., Maeda, K. and Semeluk, G. P. (1970). *Recent Developments in Mass Spectroscopy, Proceedings of Int. Conf. on Mass Spectrometry,* Kyoto, Japan, 1969, 791. (Baltimore: University Park Press)
147. Elder, F. A., Giese, C., Steiner, B. and Inghram, M. G. (1962). *J. Chem. Phys.*, **36,** 3292
148. Beck, D. and Osberghaus, O. (1960). *Z. Physik.*, **160,** 406
149. Beck, D. and Niehaus, A. (1962). *J. Chem. Phys.*, **37,** 2705
150. Beck, D. (1964). *Discuss. Faraday Soc.*, **36,** 56
151. Genzel, H. and Osberghaus, O. (1967). *Z. Naturforsch.*, **22a,** 331
152. Niehaus, A. (1967). *Z. Naturforsch.*, **22a,** 690
153. Lampe, F. W. and Niehaus, A. (1968). *J. Chem. Phys.*, **49,** 2949
154. MacNeil, K. A. G. and Thynne, J. C. J. (1970). *Int. J. Mass Spectrom. Ion Phys.*, **3,** 455
155. MacNeil, K. A. G. and Thynne, J. C. J. (1969). *Int. J. Mass Spectrom. Ion Phys.*, **2,** 1
156. MacNeil, K. A. G. and Thynne, J. C. J. (1968). *Trans. Faraday Soc.*, **64,** 2112
157. Thynne, J. C. J. and MacNeil, K. A. G. (1970). *Int. J. Mass Spectrom. Ion Phys.*, **5,** 329
158. Thynne, J. C. J. and MacNeil, K. A. G. (1970). *Int. J. Mass Spectrom. Ion Phys.*, **5,** 95
159. Thynne, J. C. J. (1969). *J. Phys. Chem.*, **73,** 1586
160. Harland, P. and Thynne, J. C. J. (1969). *J. Phys. Chem.*, **73,** 4031
161. Locht, R. and Momigny, J. (1969). *Int. J. Mass Spectrom. Ion Phys.*, **2,** 425
162. Locht, R. and Momigny, J. (1970). *Int. J. Mass Spectrom. Ion Phys.*, **4,** 379
163. Van Brunt, R. J. and Kieffer, L. J. (1970). *Phys. Rev.*, **A2,** 1899
164. Rallis, D. A. and Goodings, J. M. (1971). *Can. J. Chem.*, **49,** 1571
165. Begun, G. M. and Compton, R. N. (1969). *J. Chem. Phys.*, **51,** 2367
166. Stockdale, J. A. D., Compton, R. N., Hurst, G. S. and Reinhardt, P. W. (1969). *J. Chem. Phys.*, **50,** 2176
167. Sharp, T. E. and Dowell, J. T. (1969). *J. Chem. Phys.*, **50,** 3024

168. Collin, J. E. and Locht, R. (1970). *Int. J. Mass Spectrom. Ion Phys.*, **3,** 465
169. Chantry, P. J. (1969). *J. Chem. Phys.*, **51,** 3384
170. Naff, W. T. and Cooper, C. D. (1968). *J. Chem. Phys.*, **49,** 2784
171. Lifshitz, C. and Grajower, R. (1970). *Int. J. Mass Spectrom. Ion Phys.*, **4,** 92
172. Lifshitz, C., Peers, A. M., Grajower, R. and Weiss, M. (1970). *J. Chem. Phys.*, **53,** 4605
173. Dougherty, R. C. (1969). *J. Chem. Phys.*, **50,** 1896
174. Hastie, J. W., Bloom, H. and Morrison, J. D. (1967). *J. Chem. Phys.*, **47,** 1580
175. Hildebrand, D. L. (1970). *Int. J. Mass Spectrom. Ion Phys.*, **4,** 75
176. Uy, O. M., Muenow, D. W., Ficalora, P. J. and Margrave, J. L. (1968). *Trans. Faraday Soc.*, **64,** 2998
177. Flesch, G. D., White, R. M. and Svec, H. J. (1969). *Int. J. Mass Spectrom. Ion Phys.*, **3,** 339
178. Foffani, A., Pignataro, S., Cantone, B. and Grasso, F. (1965). *Z. Physik. (Frankfurt)*, **45,** 79
179. Winters, R. E. and Kiser, R. W. (1965). *Inorg. Chem.*, **4,** 137
180. Bidinosti, D. R. and McIntyre, N. S. (1967). *Can. J. Chem.*, **45,** 641
181. Bloom, H. and Hastie, J. (1968). *J. Chem. Phys.*, **49,** 2230
182. Ganguli, P. S. and McGee, H. A., Jr. (1969). *J. Chem. Phys.*, **50,** 4658
183. Pignataro, S., Distefano, G., Nencini, G. and Foffani, A. (1970). *Int. J. Mass Spectrom. Ion Phys.*, **3,** 479
184. Fallon, P. J., Kelly, P. and Lockhart, J. C. (1968). *Int. J. Mass Spectrom. Ion Phys.*, **1,** 133
185. Bond, S. J., Davidson, I. M. T. and Lambert, C. A. (1968). *J. Chem. Soc. A*, 2068
186. Borossay, J., Csakvari, B. and Szepes, L. (1971). *Int. J. Mass Spectrom. Ion Phys.*, **7,** 47
187. Cabaud, B., Uzan, R. and Nounou, P. (1971). *Int. J. Mass Spectrom. Ion Phys.*, **6,** 89
188. Collin, J. E. and Delwiche, J. (1967). *Can. J. Chem.*, **45,** 1883
189. Collin, J. E., Winter, R. E. and Engerholm, G. E. (1968). *J. Chem. Phys.*, **49,** 2469
190. Brion, C. E. (1969). *Chem. Phys. Lett.*, **3,** 9
191. Lossing, F. P. (1970). *Int. J. Mass Spectrom. Ion Phys.*, **5,** 190
192. Lossing, F. P. and Semeluk, G. P. (1969). *Int. J. Mass Spectrom. Ion Phys.*, **2,** 408
193. Collin, J. E. and Caprace, G. C. (1968). *Int. J. Mass Spectrom. Ion Phys.*, **1,** 213
194. Lewis, D. and Hamill, N. H. (1970). *J. Chem. Phys.*, **52,** 6348
195. Svec, J. H. and Flesch, G. D. (1971). *J. Chem. Phys.*, **54,** 2681
196. Junk, G. A. and Svec, H. J. (1970). *J. Chem. Soc. A*, 2102
197. Svec, H. J. and Junk, G. A. (1967). *J. Amer. Chem. Soc.*, **89,** 2836
198. Chambers, D. B., Coates, G. E. and Glockling, F. (1970). *J. Chem. Soc. A*, 741
199. Carrick, A. and Glockling, F. (1966). *J. Chem. Soc. A*, 623
200. Glockling, F. and Light, J. R. C. (1967). *J. Chem. Soc. A*, 623
201. Glockling, F. and Light, J. R. C. (1968). *J. Chem. Soc. A*, 717
202. Chambers, D. B. and Glockling, F. (1968). *J. Chem. Soc. A*, 735
203. Chambers, D. B., Glockling, F. and Weston, M. (1967). *J. Chem. Soc. A*, 1759
204. Chambers, D. B., Coates, G. E., Glockling, F. and Weston, M. (1969). *J. Chem. Soc. A*, 1712
205. Franklin, J. L. and Carroll, S. R. (1969). *J. Amer. Chem. Soc.*, **91,** 6564
206. Guyon, P. M. (1969). *J. Chim. Phys.*, **66,** 467
207. Ficquet-Fayard, F. and Guyon, P. M. (1966). *Mol. Phys.*, **11,** 17
208. Guyon, P. M. and Ficquet-Fayard, F. (1969). *J. Chim. Phys.*, **66,** 32
209. Blais, J. C., Cottin, M. and Gitton, B. (1970). *J. Chim. Phys.*, **66,** 1475

3
Recent Advances in Electron Spectroscopy

C. E. BRION
University of British Columbia

3.1 INTRODUCTION

Quantum states of atoms and molecules have traditionally been studied by the various methods of optical spectroscopy in which either the absorption or emission of electromagnetic radiation is monitored following initial excitation of the target species. However, in the last decade a new series of spectroscopic techniques have become available, based on the principle of accurate measurement of the energies of electrons ejected following a variety of inelastic collision processes. Electron spectroscopy may be used to study a wide variety of excitation and ionisation processes many of which are not open to investigation by conventional optical techniques either due to interfering effects or because some channels are optically disallowed[1,2].

In general the following types of process have given rise to various branches of electron spectroscopy where the velocity of the underlined electron ε is studied.

(1) $X+h\nu \rightarrow X^{+}+\underline{\varepsilon}$ (Photoelectron Spectroscopy (PES))

(2) $X+\varepsilon \rightarrow X^{*}+\underline{\varepsilon}$ (Electron-Impact Spectroscopy (EIS))

(3) $X \xrightarrow{\text{initial excitation}} X^{*} \rightarrow X^{+}+\underline{\varepsilon}$ (Autoionisation Electron Spectroscopy (AIES)

$\quad X \rightarrow X^{+*} \rightarrow X^{2+}+\underline{\varepsilon}$ (Auger Electron Spectroscopy (AES))

(4) $X+Y^{*} \rightarrow X^{+}+Y+\underline{\varepsilon}$ (Penning Ionisation Electron Spectroscopy (PIES))

Electron spectroscopy provides a great body of information relevant to studies in mass spectrometry. In particular very accurate first and inner ionisation potentials can be measured by photoelectron spectroscopy. These values, together with the excited bound and ionic states found from electron impact and Auger (and autoionisation) electron spectroscopy, are of fundamental significance in the elucidation of the complex processes occurring in the mass spectrometer ion source and this type of basic information is of great help in the understanding of fragmentation pathways. In addition the techniques of electron spectroscopy are now sufficiently developed so as to have important applications to chemical analysis. It is the purpose of this review to acquaint the reader with brief principles, potentialities and basic experimental considerations for these techniques as well as to review some of the more important developments in the literature up to the early part of 1971. In addition attention is drawn to new and developing techniques in electron spectroscopy.

To cover all the existing literature on electron spectroscopy would be an almost insuperable task in addition to which it would be an unnecessary duplication of the many fine reviews written in recent years on various aspects of this field.

The most recent comprehensive review is that published by Berry[3] in early 1969. This excellent publication covers most of the significant literature up to late 1968 for PES, EIS and PIES. Since this time a large volume of work has been carried out by PES and comprehensive PES reviews by Turner *et al.*[4], Brundle and Robin[5] and by Worley[6] have recently been published dealing mainly with the ionisation of valence shell electrons. Excellent accounts of pioneering studies of inner shells by x-ray PES (ESCA) have been given in the two volumes[7,8] by Siegbahn and his co-workers as well as more recent reviews by others [9,10,11]. Recent reviews of various aspects of EIS have been made by Trajmar, Rice and Kuppermann[12] and also by Taylor[13]. Kuyatt[14] has also discussed various aspects of EIS. More recent advances in PES and EIS will be discussed later in this review.

Auger electron spectroscopy has received much attention as a technique for studying surfaces but no attempt will be made to discuss this work since it will presumably be covered by the section on Surface Physics. However, recently gas-phase Auger and Autoionisation Studies (AES and AIES), having more relevance to chemistry, have been made and will be discussed in some detail.

The situation for PIES is that at the time of Berry's article only the pioneering studies by Cermak[15] had been carried out apart from a few brief and isolated papers. Since this time a significant number of high resolution PIES studies have been reported and a comprehensive review of PIES will be given in this publication.

3.2 GENERAL EXPERIMENTAL TECHNIQUES

3.2.1 Energy analysis of electrons

Common to all branches of electron spectroscopy is the need to energy analyse electrons of varying energies. Despite early elegant experiments such as those by Rudberg *et al.*[16], little electron spectroscopy was carried out until the late 1950s. Improvements in technology as well as an understanding of low energy electron optics have been responsible for the growth in sophistication of instrumentation. Charged particles may be momentum analysed in magnetic fields and one of the first significant experiments with a magnetic analyser was that by Nottingham[17] who studied the ionisation of mercury. Magnetic analysers have been used quite extensively for β-ray spectrometers[7,8] but in general their application for high resolution, low energy studies is complicated by the problems of fringe and stray magnetic fields affecting neighbouring parts of the experiment. This is particularly undesirable, for instance, in high resolution EIS where double analysers and more complex electron optics are usually employed. Turner and May[18] have described a high resolution magnetic photoelectron spectrometer.

Alternatively, electrostatic fields are easily produced and shielded from fringe effects. Consequently, electrostatic fields have found much wider application for electron spectroscopy. Two general types are in common use: (a) retarding analysers and (b) deflection analysers. Many retarding analysers are of the Lozier type[19] having cylindrical geometry. Retarding analysers only measure the normal component of electron velocity and hence usually exhibit a limited energy resolving power ($\sim$0.1 eV). Al-Joboury and Turner[20] used a cylindrical retarding analyser for much of their early exploratory work in PES. Vroom, Frost and McDowell[21], and also Samson[22], have described higher resolution analysers using spherical grids such that at all times electrons from a point source are normal to the retarding field. Spohr and Puttkamer[23] have described a very high resolution retarding analyser employing an Einzel lens and this device has recently been used very effectively for PIES by Hotop and Niehaus[24]. The integral stopping curves produced by retarding analysis are particularly useful for measuring relative transition probabilities[21], but are less satisfactory for accurate measurement of electron energies unless electronic curve differentiation is employed. The RPD technique[25] has found application for the production of energy-selected electron beams for EIS. Very recently[26] the modulated RPD method, in conjunction with the trapped electron technique, has been used to study electron impact spectra with a much better resolution than that obtained by previous workers. Golden and Zecca[27] have applied electron optical techniques to the RPD method and report resolutions as low as 0.01 eV. A fundamental study of the RPD gun has been made by Gordon *et al.*[28]. An automated RPD technique using a multichannel scaler is described by Chantry[29], and has recently been used by Thomas and Vogelsberg[30] to record excitation cross-section curves in a pulsed molecular beam experiment.

Analysers of the electrical deflection type have found a more widespread use in electron spectroscopy. The resulting differential spectrum is generally more amenable to spectroscopic interpretation, and where relative transition probabilities are required integration of the differential signal is readily accomplished. Several types of deflection analyser are in use and their relative merits have been investigated by Sar-el[31]. The cylindrical mirror analyser is shown to have the highest figure of merit[31, 32] but there are difficulties in designing compatible electron lens systems[33] suitable for electron impact spectrometers. An attractive feature of this analyser is its potential capability of collecting electrons at all polar angles from a point source. The cylindrical mirror analyser has been used by Mehlhorn *et al.*[34] for Auger electron studies and is the basis of several newly available commercial electron spectrometers. Magnetic field neutralisation over a large volume is particularly important for this type of analyser especially at low kinetic energies and high resolving power.

The concentric hemisphere analyser, first described by Purcell[35], has the advantage of double focusing and the theory and development of the spectrometer and its associated electron optics has received capable attention from Simpson and his co-workers[36, 37]. This device is now the basis of a sophisticated commercially available electron impact spectrometer[441, 251, 252]. Recently several important papers have been published by Read *et al.*[38–41] on aperture lenses and also for tube lenses by Heddle *et al.*[42–44], dealing with two- and three-element low-energy electron lenses. Three-element lenses, which act as simple zoom lenses, are of particular importance in minimising chromatic aberration in electron analysis. With a growing tendency towards absolute measurements these considerations are of particular significance. Concentric hemisphere analysers of advanced design have also been used by Lassettre and his co-workers[45] for an extensive programme of EIS.

An alternative form of electrostatic device is the radial 127-degree electrostatic velocity selector[46–48] latterly improved by Marmet and Kerwin[49]. Turner *et al.*[50] have used such a device for an extensive study of molecular photoelectron spectroscopy[4] where the lower electron flux does not require the use of grids[49]. Improvements to a 127-degree selector have also been discussed by Salop *et al.*[51] and Brion and Tam[52] have recently built and tested a gridless 127-degree analyser using electron lenses and virtual slits[36, 37]. It should be noted that many workers operate 127-degree selectors by scanning the voltage across the radial plates. This changes the field strength for different energy electrons and produces a resolving power which is a function of electron energy. A further disadvantage is that the electron energy must be calculated from the geometric dimensions of the analyser and this may explain the scatter of values[53] observed in PES. A more satisfactory procedure is to retard (or accelerate) all electrons to the same energy prior to their entry into the analyser which is then operated at fixed energy and resolution. For accurate intensities over a wide energy range zoom lenses[38–45] must be used.

The parallel plate electrostatic analyser[54] has been used by Eland and Danby who achieved a resolution of 0.025 eV in PES[55]. Other electrostatic analysers have been described by Siegbahn *et al.*[7, 8].

A study has been made of optimum deflection angles for 127-degree and hemispherical electrostatic spectrometers[56]. Gough[57] has described an annular curved-plate analyser with large geometric factor and high transmission.

Combined electric and magnetic fields are also suitable for electron energy analysis. Two types are in use—first the Wien filter[58] where resolutions better than 0.01 eV have been achieved. A high resolution, high transmission, double Wien filter has been described by Anderson[59] while van der Wiel[60] has used a Wien filter system as the electron energy analyser in an electron–ion coincidence experiment. A new type of crossed-field analyser, the trochoidal electron monochromator, has been developed by Stamatovic and Schulz[61, 62] who report a half-width of 0.020 V at a beam current of 10^{-9} A. With further development this simple device should find wide application, especially where an axial magnetic field in the source region is desirable.

For operation at high resolving power it is necessary to shield electron spectrometers from stray magnetic fields including that due to the Earth. This can be done by the use of field correcting coils and/or high permeability shields. Ross and Garment have reported an arrangement[63] for the servomechanical control of Helmholtz coils. This is of particular use for environments in which fluctuating magnetic fields occur.

3.2.2 Limits to resolving power

The ultimate resolution attainable in electron spectroscopy depends on several factors including analyser design and performance, shielding of stray electromagnetic fields (both varying and static), line broadening in the light source for PES and the thermal motion of the target molecules. Samson[64] has an excellent article dealing with the questions of line broadening and thermal motion and produces useful charts of doppler widths as a function of the electron energy and target mass. Siegbahn[7, 8] has reported a resolution of 0.013 V for the PES study of xenon and observes partially resolved rotational structure for H_2^+. For electron impact spectroscopy, where two analysers are used, both analysers contribute to the observed width of the electron peaks. Boersch *et al.*[65] have reported resolving powers of 0.007 V in studies at an incident energy of 30 keV and have also observed rotational levels in electronically excited *para*-hydrogen[66]. Rotational structure in the energy loss spectrum of H_2 has also been observed at lower energies[441, 67]. Simpson[36] has reported a half-width of 0.005 V using a double 180-degree hemispherical analyser system. Various efforts are being made to obtain improved resolution by such means as producing highly monochromatic electrons by photoionisation techniques. The doppler spread, which is large only in small molecules, can be minimised by the use of molecular beams of target particles[67]. The practical limit of resolution is probably of the order of one to two millivolts (8–16 cm^{-1}) which is sufficient to obtain some rotational information in small molecules. At long wavelengths electron spectroscopy cannot compete in terms of resolution with optical techniques. However proceeding to shorter wavelengths the situation becomes increasingly favourable for electrons since optical resolution is normally on a wavelength basis whereas with suitable electron-retarding devices electron spectra may be obtained at a constant energy resolution independent of the electron energy. Table 3.1 shows the 'value' of 0.01 V in angstrom units at a series of electron energies. It is obvious that as we proceed into the vacuum u.v. region of the electromagnetic spectrum, and beyond, that a distinct advantage exists in the use of electron spectroscopy.

Table 3.1

Electron energy/eV	*Equivalent wavelength*/Å	0.01 eV *in* Å
1	12398.0	110.0
10	1239.8	1.2
20	619.9	0.3
30	413.3	0.14
40	310.0	0.08
50	248.0	0.05
100	124.0	0.013

Ripple problems are frequently encountered when most commercial power supplies are operated in a floating mode above ground potential. Although capacitive loading minimises this effect it also unfortunately produces severe damping effects when voltages are scanned. A.C. ripple and pick-up seriously degenerates resolving power. The use of batteries is a simple and inexpensive solution particularly with the availability of long life mercury and alkaline cells as well as rechargeable batteries.

3.2.3 Signal detection

Frequently a practical limit to performance in electron spectroscopy is the intensity of the signal although a growing use of signal averaging techniques and multi-channel scaling makes many new measurements possible. Various types of electron multiplier have been used[14] mainly in the pulse counting mode. The continuous channel capillary electron multiplier[68, 69] has found wide application due to its simplicity and stable surface properties with regard to air exposure. However with the rapid growth of electron spectroscopy for chemical applications multipliers are needed which are resistant to chemical attack particularly by many inorganic compounds. Spectrometers for such applications are best designed with very efficient differential pumping between source and detector.

In many aspects of electron spectroscopy problems exist due to background function and impurity gases. In such cases it would be advantageous to use modulated sample beams and phase sensitive detection for electron currents. Conventional analogue 'lock-in' techniques will probably not be sensitive enough for many of these applications but recent developments in digital 'lock-in' methods[446] with counting systems should prove to have a wide application.

3.3 PHOTOELECTRON SPECTROSCOPY (PES)

3.3.1 Introduction

In photoelectron spectroscopy the kinetic energy of the ejected electron is determined for the process.

$$X + h\nu \rightarrow X^+ + \varepsilon$$

Due to the large disparity in mass between the electron and the product ion the electron can be considered for all normal purposes to possess an energy given by.

$$E_\varepsilon = h\nu - I(X)$$

where E_ε = energy of electron
$h\nu$ = energy of incident photon
$I(X)$ = ionisation energy of the target X.

In general, for molecules, $I(X)$ includes not only electronic energy but also vibrational and rotational contributions:

$$I(X) = E_{el} + E_{vib} + E_{rot}$$

It can be seen therefore that photoelectron spectroscopy is a method of determining ionisation potentials and therefore, to a first approximation, orbital energies if the independent electron approximation is assumed. A particularly significant feature is that PES is not a threshold technique in that a fixed light quantum (i.e. a single wavelength) is normally used and hence excitation to super-excited neutral states cannot generally occur (except at the exciting wavelength[5, 70]). Consequently, in contrast to techniques of optical spectroscopy and both electron and photon impact mass spectrometry[71], PES is not complicated by autoionisation processes. PES is a detector of direct photoionisation and as a result has proved to be the best and most prolific source of both inner and outer ionisation potentials.

Historically PES has become divided into two general classifications—

those dealing with outer valence shells called Molecular Photoelectron Spectroscopy and studies of inner shells using x-ray photons (ESCA). Each will be discussed in turn. Berry[3] has given an extensive review of the field up to the end of 1968.

PES may now be said to have 'come of age' with a large and varied body of work published and with commercial instruments now available. The method is beginning to find application as an analytical tool in organic and inorganic chemistry as well as for continuing basic studies of molecular structure.

3.3.2 Molecular photoelectron spectroscopy

This technique utilises photons in the vacuum ultraviolet to photoionise valence shell electrons and the majority of published work has been performed using the He resonance line at 584 Å (21.21 eV). This intense radiation may be produced in a D.C. or microwave discharge in low-pressure helium using a windowless system and differential pumping. The radiation is usually used in an undispersed fashion (i.e. without a vacuum monochromator) in which case care must be taken to avoid impurities giving rise to spurious wavelengths[72, 73]. Helium 584 Å sources contain a minor contribution from 537 Å [74] and, not infrequently, small amounts of impurity hydrogen giving Lyα(1216 Å = 10.2 eV) radiation. Resonance lines from the other rare gases have also been used for PES, as well as hydrogen Lymanα. The other rare gas lines are generally unsatisfactory as undispersed sources since they contain several different wavelengths which complicate the photoelectron spectrum. Several groups have used vacuum monochromators for wavelength selection to avoid these spurious affects[75, 76]. Until recently most PES work has been with 584 Å helium light which restricts the measurement of orbital energies to those less than 21.21 eV. However, if the 2p excited state of the helium ion can be populated with sufficient intensity, significant amounts of He II resonance radiation can be produced at 304 Å (40.8 eV) [77–79].

Following the first isolated experiments in the Soviet Union[80], the technique was highly developed and exploited by Turner and his co-workers. A full account of this and other related work up to late 1969 has been recently published[4] along with much relevant information on the technique. A comprehensive documentation is given of the photoelectron spectra of a wide range of organic and inorganic molecules. Berry[3] has given a detailed discussion and critique of molecular photoelectron spectroscopy illustrating its potentialities, achievements and shortcomings and most of the significant publications up to late 1968 are discussed. Subsequently Brundle and Robin[5] have published a fairly comprehensive report covering the period up to early 1970. Following a brief introduction tracing the historical development of PES and the instrumentation available Brundle and Robin give a useful discussion of the relative merits of different methods for determining ionisation potentials. A consideration of the relationship between PES and the electronic structure of molecules is followed by a useful section describing complicating features occurring in photoelectron spectra due to inter-electron effects. These include exchange, multiplet, Jahn–Teller and spin–orbit splitting. The remainder of the review is given to a discussion of recent developments of PES with special reference to organic chemistry. A short discussion is given of the possible uses of photoelectron intensities but no mention is made of instrumental effects such as collection efficiency or chromatic aberration. If relative intensities and absolute measurements of individual differential photoionisation cross-sections are to be made with precision, designers of spectrometers will have to give more serious consideration to the use of improved electron optics and zoom lenses[38–45] (see earlier section on instrumentation). Brundle and Robin finally discuss the correlation of spectra in related series of molecules. In their conclusions they draw attention to the significant relation between the experimental findings of PES and the results of molecular structure calculations.

Another excellent recent review by Worley[6] of polyatomic molecules discusses important developments in wide areas of organic and inorganic chemistry, with special emphasis on the interpretation of PES using quantum molecular orbital procedures. Following a discussion of the theoretical methods available for calculating electron orbital energies, an excellent account is given of the PES of the following classes of molecules and their related compounds – alkanes, alkenes, alkynes, aromatic and other cyclic hydrocarbons, azabenzenes and naphthalenes, metal carbonyl compounds, furans, pyroles and thiophenes, amines, hydrazines, alcohols, aldehydes, ketones, acids, amides, ethers and oxides, compounds containing boron, phosphorus, arsenic, sulphur, silicon and mercury. Dewar and Worley have also published a useful tabulation of the ionisation potentials of 67 organic compounds[81].

The reader is referred to these other publications[3–6] which cover most of the important work until early 1970. The present reviewer will seek only to comment on new work and earlier papers not discussed in these other publications.

Useful discussions of various aspects of PES have recently appeared[82, 447] and Baker[83] has published a short general review of PES illustrated by the spectra of some simple molecules. A considerable number of papers have recently appeared dealing with the various physical chemical aspects of the photoelectron spectra of small molecules. Lindholm[84] has discussed the analysis of Rydberg series of small molecules and finds that it can be simplified using recent results of PES and electron impact energy loss spectra. It is shown that the quantum defect depends mainly on the nature of the Rydberg orbital and is approximately independent of the nature of the molecule and of the initial orbital of the electron. These findings are applied to CO [85], N_2 [86], O_2 [87], CO_2 [88], N_2O [89] and NO [90].

Cornford *et al.*[91] have re-examined the spectra of H_2^+, HD^+ and D_2^+ at 584 Å under high resolution using a 180-degree hemispherical analyser[73]. The vibrational intervals in the $^2\Sigma_g^+$ states have been measured leading to the derivation of the vibrational constants which compare favourably with theoretical predictions[92].

Asbrink has made a study of H_2 at high resolution[93]. Some individual rotational lines are resolved and accuracy of ionisation energies comparable with the best spectroscopic measurements is claimed. Rotational and vibrational constants are in good agreement with theoretical values. The published spectra show many peaks ascribed to impurities and this work well illustrates the increasing need for great care and caution in the running and interpretation of photoelectron spectra at high resolution. It would seem highly desirable to monitor simultaneously gas samples by mass spectrometry to identify possible impurities. It should be stressed that a low percentage impurity will nevertheless make a significant contribution if it has a high photoionisation cross-section.

High resolution studies of HF and DF [94] have led to values for the dissociation limits. The data are compared with previous results and suggest that recent mass spectrometric values of D_0^0 (HF) and D_0^0 (F_2) may be in error.

The molecules N_2, CO and O_2 have been further studied at low resolution by Collin and Natalis[95] at 584 Å and also with 736–744 Å neon radiation. At the longer wavelengths spurious vibrational excitation occurs[70] presumably due to the overlapping of autoionising levels. Lindholm and his co-workers[96] have also examined N_2, CO and O_2. In particular the previously undetected $A^2\Pi_u$ state of O_2^+ has been partially resolved. A high resolution spectrum at 304 Å [97] not only convincingly demonstrates the $A^2\Pi_u$ state (with a new value of the I.P. = 17.045 eV) but also shows the $C^4\Sigma_u^-$ and $^2\Pi_u$ states between 24 and 25 eV. Relative intensities for the $^2\Pi_u$ states are in good agreement with the predictions of Dixon and Hull[98]. The spin–orbit splitting in the $X^2\Pi_g$ ground state of O_2^+ is also clearly shown. Bahr *et al.*[99] have studied the photoionisation of CO_2 by photoelectron spectroscopy throughout the wavelength range 584–720 Å. Partial cross-sections have been obtained for the production of each of the accessible states of CO_2^+ as a function of wavelength. Two studies[100, 101] of NH_3 and ND_3 at high resolution are in fair agreement showing a small isotope effect. The first band ($^2A_1$) in the spectrum of each compound shows a well-resolved vibrational progression which is assigned to the totally symmetric out of plane bending mode, v_2. The second band ^2E is not clearly resolved. Weiss *et al.* have also studied HCl and DCl [102] using 584 Å and 736–744 Å radiation. Ionisation potentials and transition probabilities for the $X^2\Pi$ and $A^2\Sigma^+$ states were in good agreement with spectroscopic values. Branton *et al.*[74] have reported five ionisation potentials in ethylene and ethylene-d_4. Vibrational structure in the first and fourth bands was resolved leading to the assignment of vibrational modes. These authors also report the photoelectron spectrum of neon due to the 537 Å minor component (2 %) in the helium lamp. The neon spin–orbit doublet is clearly resolved (0.096 eV). Nitrous oxide (N_2O) has been studied by Natalis and Collin[103]. Six bands are observed including two (at 14.3 and 19 eV) not previously reported. However the existence of structure at these energies in the reported curves is highly speculative and any definite interpretation must await further study. The high resolution spectrum reported by Turner[4] does not show significant structure at these energies.

The bent molecule, nitrogen dioxide (NO_2), has been the subject of much recent study. The ground state of NO_2^+ is probably linear like the isoelectronic species CO_2. Due to the change in inter-nuclear distance the vertical ionisation potential is expected to involve a high vibrational level resulting in little intensity to the zeroth level. Consequently there has been a wide range of values quoted for the first I.P. of NO_2 – due in part to differing instrumental sensitivities. A careful high resolution study by Brundle[104] gives a value of 10.0 eV for the adiabatic I.P. and a frequency of 650 cm^{-1} for the bending mode v_2. An earlier lower value of 8.8 eV [105] is discredited by Brundle who suggests it is due to impurity lines in the argon resonance lamp. Using 304 Å radiation the high resolution photoelectron spectrum of NO_2 has been independently studied by Brundle *et al.*[106] and by Edquist *et al.*[107]. Adiabatic ionisation potentials are reported by Brundle *et al.* at $\geqslant$9.75, 12.85, 13.6, 16.07, 14.37, 16.99, 18.86, ~20.8 and 21.26 eV and by Edquist *et al.* at ~10.3, 12.86, 13.6, 14.07, 14.45, 17.07, 17.265, ~18, 18.86, 20.7, 21.26 and 23.2 eV. Brundle[104] also shows the first band of nitric oxide (NO) with an adiabatic I.P. of 9.26 eV confirming the conclusions of Turner in view of an earlier controversy concerning this I.P.[4]. Samson[108] has reported higher ionisation potentials of NO at 21.72 and 23.1 eV using a u.v. monochromator together with a spherical retarding analyser. The weak first band and the possibility of NO impurity complicates the interpretation of the first band. A detailed discussion of the spectrum is given in both papers. Two important studies[109,110] of various halomethanes have recently been fully discussed by Worley[6]. Subsequently Brundle *et al.*[111] have published an extensive article on CH_4 and the fluoromethanes as well as deuterated analogues using both 304 Å and 584 Å radiation. The results have been compared with Hartree–Fock calculations with generally good agreement. Correlations are demonstrated between the compositions of the orbitals from which electrons are ejected and the characters of the photoelectron bands. Jahn–Teller effects are evident in CH_4 and CH_3F. Hashmall and Heilbronner[112] have reported ionisation potentials for alkyl bromides. A comparison has been made between PES values and the calculated I.P. values of linear alkanes[113]. Brundle *et al.*[114] have compared the spectra of the isoelectronic molecules C_2H_4 and B_2H_6 with the results of Hartree–Fock calculations. Good agreement is obtained lending support to theories concerning the electronic structure of diborane. Baker *et al.*[115] have discussed the use of electronegativity values in the interpretation of photoelectron spectra with special reference to hydrogen halides and halobenzenes. Excellent correlation is reported between the electronegativity of the halogen atom and I.P. values from orbitals of the same symmetry.

A high resolution study of benzene and pyrazine by Asbrink *et al.*[116] reveals vibrational structure which is used to identify various degenerate and non-degenerate orbitals in benzene. Pyridine has been studied using both PES and photoionisation[117,118]. By using the Ar and Ne resonance lines for PES two I.P. values (10.72 and 11.44 eV) are reported which are not found using the usual 584 Å line. It is suggested that these results together with others on benzene and water[119] may be due to (a) low photoionisation cross-sections at threshold (and consequently smaller at the chosen energy) and (b) anisotropy in the angular distributions of some groups of electrons. The authors point out the consequent need to make PES studies over a wide range of wavelength and with various geometries for the photoelectron analyser.

Other recent studies of organic molecules include amides and carboxylic acids[120], cyclic alkenes[120,121] and alkanes[122], hexafluorobutadiene[123] and bicyclic and exo-cyclic olefins[124].

Most of the early work in PES was concerned with atoms and simple polyatomic molecules. This was followed by studies of a wide variety of organic compounds primarily due to the interest in the comparisons that can be made with the results of molecular orbital calculations and, in particular, π-electron energies. More recently PES has expanded into the field of inorganic chemistry. A study of xenon difluoride (XeF_2)[125] at 584 Å and 304 Å reveals eight vertical I.P. values which compare favourably with calculation. Delwiche *et al.*[126, 127] report adiabatic I.P. values for H_2S and H_2Se as well as vibrational frequencies. Loss of vibrational structure in the second band is interpreted as being due to predissociation leading to S^+ and Se^+ formation. Bassett and Lloyd[128] discuss correlations between the photoelectron spectra of NF_3 and ONF_3 and MO calculations. Other inorganic molecules featured in recent publications include pentacarbonyl manganese derivatives[129], transition metal carbonyl derivatives[130], silane and germane[109], diborane[114], vanadium tetrachloride[131] and phosphonitrilic fluorides[132].

3.3.3 Free radicals and transient species

A new application of photoelectron spectroscopy is the study of transient species and free radicals. The first reported work of this type by Jonathan *et al.*[133] was concerned with atomic species of H, O and N. Spectra were recorded using a 180-degree electrostatic hemispherical analyser and 584 Å radiation, and the atoms were produced in each case by an electrodeless 2450 MHz microwave discharge of the parent gas in a flow tube. The following transitions were observed; hydrogen $H(^2S) \rightarrow H^+(^1S)$, oxygen $O(^3P) \rightarrow O^+(^4S)$, $O(^3P) \rightarrow O^+(^2D)$ and nitrogen $N(^4S) \rightarrow N^+(^3P)$. In addition to the study of the spectroscopy of atoms these studies give a new dimension to vacuum ultraviolet photoelectron spectroscopy as a tool for qualitative and quantitative gas kinetic studies. The method can presumably be extended to other atoms such as the halogens. It is possible that the technique might be useful for the determination of reaction rates in flow systems. The authors also suggest the possibility of studying atom–molecule exchange reactions such as

$$H + X_2 \rightarrow HX + X$$

where in addition the electronic state of the halogen atom ($^2P_{\frac{1}{2}}$, $^2P_{\frac{3}{2}}$) might be determined. Subsequently Jonathan, Smith and Ross[134] have reported the photoelectron spectrum of the transient species O_2 ($^1\Delta_g$) produced by microwave discharge in an oxygen stream. The ionisation potential of O_2 ($^1\Delta_g$) is found to be 11.09 ± 0.005 eV and the mean separation of the vibrational levels is 0.22 eV. The photoelectron spectrum for SO ($^3\Sigma^-$) has also been studied[135].

Cornford *et al.* have recently reported[136] the 584 Å photoelectron spectrum of the radical NF_2. The source of NF_2 was the pyrolysis of tetrafluorohydrazine in an electrically heated furnace adjacent to the ionisation chamber of the photoelectron spectrometer. A very high percentage pyrolysis to NF_2 was achieved. The photoelectron spectrum of NF_2 shows evidence for four ionisation potentials with vertical values of 12.10, 14.60, 16.38 and ~17.6 eV. The adiabatic values of the first two I.P. values are 11.62 and 14.60 eV respectively. The experimental values are interpreted with the aid of LCAO–INDO calculations on NF_2 and NF_2^+.

Further applications of photoelectron spectroscopy to the study free radicals and transient species can be anticipated in the near future.

3.3.4 X-ray photoelectron spectroscopy (ESCA)

Conventional molecular photoelectron spectroscopy is concerned with valence shell orbitals due to the use of ionising radiation in the vacuum ultraviolet. However, with the use of x-rays ($\sim$1000 eV), it is also possible to remove inner-shell electrons and historically such studies have been separately classified as x-ray PES or ESCA (Electron Spectroscopy for Chemical Analysis) studies. A major pioneering contribution to ESCA studies has been made by Siegbahn and his numerous co-workers at the University of Uppsala. Siegbahn *et al.* have written two outstanding books[7,8] describing the various aspects of ESCA studies for both solids and free molecules. In the face of this it would be presumptous to attempt to re-review the field of ESCA studies and the reader is referred to these works[7,8] in which a detailed account of both techniques and experimental results are given. Siegbahn has also published a shorter report[137] demonstrating the scope of ESCA studies for analysis and molecular structure. Several other recent reviews of ESCA studies have been published[5,9,11,138]. This present review will be simply confined to a few general comments on the method and a discussion of some of the most recent literature.

The techniques of ESCA are in most ways similar to those used in molecular photoelectron spectroscopy with the obvious exception that x-rays are used in place of vacuum u.v. radiation. Most commonly the Mg and Al K_α lines at 1253.6 and 1486.6 eV have been used. An inherent limitation to the resolution in ESCA spectra is caused by the natural linewidths of the x-rays ($\sim$1 eV) although efforts are now being made to monochromate x-rays[8]. The interest in ESCA, and the drive to improve resolution, centre in the fact that inner shell binding energies are found to be a sensitive probe of the molecular environment in which an atom finds itself. Atoms of a given species in differing molecular environments exhibit characteristic chemical shifts (so called by loose analogy with n.m.r.). Studies in both gas and solid phases have been made, although gas-phase spectra are more easy to interpret quantitatively. Problems with electrical charging can occur with the use of solid samples in the ionisation region. An additional problem from the standpoint of chemical interest is that with solids the ejected photoelectrons must come from regions close to the surface and thus surface contamination may obscure the bulk chemical properties of the molecular environment. Conversely, of course, the technique has excellent possibilities for surface studies.

The utility of the ESCA method is well illustrated by consideration of the characteristic K shell binding energies of first row elements (in eV), H (14), He (25), Li (55), Be (111), B (188), C (284), N (399), O (532) and F (686). Remembering that the inner shells in molecules retain a very high degree of atomic character and that the environment produces a characteristic 'chemical shift', the attractive features of the ESCA method are obvious. Typical shifts in, for example carbon compounds, are of the order of a few eV (e.g. ~10 eV for C (1s) in CH_4 and CF_4). Despite the natural x-ray line width the positions of ESCA peaks can be determined with good precision but some improvement in resolution would afford clearer interpretation in many cases. A further inherent limiting factor may be the short lifetime of inner-shell ionised states. From Auger studies Siegbahn *et al.*[8] have shown, for example, that the natural width of the C (1s) state in CO is <0.15 eV. An additional complication can occur with Auger electrons resulting from inner-shell vacancies produced by the x-rays (see Section 3.5). However, such Auger electrons can be distinguished by the fact that their energy is independent of the x-ray wavelength.

The observed experimental chemical shifts of ESCA lines have naturally stimulated much theoretical work and various correlations have been made between ESCA shifts and quantities indicative of atomic charge density. For example, there appears to be a close parallel between electronegativity and chemical shift[7,8,11]. Various other semi-empirical and *ab initio* calculations of charge density have also been used in attempts to correlate quantitatively ESCA chemical shifts with known features of molecular structure and chemical bonding. The results are most encouraging and indicate a bright future for x-ray photoelectron spectroscopy. There has been a consequent rapid increase in the growth of the literature and several ESCA machines are now commercially available.

Helmer and Weickert[139] have discussed enhancement of sensitivity in ESCA spectrometers by retarding source electrons and at the same time matching the design of the spectrometer to the retarded source. Fadley, Miner and Hollander have described an efficient magnetic spectrometer[140] with double focusing and a multichannel detector. A computer-controlled ESCA spectrometer system has been described by Fahlman *et al.*[141].

A wide variety of ESCA studies has been featured in recent publications by Siegbahn's group. The spectra of the paramagnetic molecules O_2 and NO [142] show a splitting of core electron levels due to interaction between the emitted core electron and the unpaired valence electrons. No such effect is observed with the nitrogen molecule which is diamagnetic. Binding energies of L, M and N electrons in promethium have been reported[143] showing the ability of the ESCA technique to study radioactive samples. Band structure in a wide selection of transition metals has been observed[144]. The electron spectra of an extensive series of sulphur compounds have been studied[145] and the influence of structure on the electron binding energies is discussed in terms of a calculated atomic charge based on the concepts of electronegativity and partial ionic character of bonds. The results are applied particularly to a discussion of the sulphur–oxygen bond. A series of carbon compounds consisting of saturated aliphatics, carbonyl compounds and some aromatics have also been studied[146] and the shifts in carbon 1s energies are correlated with a charge parameter obtained from electronegativity considerations. The shifts are also analysed in terms of group shifts from which group electronegativities are derived. A comparison is made between shifts in solid and gaseous samples and it shown that solid state effects are small for non-ionic compounds. Comparisons are also made with semi-empirical and *ab initio* molecular orbital calculations.

A study of simple hydrocarbons has been reported by Thomas[147]. Ionisation potentials for C (1s) are found to be about 5% smaller than values of orbital energies calculated by using Koopman's theorem. The ionisation potentials for the electrons of carbon monoxide have also been measured[148] with the shifts indicating a positive charge on the carbon and a negative charge on the oxygen in agreement with the relative electronegativities. A comparison has been made[149] of core-level binding-energy shifts in small organic molecules with predictions based on SCF–MO calculations. It is concluded that SCF 'frozen orbital' values can be used to give estimates of the binding energy shifts in these molecules to within about 1 eV. Fadley and Shirley[150] have studied multiplet splitting of metal atom electron binding energies in both inorganic solids and gases. Chemical shifts in arsenic and bromine compounds have been reported by Hulett and Carlson[448].

Various theoretical papers have been published[157–163] exploring the correlations with experimental chemical shifts. In general, good agreement is obtained even with simple predictions based on Koopman's theorem.

3.3.5 Angular distributions of photoelectrons

The angular distribution of photoelectrons can in principle provide some useful information concerning the nature of the orbital from which an electron has been removed and several experimental[164–167] and theoretical[168–173] investigations have been made. Berry[3] has discussed much of the earlier work in some detail. For polarised light the differential angular distribution is of the form[168, 169, 171]:

$$\frac{d\sigma}{d\omega} \propto (1+\beta P_2(\cos\Theta))$$

where β is an assymmetry parameter and Θ is the angle the outgoing electron makes with the polarisation direction of the incoming photon. Chaffee[164] and also Hall and Siegel[167] have given experimental proof of this relationship for certain species using polarised light.

However in most experiments the direction of polarisation cannot be defined and the angular distribution expression is replaced by

$$\frac{d\sigma}{d\omega} \propto (1+\beta/2(\tfrac{3}{2}\sin^2\theta - 1))$$

where θ measures the angle between the direction of the ejected electron and the propagation direction of the incident light. This expression has received experimental confirmation in several cases[165, 166].

In more recent work Vroom, Comeaux and McGowan[174, 175] have studied angular distributions of photoelectrons at 584 Å and 744/736 Å using a segmented spherical retarding analyser. At the higher energy the distributions obtained for argon[174] are in general agreement with theoretical predictions, but at the lower photon energies (nearer the photoionisation threshold), anomalous distributions were obtained. Broadening was observed as well as the appearance of some structure. Further studies of Ar, Kr, Xe, H_2, N_2 and O_2 [175] are in general agreement with theory but show anomalies for Ar and Kr at 744/736 Å. The anomalies are tentatively ascribed to competition for formation of the doublet states of the rare gas ion. No such anomaly was observed in hydrogen and nitrogen thus seemingly ruling out instrumental effects. The angular distributions in argon at 584 Å and 744/736 Å have been re-examined by Morgenstern, Niehaus and Ruf[176] who report no anomalous features. They obtain smoothly varying functions in agreement with general theoretical predictions. It is pointed out that at 744/736 Å the photoelectrons are comparatively slow (~ 1 eV) and would be sensitive to experimental interferences.

Angular distributions and partial photoionisation cross-sections for Ar, N_2 and O_2 have been measured by Samson[177]. Samson points out that the measured value of partial photoionisation cross-section (i.e. the probability that an incident photon will eject an electron from a particular orbital and hence leave the ion in a known state) will be affected due to (a) the effects of autoionisation and (b) the fact that electrons are ejected in preferred directions with respect to the direction of the incident light. It is usually impractical to collect photoelectrons over the total 4π geometry and it is shown that at an angle of 54 degrees 44 minutes with the direction of the electric vector of the photon beam the electron distribution is independent of the degree of polarisation and of the assymmetry parameter β. It is suggested that all future measurements are made at this angle. The interfering effects of autoionisation in O_2 and N_2 at certain wavelengths are demonstrated. Harrison[178] has studied angular distributions of photoelectrons from cadmium and zinc at 584 and 1048 Å.

Sichel[179] and Buckingham *et al.*[180] have developed the theory of angular distributions to consider the potential information contained in rotational fine structure in photoelectron spectra but confirmation must await the necessary improvements in experimental technique outlined earlier in this review. Angular distributions of the photoelectrons from outer shells of noble gases have also been investigated[181, 182].

3.3.6 Special techniques in photoelectron spectroscopy

In conventional molecular photoelectron spectroscopy photoionisation is effected at an energy (usually 21.21 eV) which is considerably in excess of the threshold energy for most valence orbital electrons. Due to the general decrease observed in photoionisation cross-section as the incident energy is increased above the threshold[21] it is possible that some processes may not be observed at 21.21 eV. It is also apparent that the differing detection sensitivity of various spectrometers will assume a significant role in such cases. This may, for instance, explain the variation in values reported for the first ionisation potential of methane. In an attempt to resolve these difficulties Villarejo *et al.*[183–185] have developed a threshold photoelectron spectrometer employing a scanning u.v. monochromator and a 127-degree cylindrical electrostatic analyser. An upper bound of 12.75 eV for the first ionisation potential of methane is reported, in better agreement with photoionisation studies for positive ions[186] than with other photoelectron work[187, 188].

It should be pointed out that accurate Franck–Condon factors can only be directly obtained from *threshold* photoelectron spectra, and so the development of such techniques is of importance in obtaining data suitable for comparison with theoretical calculations. Conventional photoelectron spectra of vibrational bands will lead to erroneous Franck–Condon factors, due to the variable 'drop-off' of photoionisation cross-section for each vibrational level[21].

Peatman, Borne and Schlag[189] have developed an alternative threshold technique employing a steradiancy analyser for photoionisation resonance studies. The method has been applied to NO and C_6H_6 [189] and also to CH_3I and NO [190]. The authors claim that the technique offers several important advantages including increased flux, reduced effect of magnetic fields, diminished importance of contact potentials and simplicity of design.

Berkowitz and Chupka[191] have studied the photoelectrons arising from autoionising states of hydrogen and nitrogen in an attempt to further understand the process of autoionisation. Another interesting development has been the application by Puttkamer[192] of coincidence techniques to photoelectron spectroscopy. Coincidence spectra are measured at 21.21 eV for photoions and photoelectrons in H_2O, NH_3, CH_4, CD_4, C_2H_4, HCOOH and CH_3OH. The coincidence curves show the fragmentation of the molecular ion as a function of the internal energy of that ion. It is also observed that fragmentation does not always occur completely, even though the internal energy of the ion is sufficient.

It is anticipated that the further development and application of coincidence techniques will play a major role in the understanding of the mechanism of ionisation and dissociation (see also coincidence experiments in Section 3.4 on electron impact).

3.4 ELECTRON IMPACT SPECTROSCOPY (EIS)

3.4.1 Introduction, background and methods

In electron impact spectroscopy the kinetic energy of the ejected electron ε_2 is determined for the process

$$X+\varepsilon_1 \rightarrow X^* + \varepsilon_2$$

For this process.

$$E_2 = E_1 - E^*$$

where

E_1 = kinetic energy of incident electron ε_1
E_2 = kinetic energy of ejected electron ε_2
E^* = energy of excited state X^*

In general E^* will have contributions from electronic, vibrational and rotational excitation. The method is a means of studying quantum states of neutrals (i.e. bound states) and is complementary to PES where ionic states are studied. Lindholm[84–90] has shown how these two types of measurement may be used together to elucidate new Rydberg series and to determine orbital symmetries from quantum defects. It is possible to use EIS for the study of direct ionisation processes of the type

$$X+\varepsilon_1 \rightarrow X^{n+}+(n+1)\varepsilon$$

if coincidence techniques are used to uniquely determine the distribution of energy between the electrons.

With improvements in technology and, in particular, low energy electron optics, there has been phenomenal growth in the field of electron impact spectroscopy and it may be said to have 'come of age' in that the reliability of the technique permits detailed studies of atomic and molecular structure to be made. A comprehensive account and literature review of EIS up to the end of 1968 is contained in the review by Berry[3] and this is highly recommended as a detailed source of information to that date. Several other reviews of electron impact spectroscopy and associated topics may be found in the literature. Kerwin, Marmet and Carette have discussed the production and application of high resolution electron beams[193]. Kuyatt[14] and Simpson[194, 195] have given extensive accounts of various aspects of electron scattering from a static gas target. The electron impact excitation of atoms is discussed by Moiseiwitsch and Smith[196] and by Rudge[197]. Kieffer has assembled compilations of data on ionisation, dissociation and vibrational excitation[198] as well as electronic excitation cross-sections[199]. Graphic displays of selected experimental data on cross-sections as a function of electron energy are presented for a large variety of molecular and atomic targets and criteria for data selection are discussed. Scattering experiments for three-body atomic systems, dealing with resonance and threshold behaviour are discussed by McGowan[200] in relation to existing theoretical work. Electron excitation functions are the subject of a review by Heddle and Keesing[201]. Excellent articles on various aspects of energy-loss spectroscopy have been published by Lassettre *et al.*[45, 202] and by Trajmar, Rice and Kuppermann[12, 203].

Several reviews of various theoretical aspects of electron-impact spectroscopy are also to be found in the literature. Models, interpretations and calculations concerning resonant electron scattering processes are discussed by Taylor[204]. Particular attention is given to physical models of resonance states that are of use in the interpretation of experimental data for hydrogen, nitrogen and carbon monoxide as well as atomic systems. The calculation of resonant energies and widths is also discussed. A qualitative description of the various types of resonant states is given in an earlier paper by Taylor *et al.*[205]. Resonances (or compound states) have also been discussed by Burke[206], Fano[207, 208], Bardsley and Mandl[209] and by Chen[443]. Fano and Cooper[210] have surveyed information on the oscillator strengths for neutral atoms and this article is of particular significance with the growing emphasis on the determination of oscillator strengths from electron energy-loss measurements. A book by Bonham[211] deals with the theory of electron impact spectroscopy.

There are some important differences between optical and electron impact spectroscopy. From an experimental standpoint (see Section 3.2) there are advantages for EIS in the vacuum u.v. with regard to resolution, intensity and the availability of a continuum of energies. A more fundamental difference occurs with regard to selection rules for transitions. The probabilities for optical transitions are governed by fairly rigid selection rules, which are also found to be closely followed for transitions induced by high energy electrons at small scattering angles. Consequently it is possible to derive oscillator strengths and optical absorption coefficients under conditions of small momentum transfer (high energy). In short it is possible to carry out 'optical' absorption spectroscopy with electron beams at high energy both qualitatively and quantitatively. Using the concept of generalised oscillator strength ntroduced by Bethe[212] in 1930 it can be shown[212, 213, 214] that the differential cross-section for scattering by fast electrons (i.e. where the Born approximaion holds) is given by

$$\sigma(\theta,E) = \frac{2}{E} \quad \frac{1}{K^2}\frac{\mathrm{d}f(K)}{\mathrm{d}E}$$

where θ is the scattering angle, E the energy loss, k_o and k_n are the magnitudes of the momenta of the primary electron before and after the collision, K is he magnitude of the momentum transfer ($K = k_o - k_n$) and $\mathrm{d}f(K)/\mathrm{d}E$ is the eneralised oscillator strength. The latter quantity may be expanded in erms of K^2

$$\frac{\mathrm{d}f(K)}{\mathrm{d}E} = \frac{\mathrm{d}f}{\mathrm{d}E} + a\,K^2 + b\,K^4 + \ldots\ldots$$

where $\mathrm{d}f/\mathrm{d}E$ is the optical oscillator strength. It can be seen that in the limit s $K \rightarrow 0$ (as $E \rightarrow \infty$) the generalised oscillator strength will approach the ptical oscillator strength. Thus by the extrapolation of electron impact ata to zero momentum transfer, a value for the optical oscillator strength s obtained. In general, only relative cross-sections are measured in electron npact work leading to relative generalised and optical oscillator strengths. Normalised procedures must therefore generally be used. Some examples of he use of these ideas are to be found in the recent literature[60, 244, 257, 261, 283]. Photoionisation cross-sections may be derived from electron impact data sing the further relationship $\sigma(\text{photoionisation}) = (mc/\pi e^2 h)\,\mathrm{d}f/\mathrm{d}E$[244].

At lower incident electron energies and small scattering angles however it found that optical dipole selection rules are violated, often with high robability. These violations are well illustrated for the differential scattering om helium by low energy electrons[203]. Dipole forbidden, but electric uadrupole allowed, transitions (e.g. $1^1\mathrm{S} \rightarrow 2^1\mathrm{S}$) in helium as well as spin orbidden transitions (e.g. $1^1\mathrm{S} \rightarrow 2^3\mathrm{S}$, $1^1\mathrm{S} \rightarrow 2^3\mathrm{P}$) show increasing probability nd characteristic angular dependence[12, 203]. At threshold energies these fects are often dominant as, for example in the threshold electron impact pectrum of helium[215, 216], where the $2^1\mathrm{S}$ and $2^3\mathrm{S}$ peaks are the most probable ansitions while the optically allowed $2^1\mathrm{P}$ adopts a relatively minor role. ery low energy electron impact techniques are consequently of importance r the study of triplet and other optically inaccessible states. Further exnples of forbidden transitions observed by electron impact are the electric uadrupole allowed transitions in argon and nitrogen which have been udied at high resolution by Lassettre *et al.*[45, 202].

Vibrational excitation by electron impact was first observed by Haas[217] nd later by Schulz[218–220] who observed forbidden vibrations in homonuclear diatomics at low energies (2–3 eV). From a study of the vibrational structure and the nature of the cross-sections to individual vibrational states it was concluded that resonant states[209] (temporary negative ions) are involved. More recently, Lassettre and his co-workers report the observation of direct excitation of infrared-inactive vibrational modes in O_2, N_2, H_2O and CO_2 [221, 222]. The rotational and vibrational excitation of molecules by low energy electrons has been discussed by Phelps[223]. Geiger and Wittmaach[22 have observed dipole-allowed vibrations using high energy electrons (33 keV) Garstang[1] has discussed at length the various types of forbidden transition while the validity of selection rules in excitation and ionisation collision processes is the subject of an article by Collin[2].

In practice, electron impact spectroscopy may be studied in several ways The bombarding energy E_1 may be kept constant, while the intensity of the scattered electrons ε_2 is measured as a function of the energy they have lost $(E_1 - E_2)$ and the scattering angle θ. This type of measurement, analogous to an optical u.v. absorption spectrum, is known as an Energy-Loss Spectrum Alternatively the energy loss $E_1 - E_2$ may be maintained constant while the incident energy E_1 or the angle θ is varied. This type of measurement pro duces an excitation function for a given process and has been less commonly studied than energy-loss spectra presumably due to the problems of chro matic aberration in low energy electron optics (see Section 3.2). A third type of operation is also possible in which the incident energy E_1 is varied while the scattered electrons ε_2, which have a fixed amount of energy, are collected Most work of this type has been carried out at energies close to the threshold values for excitation processes where forbidden transitions are frequently dominant.

In most electron impact work it is necessary to monochromate the inciden electron beam prior to energy analysis of the scattered electrons. This is du to the large energy bandwidth of an electron beam produced by thermioni emission. A narrow 'slice' suitable for spectroscopic studies may be selecte from the Maxwell–Boltzmann distribution by one of the methods referre to in Section 3.2 of this review. A second analyser is necessary to monito the scattered beam. High resolution electron spectrometers of the Wie filter type have been described by Boersch *et al.*[58] and by Anderson an LePoole[59], but so far have not received widespread application other tha the continuing work of Boersch *et al.* Perhaps the most widely employe type of electron spectrometer is the spherical analyser[35], and high resolutio electron impact spectrometers based on this principle have been describe by Simpson and Kuyatt[36, 37] and Lassettre[45]. The majority of intermediat energy ($\sim$20–300 eV) electron impact spectroscopy has gained its impetu from the work of these two groups who have extensively applied electro optical principles to the analysis and focusing of low energy electron beam (see Section 3.2). Simpson and Kuyatt[36, 37] have also developed and used th idea of virtual apertures in which a narrow object aperture at higher energ is electron optically imaged on to the focal plane of the energy analyser b using a retarding lens. This technique eliminates scattering from slit edges i the analyser and offers improvements in space charge conditions since th object slit is at a higher energy than in a normal analyser. The 127-degre radial electron selector has also been used quite extensively in electro npact spectroscopy. Most of the successful work with this type of device as been with selectors of the 'gridded' type developed by Marmet and .erwin[49]. Concentric deflector grids and electron catcher electrodes are used ithin the selectors to minimise effects of space charge and reflected electrons. nteresting applications of this device to EIS are typified by the work of chulz[225], Ehrhardt *et al.*[226] and by Dolder[227]. More recently the principles virtual apertures have been successfully applied to the slit geometry of ne 127-degree selector and it is found that grids are not required for efficient peration[52].

Retarding analysers based on the RPD technique[25] have found certain pecific applications in EIS such as in the trapped electron technique de- eloped by Schulz[228, 248] and recently improved by Hall *et al.*[26]. By applying ne principles of electron lenses and low energy electron optics Golden and ecca[27] have produced a superior RPD spectrometer without the need for ne magnetic field used in earlier RPD devices.

The mathematical process of deconvolution offers an analytical alternative the process of electron beam monochromation and analysis. Morrison[229, 230] as proposed a method of deconvolution which has been applied with some uccess to the study of fine structure in ionisation efficiency curves[229, 231] nd to the improvement of negative ion data[232]. Cowperthwaite and Myers[233] ave also discussed the convolution analysis of electron impact data. A ariation of the deconvolution method is the energy distribution difference EDD) method proposed by Winters *et al.*[234–235] which is a mathematical nalogue of the experimental RPD method[25]. The techniques of deconvolu- on have been little applied to electron impact spectroscopy. A limited nprovement in threshold spectra by deconvolution of data has been re- orted by Hubin-Franskin and Collin[340], but the results are still inferior to ose obtained using energy selection of electron beams. Although decon- olution methods show some promise, the problems and complexity of their pplication have not so far permitted widespread reliable application and rect experiment is to be preferred.

For purposes of this review EIS can conveniently be subdivided into three tegories, (1) high energy (>500 eV), (2) intermediate energy 20–500 eV nd (3) low energy and threshold studies. The recent literature will be dis- ussed under these classifications.

4.2 High energy studies (>500 eV)

oersch, Geiger and their collaborators at the Technical University of erlin continue to be the only major contributor to high resolution, high nergy electron impact spectroscopy although recently a new high energy, gh resolution Wien filter has been described by Andersen and LePoole[59]. oersch, Geiger and Topschowsky[236] have reported the angular distributions nd energy-loss spectra of 25 keV electrons by ions, electrons and excited oms produced in a discharge tube. For argon an increase in small-angle attering is observed in the presence of a discharge, due to Rutherford attering on the ions and to free electrons in the discharge. Energy-loss ectra are also observed from excited helium atoms (2^3S, 2^1S, 2^3P, 2^1P)

formed in the discharge. Super-elastic collisions resulting in an energy gai of 1.14 eV were due to the de-excitation $2^3P \rightarrow 2^3S$. Geiger and Schmoranzer[237] have studied the dependence of the electronic transition moment o the nuclear distance for the Lyman and Werner bands of hydrogen usin H_2, HD and D_2 at an electron energy of 34 keV and a resolution of 0.01 eV Total vibrational band intensities obtained by integration over the individua rotational–vibrational bands showed a considerable isotope effect in agreement with theory. The electronic transition moment is found to increase wit increasing nuclear distance. The high resolution electron impact spectrum c helium has also been reported[238] in the region 20–66 eV. This region of th spectrum is difficult to investigate optically other than with the use c synchrotron radiation. The Rydberg series $1^1S \rightarrow n^1P$ is investigated quantitatively and is in fair agreement with theory[239]. The double electron excitations $1s^2$ (^1S) $\rightarrow$ 2s2p(^1P) are also observed. Oscillator strengths of the lo lying resonance lines of rare gas atoms have been measured[240] using 36 ke electrons. They are in general agreement with optical measurements, excep in the case of xenon. Geiger and Bonham[241] have discussed the differentia cross-sections for the electron impact excitation of vibrational levels of molecules.

A unique experiment involving energy loss of fast electrons at moderat resolution has been described by van der Wiel[60, 242–244]. Electrons (from a oxide cathode) of 10 keV primary energy, scattered by gaseous targets throug a small angle (0–10^{-2} radians), are energy analysed in a double Wien filte and detected in coincidence with the ions formed. Studies have been made t energy losses in excess of 500 eV providing a powerful technique for investgating inner-shell processes. The coincidence technique not only serves t determine the mass and the charge of the ions produced by the scattere electrons, but is also helpful in defining the effective interaction region whic is usually a problem for such small angles. Since the target gas density an the overall detection efficiency for electron energy analysis are unknow absolute intensities of scattering are not measured. Optical oscillator strength are derived for He, Ne and Ar[60, 242] by including normalisation on know values. However, unlike most other electron impact determinations of oscilator strengths it does not involve an extrapolation to zero momentu transfer. A comparison of the results for singly charged ions with photo ionisation cross-section data reveals outstanding agreement on a quantitative basis over a wide energy range. This experiment (being sensitive to th charge on the ion unlike total photo-absorption experiments) also give information on the relative importance of multiple ionisation in the photo experiments. This is convincingly demonstrated by a detailed study of mutiple ionisation in argon[244]. A variety of processes resulting from vacancie and excitations of M- and L-shell electrons have been studied. Oscillato strengths for the individual charge states Ar^+–Ar^{4+} are derived from th electron–ion coincidence measurements. An application of the sum rul (following normalisation at a single energy for Ar^+ alone) over a range 400 eV gives outstanding agreement with the results of photo-absorptio A detailed examination of the oscillator strength spectra reveals that a significant contribution to double ionisation in the M shell arises from doub excited resonance states. The K shell energy-loss spectra of the isoelectron molecules N_2 and CO have been studied by van der Wiel, El-Sherbini and Brion[245]. In addition to the K-shell ionisation continua, discrete transitions at 400.8 eV for N_2 and at 287.7 eV and 534.4 eV for CO are observed corresponding to excitation of the atomic K shells of N, C and O respectively. Electron–ion coincidence studies provide specific information on the decay processes following K-shell excitation and ionisation. These studies complement the results of electron synchrotron[246] and gas-phase Auger electron experiments[8, 247] (See Section 3.5). The coincidence experiment has proven to be a powerful technique for investigating ionisation processes and obtaining quantitative data. The extension of these methods to obtain absolute data (without normalisation) should be very fruitful not only for electron impact measurements but also for deriving optical quantities by a non-optical method.

3.4.3 Intermediate energy studies (20–500 eV)

The technique of electron energy-loss spectroscopy is now well developed in the intermediate energy range following the classic work of Simpson and Kuyatt[36, 37], and Lassettre[45, 214] and their various co-workers. In a recent review[202] Lassettre has given a comprehensive discussion of the inelastic scattering of electrons by atmospheric gases and essentially all of the significant work to mid 1968 is covered. So far the technique has largely been applied both qualitatively and quantitatively to the study of atoms and small molecules, but now it can be anticipated that electron impact spectroscopy will find an increasing application to problems of applied chemical interest. Simpson[249] has discussed applications to gas-phase chemical analysis. In particular the technique is compared with mass spectrometry and optical absorption spectroscopy. Inelastic scattering from organic compounds is discussed by Gerber[250] with a view to its relevance to the investigation of high temperature organic superconductivity. Rendina and Grojean[441, 251, 252] have described an electron impact spectrometer specifically designed for qualitative and quantitative gas analysis.

Significant contributions from Lassettre and his co-workers have recently appeared in the literature. Relative generalised oscillator strengths for the $1^1S \rightarrow 2^1P$ transition of helium have been determined[253] as a function of the momentum of the incident electron. Following extrapolation to zero momentum change, the result was normalised to the theoretical oscillator strength. Theoretical investigation of the normalising procedure shows that the limit of the generalised oscillator strength at zero momentum change is equal to the optical oscillator strength for any atom or molecule at any incident energy, regardless of whether the first Born approximation holds[253]. The ratio of elastic to inelastic cross-sections was also determined and the elastic cross-sections calculated from the data. In the meantime Bromberg[254–256] (working in the same laboratory) determined the absolute elastic collision cross-sections for helium. Subsequently Lassettre *et al.*[257] have re-normalised their data[253] using Bromberg's results[254] and the oscillator strength compares favourably with other experimental results and theory. In particular there is agreement, within experimental error, with the recent electron impact results of Chamberlain, Mielczarek and Kuyatt[258]. The absolute cross-section measurements have been made possible by a new technique used by Lassettre *et al.* This involves firstly the measurement of absolute elastic collision cross-sections (see Bromberg[254–256]) and secondly the measurement of the ratio of inelastic to elastic peak areas using a high resolution spectrometer[45]. The technique has been used to obtain absolute cross-sections for helium[257], mercury[260], nitrogen[261] and carbon monoxide[262]. Kessler[444] has discussed the determination of absolute differential cross-sections for electron scattering at intermediate energies[257, 258]. Energy-loss spectra[259] and generalised oscillator strengths[260] have been reported for mercury and the limiting oscillator strength is found to be in good agreement with the optical value. Absolute differential electron collision cross-sections and generalised oscillator strengths have been measured for two forbidden transitions $X^1\Sigma_g^+ \rightarrow a^1\Pi_g$ and $X^1\Sigma_g^+ \rightarrow a''^1\Sigma_g^+$ in N_2 [261]. The Born approximation apparently holds for the first transition but not the second. However generalised oscillator strengths extrapolate to the optical oscillator strength (zero) in both cases. Studies of absolute generalised oscillator strengths in CO [262] give values in generally good agreement with calculations based on lifetime measurements. However in the fourth positive bands experiment shows a large discrepancy (a factor of 2). An independent experiment by Meyer and Lassettre[263] confirms the electron impact result[262]. Following experimental observations[261], Lassettre[264] has discussed further selection rules for excitation by electron impact at small momentum changes.

Simpson[265], studying the angular dependence of inelastically scattered electrons (2^1P, helium), concludes that, at least below 500 eV impact energy the Born approximation does not adequately describe the angular dependence of the differential cross-section. Chamberlain, Mielczarek and Kuyatt[258] have measured absolute differential cross-sections for electron scattering in helium. The method is based solely on experimental parameters and is not dependent on any normalising procedure such as used by Lassettre[257]. Cross-section values for helium 2^1P excitation are found to be less than predicted by the Born approximation whereas elastic scattering agrees well with recent theoretical calculations. Krauss and Mielczarek have re-investigated the electron impact spectrum of ethylene[266]. The characteristic minimum which occurs in the generalised oscillator strength, as a function of momentum transfer for the electron impact excitation of low lying Rydberg states, is used as a probe of the Rydberg character of four transitions in C_2H_4. Miller[267] has made a related theoretical study of this problem.

In a series of papers Hertel and Ross[268–271] have reported the energy–loss spectra of the alkali metals at a series of energies. Absolute generalised oscillator strengths have been obtained by normalising the resonance transition to the optical oscillator strength. From an analysis of the results of rubidium and caesium[270] and using the series expansion[272] for the generalised oscillator strength Hertel and Ross find evidence for electric octopole transitions[271].

Following a preliminary study of low energy, large angle electron impact spectra in He, N_2, C_2H_4 and C_6H_6 [273] Doering and Williams[274] have reported an improved electron spectrometer with a useful primary energy range of 6–100 eV, variable scattering angle of 0–100 degrees and a useable resolution of about 0.04 eV. A study of the nitrogen energy-loss spectrum[274, 275] in the

12–14 eV region shows different angular behaviour for excitation to the $b^1\Pi_u$ and $p^1\Sigma_u^+$ states and a possible triplet state at 13.2 eV is also reported. Further studies have been made of benzene[276] at 13.6 and 20 eV and at scattering angles from 9 to 80 degrees. Three singlet–triplet transitions at 5.0, 6.2 and 6.9 eV have been identified in agreement with optical work. Three other possible triplet states at 3.9, 4.7 and 5.6 eV are reported.

A preliminary study of low energy, high angle electron scattering in He, H_2, N_2, CO and C_2H_2 by Kuppermann, Rice and Trajmar[203] has given a convincing overview of the significant behaviour of forbidden transitions in EIS. A recent review[12] gives a detailed account of studies of these molecules and in addition, ethylene. A useful general discussion of the electron impact method is also given. The differential and integral cross-sections for excitation of the 2^1P state of helium are discussed at length[277] and the normalised experimental cross-sections are compared with the results predicted by the Born and several other first approximations. In these, direct excitation is calculated in the Born approximation and exchange scattering by approximations of the Ochkur type. These theories are shown to be poor near threshold, and exchange is important at high scattering angles for all energies. The use of good SCF functions yields differential cross-sections in quantitative (20 %) disagreement with the results, although the energy and angle dependences are predicted qualitatively correctly. A theoretical study by Steelhammer and Lipsky[278, 279] of this work[277] and also the He $1^1S \rightarrow 2^3S$ transition[12] confirms the breakdown of the theory. Truhlar *et al.*[280–282] report an extensive theoretical and experimental study of electron scattering by H_2. The use of a polarised Born–Ochkur–Rudge approximation gives better agreement with experiment than any previous calculation for elastic scattering. It is found that the inclusion of polarisation is necessary to obtain accurate cross-sections at low scattering angles. Less good agreement is found for inelastic scattering. Brinkmann and Trajmar[283] have obtained differential electron energy-loss spectra for nitrogen at scattering angles of 0–80 degrees and at incident energies of 15–80 eV. The cross-sections have been made absolute by normalising to known cross-sections. Trajmar, Williams and Kupperman have used angular dependences to detect and identify two singlet–triplet transitions in H_2O at 4.5 eV and 9.81 eV [284] of which the level at 9.81 eV was previously unknown. Claydon, Segal and Taylor[285] have given a theoretical interpretation of the optical and electron scattering spectra of H_2O. Cartwright[286] reports total cross-sections for the electron impact excitation of the seven lowest triplet states of N_2 together with quantum mechanical calculations which are in fair agreement with the observed values. The relative magnitudes of the excitation cross-sections are used to predict N_2 processes in the upper atmosphere. Calculations of electron impact excitation cross-sections of the lowest triplet states ($^3B_{1u}$, $^3E_{1u}$ and $^3B_{2u}$) of benzene indicate a peak in the differential cross-section at a large scattering angle in the low energy region[287, 288]. These and other predictions are consistent with the experimental work of Lassettre *et al.*[45] and of Doering and Williams[273, 276]. The energy-loss spectrum of oxygen has been studied by Konishi *et al.*[289] at energies of 20–70 eV and variable scattering angle. Absolute cross-sections are obtained for excitation to the $a^1\Delta_g$, $b^1\Sigma_g^+$ and $A^3\Sigma_u^+ + C^3\Delta_u$ states in O_2 by a normalising method based on measurements of the differential elastic and inelastic scattering. Foo, Brion and Hasted[290], have studied differential energy-loss spectra for the triatomic molecules CO_2, COS, CS_2, N_2O and SO_2. Rydberg series are interpreted using the ideas discussed by Lindholm[84–90]. Weiss *et al.*[445] have also reported an energy-loss spectrum for N_2O.

The ionisation of helium has been studied by Ehrhardt *et al.*[291–292] in an elegant experiment in which the two outgoing electrons are detected in coincidence as a function of the angle between them. At the same time the energies of the two electrons are determined. The angular correlation distributions are found to be a very sensitive test of theory compared with measurements of energy loss, total or differential cross-sections. Vriens[293–295] has given a theoretical discussion of angular distributions of scattered electrons in ionising collisions. The angular distributions of electrons scattered elastically and inelastically (energy loss 6.7 eV) from mercury have been measured by Gronemeier[296].

3.4.4 Low-energy and threshold studies

In addition to the studies at intermediate energies outlined above, there has been a continuing interest in various other types of low-energy scattering experiments which provide information on elastic and inelastic cross-sections and temporary negative ion resonance states. Of particular interest are scattering experiments involving 'simple' atomic systems, dealing with resonance and threshold behaviour, which provide tests of theory. A lucid account of studies for the three-body atomic systems He^+, H^- and H_2^+ has been given by McGowan[200] in an article which covers most of the significant work reported to date. McGowan, Williams and Curley[297] have recently also studied resonance in the atomic hydrogen 2p excitation channel. Recent experimental progress in resonance electron scattering for simple atoms has also been discussed by Ehrhardt[298]. Gibson and Dolder[227] have investigated the resonant differential scattering of electrons by helium at energies close to 19.3 eV where the well-known He^- resonance (1s $2s^2$, $^2S_{\frac{1}{2}}$) occurs. Using a data logging technique the resonance profile was carefully examined over a wide range of scattering angles. Additional structure observed at 19.45 eV is found to have the properties of a p-wave resonance of He^- (1s 2s 2p, $^2P_0$). Structures in the low energy electron–helium scattering cross-section have also been observed by Golden and Zecca[299] using an improved RPD spectrometer[27]. The p-wave structure at 19.45 eV reported by Gibson and Dolder[227] is confirmed together with numerous other possible states of He^-. 24 structures have been observed between the 19.3 eV resonance and the He^+ onset at 24.6 eV (only 11 of these have been previously reported). Assignments are made on the basis of theoretical predictions[300]. Using the trapped electron technique Burrow and Schulz[301] report evidence for the observation of the decay of a compound state, He^- ($2s^2$ 2p, ^2P) into $He^+ + 2e$, by the simultaneous emission of two electrons. Although inelastic decay into other channels by single-electron emission (into He 2^3S, 2^1S and 2^1P) has been previously observed this is the first reported two-electron decay in an atomic system. Subsequently Grissom, Garrett and Compton have reported similar resonance structure in the trapped electron spectrum of helium[302] and neon[303].

There appears to be a small energy-scale discrepancy ($\sim$0.3 eV) between the two results for helium[301, 302]. Similar resonant structure has been observed by Brion and Olsen in the threshold electron impact spectra of krypton and xenon[216]. Burrow[304] has studied the formation of the lowest doubly excited state of He ($2p^2$, ^3P) by electron impact. An energy of 59.64 eV is found in good agreement with theory and an estimate of the cross-section at threshold gives a value of 4×10^{-20} cm^2/eV. The trapped electron spectrum of mercury has been measured by Borst[305]. An absolute cross-section scale was established by monitoring slow inelastically scattered electrons near the threshold for Hg(6^3P_2). Hall *et al.*[306] using an improved trapped electron tube[26] have studied excitation of H_2 in the 11–13.5 eV range from threshold to 0.3 eV above threshold. Above 0.2 eV, super-imposed on the threshold spectrum, new structure is observed and is interpreted as being due to resonances in the excitation cross-sections of H_2 ($B^1\Sigma_u^+$ and $C^3\Pi_u$). Using 127-degree selectors and a molecular beam, Weingartshofer *et al.*[307] have measured energy and angular dependences of the elastic and inelastic scattering from H_2 in the energy range 10–16 eV from 10 to 120 degrees. The formation and decay of three series of H_2^- resonances has been investigated and absolute cross-sections derived by normalising procedures. Branching ratios for the decay of resonance series into various electronic states are obtained, and from the observed angular dependence the configuration of the resonances has been determined for comparison with theoretical predictions. Hall *et al.*[308] have made a high resolution study of the trapped electron spectra of H_2 and D_2. Structure between 11.7 and 14 eV has been positively identified as being due to the vibrational series of the $C^3\Pi_u$ state. Spence and Schulz have studied vibrational excitation in O_2 using the trapped electron technique[309] with 0–1 eV electrons. Structure is associated with vibrational levels of O_2^- ($^2\Pi_g$) A further study of O_2 by Boness and Schulz[310] using 127-degree energy analysers provides measurements of the differential cross-section for elastic scattering at 50 degrees over the energy range 0–1.5 eV. Structure due to the compound state O_2^- ($^2\Pi_g$) is used to calculate $\omega_e(O_2^-) = 1089\ \text{cm}^{-1}$ and $r_e(O_2^-) = 1.377$ Å. A portion of the O_2^- Morse function potential energy curve is plotted using the derived constants. Theoretical studies of the angular distributions for resonant scattering of electrons by molecules have been made by Read[311, 312] and by Bardsley and Read[313]. The trapped electron threshold spectrum of nitrogen obtained by Hall *et al.*[26] is characterised by excitation of triplet states. The shape of the spectrum between 9.6 and 11.6 eV suggests the excitation of a repulsive or slightly stable state which is tentatively identified as being due to the $^5\Sigma_g^+$ state or some presently unknown state. The large peak at 11.87 eV is shown by well-depth studies to be resonant at the threshold energy in agreement with the findings of scavenging studies[215]. N_2 has also been studied by Bensimon *et al.*[314]. Unusual structure, attributed to resonances, is observed but it seems possible that the effects are instrumental. A trapped electron spectrum of CO reported by Rempt[315] shows an apparent anomaly in the Franck–Condon factors for $v = 0$ and $v = 1$ of the $a^3\Pi$ band when compared with results of electron scavenging[316]. The same effect appears in the trapped electron work by Brongersma and Oosterhoff[317]. Subsequently Brongersma[318] has shown that the effect is due to an apparatus function in his trapped electron tube and that, when account is taken of this, correct Franck–Condon factors are obtained. Hasted and his co-workers[319,320] have studied electron scattering in N_2, CO, O_2, NO, CO_2, N_2O and NO_2. Structure is interpreted in terms of compound negative ion states. Further study of N_2O and CO_2 by Larkin and Hasted[321] at zero scattering angle shows vibrational excitation at energy losses within 1 eV of the elastic peak indicating a very clean 'cut off' of the primary electron beam. The vibrational excitation for CO_2 is in good agreement with the earlier work of Andrick, Danner and Ehrhardt who show excitation of the 010, 001 and 100 modes[322]. The optically active (010, 001) modes show a strong forward scattering whereas the 100 mode has low intensity and is almost isotropic at low energy. This result is interpreted as meaning that dipole interaction, rather than the usual resonance mechanism, is responsible for the strong forward scattering. An extensive theoretical interpretation of the electron scattering spectrum of CO_2 has been given by Claydon, Segal and Taylor[323].

The trapped electron technique has been used by Brongersma *et al.*[317,324] for a series of studies of threshold excitation in atoms and small molecules. Total cross-sections have been obtained for electronic excitation and ionisation processes in He, CO, N_2, O_2, CO_2, C_2H_4 and C_6H_6 [325]. Threshold spectra and excitation functions near threshold were obtained. By observing the production of low energy ions as well as the production of low energy electrons, resulting from an inelastic collision, the excitation cross-sections were normalised on the well-known ionisation cross-sections. Brongersma and Oosterhoff[326] have also studied the threshold spectra of methane, ethane, cyclopropane and cyclo-octane. A comparison of the threshold spectra with the results of corresponding vacuum u.v. absorption spectra shows significant differences which are interpreted as being due to intense singlet–triplet transitions. This work has been continued and developed by Knoop *et al.*[327] who report threshold spectra for 1,3,5-cycloheptatriene (CHT) and 1,3,5,7-cyclo-octatetraene (COT). In such conjugated systems it is of interest to compare the measuring excitation potentials with theoretical values resulting from quantum chemical calculations. Theory predicts the first triplet state for CHT at about 2 eV in rather good agreement with a peak observed at 2.1 eV. Comparisons for other excited states are also made both for CHT and COT with a reasonable degree of correlation being observed. Recently an interesting development of the trapped electron technique has been reported by Knoop, Brongersma and Boerboom[328]. By applying a small modulation to the electron-trap well-depth and using a double-phase sensitive detection sequence it is possible to obtain excitation spectra of energies well above threshold while still retaining the resolution advantage of a small effective well-depth. Further development of this technique[328] should provide a very useful tool for low energy electron impact spectroscopy.

The trapped-electron technique has proved to be very useful for the study of threshold electron impact excitation. There are, however, some limitations to the method such as the possibility of negative ion contamination of the electron spectra. In addition, for reasons of geometry as well as field penetration it is difficult to use electron multipliers for electron detection. Furthermore, the technique is dependent on the use of an axial collimating magnetic field to confine the electron beam and the higher energy inelastically scattered electrons. This magnetic field requirement prohibits the use of electrostatic electron monochromators of the deflection type which have the capability of higher electron energy resolution than the RPD gun.

A new approach to the study of threshold electron impact spectra was demonstrated by Curran[329] who used sulphur hexafluoride to scavenge the zero-energy ($E = 0$) electrons ejected at electron impact excitation thresholds. SF_6 has a very high cross-section (about 10^{-15} cm^2) for electron attachment, and the resonance has been shown to have a width of less than 20 mV[61]. Using the RPD gun in a conventional mass spectrometer Curran obtained a threshold excitation spectrum for N_2 by recording the SF_6^- ion current as a function of electron energy in a mixture of SF_6 and N_2. The general scheme for electron scavenging is

$$X + \varepsilon(E = E^*) \longrightarrow X^* + \varepsilon(E = 0) \xrightarrow{SF_6} SF_6^-$$

Subsequently Jacobs and Henglein[330], using an unselected electron beam, made low-resolution electron scavenging studies of atoms and small molecules using SF_6^- and I^- as detectors. More recently Compton *et al.*[215] have further exploited the method using a time-of-flight mass spectrometer equipped with an RPD gun. Threshold excitation spectra were reported for He, N_2, HCl, H_2O, D_2O, C_6H_6, toluene, a series of halogenated benzenes and naphthalene. A large number of high intensity forbidden transitions are observed as well as optically allowed processes and temporary negative ion resonances. Compton *et al.*[215] have also shown how electronic excitation by electrons with a wider range of incident energies above threshold ($\sim$0.5 eV) may also be investigated by recording the energy dependence of the SF_5^- ion current. The method is used to investigate the energy dependence of temporary negative ion resonances in nitrogen. In this manner detection of the SF_5^- ion current is analogous to increasing the well-depth in the trapped electron technique. It has been found[215] that energy-scale calibration using the SF_6^- primary peak is inaccurate, presumably due to the convolution of the SF_6^- cross-section with the rapidly rising electron current at threshold. The nitrogen spectrum energy scale is calibrated with SF_6^- in Compton's paper[215] and appears to be in error by $\sim$0.4 eV. For the other molecules the Cl^-/HCl peak at 0.82 eV has been satisfactorily used for calibration of the SF_6^- scavenging spectrum.

Huebner, Compton and Christophorou have investigated[331] two temporary negative ion states in pyridine using both SF_6^- and Cl^-/HCl as detectors. The two states are believed to arise from the two lowest unfilled π-orbitals which are observed as being degenerate in benzene. The degeneracy is removed by the replacement of CH by N in the ring. A study of NH_3 and ND_3 using both the trapped electron and scavenger techniques has been reported by Compton, Stockdale and Reinhardt[332]. Both methods show four well-defined peaks between 6 and 12 eV and negative ion formation in ammonia is also studied. Begun and Compton[333] have measured the threshold spectrum for XeF_6 and XeF_4 using SF_6^- scavenging. No low-lying electronic states are found in the region 1–3 eV. Comparison with u.v. spectra reveals extra structures in the electron impact curves which are attributed to differences in resolution and the possibility of electric quadrupole and singlet–triplet transitions.

Some improvement in the scavenger technique has been obtained by Brion *et al.*[216, 316, 334–336] by using a 127-degree electron monochromator for the incident electron beam and a monopole mass spectrometer[337] to monitor the negative ions. Studies of the rare gases He, Ne, Ar, Kr and Xe [216, 335, 336] show a variety of spin and parity forbidden transitions for excitation of both p and s outer-shell electrons. The shape of the SF_6^- spectrum has been measured above the helium ionisation potential and gives a measure of the distribution of zero energy electrons among the two electrons ejected in the ionisation process. Using theory developed by Temkin[338], the curve shape provides further information on the nature of the threshold law for ionisation by electron impact. Studies of the excitation of the outer s electrons in Ar, Kr and Xe [216] show significant structure superimposed on the p electron 'ionisation continuum'. The peak positions correlate well with known and calculated levels. Structures beyond the M_1 edge in argon correlate well with doubly excited states observed by Madden, Ederer and Codling using electron synchrotron radiation[339]. Similar structures are observed for Kr and Xe [216]. Brion and Olsen[316] have investigated the vibrational band structure for the optically forbidden excitation to the $a^3\Pi$ state of CO.

Threshold electron impact excitation by the SF_6 scavenger technique has been used by Hubin-Franskin and Collin to study N_2 [340], C_6H_6 and C_2H_4 [341] and CO [342]. An analytical mass spectrometer was used without monochromation of the electron beam. Some improvement of the electron energy resolution was obtained by application of a smoothing and deconvolution method but vibrational structure is not resolved. Forbidden transitions and temporary negative ion states have been observed. Lifshitz and Grajower report temporary negative ion resonances in perfluorocarbons[343]. Using the trochoidal monochromator[61, 62], Stamatovic and Schulz have studied the excitation of vibrational modes near threshold in CO_2 and N_2O by means of SF_6 scavenging[344]. In general, the results confirm other experimental[322] and theoretical[323] findings for the mechanism of vibrational excitation at low energy. Ion cyclotron resonance (ICR) mass spectrometers have been used in two studies of threshold electron impact excitation using SF_6^- [345] and Cl^-/CCl_4 [346], respectively, as electron scavengers. The high sensitivity of the ICR mass spectrometer makes it a useful device for the study of these collision processes but further application requires electron beam monochromation which could best be achieved by a modulated RPD technique due to the presence of the high magnetic field.

3.4.5 Special techniques in electron impact spectroscopy

The use of coincidence techniques in electron impact spectroscopy has already been discussed in the preceding sections for electron–ion coincidence[60, 242–245] and electron–electron coincidence[291, 292]. Imhof and Read[347] have described an ingenious method for measuring atomic and molecular lifetimes based on an electron–photon coincidence experiment. Using a double hemispherical electron energy-loss spectrometer and a beam of helium atoms, photons and energy-loss electrons from the 4^1S transition are detected in delayed coincidence. A mean lifetime of 73 ± 2 ns for the 4^1S state is derived which compares favourably with measurements by other techniques.

Hamill *et al.*[348, 349] have developed a technique involving low energy electron reflection spectrometry (LEER) from thin films of molecules. Electrons reflected from thin films, frozen on to a kovar block, are detected (at low resolution) with a retarding analyser. The resulting curves exhibit structure which correlates with known molecular levels. Triplet states are detected in benzene and water. A further study of aromatic and aliphatic molecules[350, 351] reveals low lying bands not observed optically and which may be due to triplet states. Despite problems of surface charging and destruction by the electron beam, this technique (LEER) shows great promise for providing further information on molecular energy levels. Further development, using energy-selected incident electron beams should be very fruitful.

3.5 AUGER (AES) AND AUTOIONISATION (AIES) ELECTRON SPECTROSCOPY

3.5.1 Introduction

If a super-excited state X** (i.e. a neutral or ionic excited state at an energy in excess of the lowest ionisation potential of X) is formed, its subsequent autoionisation (a radiationless decomposition) may be studied by gas-phase Auger electron spectroscopy.

Following the discovery of the Auger effect in 1925[352] a large amount of Auger electron spectroscopy has been carried out, mainly involving solids and the study of surface physics[353]. However, in recent years gas-phase studies have been made and provide information, such as electronic binding energies in free atoms and molecules, which is of more direct interest to chemists. The study of the effects resulting from the formation of inner 'holes' in atoms and molecules is of particular relevance to subjects such as radiation damage. Consider the following sequence for neutral excited states.

$$\mathrm{X} + \varepsilon_1 \longrightarrow \mathrm{X}^{**} + \varepsilon_2 \qquad \text{(a)}$$

$$\mathrm{X}^{**} \xrightarrow{\text{autoionisation}} \mathrm{X}^{+} + \varepsilon_3 \qquad \text{(b)}$$

The initial excitation (a) may also be effected by ion impact or by photons. For the present purposes the process of autoionisation (b) may be considered as a separate entity for which

$$E_3 = E^{**} - I(\mathrm{X})$$

where E_3 = kinetic energy of electron ε_3
E^{**} = energy of super-excited state X**
$I(\mathrm{X})$ = ionisation energy of X to the final ionic state X^+

The measurement of E_3 give rise to the method of Autoionisation Electron Spectroscopy (AIES). Since a discrete state X** is the 'reactant' for process (b), energy E_3 will be independent of the mode or conditions of the formation reaction (a). A significant consequence is that high intensity unmonochromated electron beams may be used for ε_1 in process (a).

Auger processes will occur when an inner hole state X^{+*} is created by electron, ion or photon impact. The excited ion may undergo an Auger transition to a multiply charged ion state. In the case of doubly charged ion formation.

$$X^{+*} \longrightarrow X^{++} + \varepsilon_4$$

for which the energetics can be simply represented by

$$E_4 = E^{+*} - E_{X^{++}}$$

where E_4 = kinetic energy of electron ε_4
E^{**} = energy of inner hole state X^{+*}
$E_{X^{++}}$ = ionisation energy of X to the final state X^{++}

Auger Electron Spectroscopy (AES) involves the measurement of E_4. When a vacancy is created in the inner shell of an atom it may be filled (rapidly) either by a radiative transition or by an Auger process. New vacancies created in this transition may in turn be filled by further transitions and this continues until all vacancies reach the outermost shell (vacancy cascade). In lighter atoms Auger processes are generally much more probable than radiative transitions. The vacancy cascade can give rise to highly charged ions[244]. It is also possible for X^{**} to autoionise to a multiply charged ion state[244, 245, 354]. Following formation of the initial vacancy, a variety of electron orbital rearrangements may occur including further subsequent ionisation (shake-off) and excitation (shake-up)[8]. It can be seen that Auger processes and autoionisation are similar types of processes and they are therefore discussed together in this section.

Since the decompositions involving autoionisation and Auger transitions occur between two 'excited' states a process is not uniquely defined by a measurement of E_3 (or E_4). Other evidence must be found to establish unequivocally the identity of one or other of the two states involved. A unique elucidation becomes possible when coincidence studies[244, 245] are made involving two or more of E_2 (see equation (a)), E_3 (or E_4), X^+ (or X^{n+}) and any photons which may be ejected from the excited states in competition with the radiationless decomposition.

3.5.2 Recent studies

Gas-phase Auger and autoionisation studies have received considerable impetus following the early work[34] and continuing contributions[247, 355–366] by Mehlhorn and his co-workers. Extensive contributions have also been made by Carlson *et al.*[354, 367–380]. The theory of the Auger effect received earlier attention[383–384, 442] but has come under more recent scrutiny with the availability of high quality experimental data.

In an early study, Mehlhorn[34] studied the Auger spectra of Ar, N_2, O_2 and CH_4 using x-ray irradiation to create the initial hole states. This work demonstrates that Auger spectra could lead to the derivation of inner shell binding energies. A value of 409.4 eV obtained for the K-shell in N_2 agrees favourably with the later results of ESCA studies[7,8]. This work was later extended to a study of neon[355,358], argon[358] and krypton[356,357] and also led to the derivation of K-shell binding energies. Electron bombardment[358] was introduced to create the initial hole states and offers advantages over the use of x-rays with regard to intensity. The use of x-ray photons is also restricted of course if autoionisation studies of neutral (bound) states are to be made. More recently, Carlson *et al.*[377] have used x-rays and electron bombardment to study the KLL Auger spectrum of N_2 showing that the use of both techniques permits a more definitive interpretation of the spectrum. Using improved theory, Mehlhorn and Asaad[359] were able to largely resolve the earlier differences between theory and experiment for Auger transitions involving elements of low atomic number ($10 \leqslant Z \leqslant 36$). A study of the autoionising doubly excited states of helium[360] shows the familiar assymmetric Fano resonance profiles in agreement with the earlier reports of this phenomenon using electron scattering[385,386], electron synchrotron radiation[387] and ion impact[388,389]. Siegbahn *et al.*[8,381] have recently reported similar studies of the autoionisation spectra of helium at higher resolution. Glupe and Mehlhorn[361] have obtained relative and absolute electron impact ionisation cross-sections of the K-shell electrons in Ne, N_2, O_2 and for C in CH_4. The most recent Auger studies by Mehlhorn *et al.* have been detailed examinations of the spectra of argon[363–365], neon[366] and N_2 [247]. A new binding energy of 326.5 eV was obtained for the L_1-shell in argon[363] and this is confirmed by ESCA studies[8]. The study of nitrogen[247] provides significant new information on the excited states of the doubly charged nitrogen molecular ion and the energies of six states of N_2^+ were obtained from the K Auger spectra. The Auger method is of particular significance as a tool for studying multiply charged ion states since only very limited alternative methods exist. An Auger line at 383.7 eV[247] can only be explained by an autoionising transition from a highly excited neutral state to $N_2^+(A^2\Pi_u)$. Such a neutral state has been reported in the absorption spectrum (at 400.24 eV) by Nakamura *et al.*[246] and its existence and the autoionising mechanism have been confirmed by electron energy-loss studies in the region of K-shell excitation[245]. Combined x-ray and electron impact Auger studies are also in accord with this mechanism[377]. Higher resolution Auger studies[8] confirm the findings by Mehlhorn *et al.*[247]. Ion–electron coincidence studies[245] have also shown that dissociative autoionisation makes a significant contribution to the decomposition of K-shell excited states.

Carlson, Krause and co-workers have made a series of experiments on the Auger spectra of atoms and also related mass spectrometric studies of the charge spectra of the ions resulting from inner-shell vacancies. Following an early electron–ion coincidence study of neon[367] Auger electron and ion charge spectra were obtained for neon[368,371,374,375,378,379], argon[354,368,374–376,378], krypton[370,372], xenon and mercury[373]. Carlson *et al.* also made early observations of double electron emission in neon[368] and later in krypton[372] and argon[354,376]. This phenomenon, as well as more highly charged states of argon, has been investigated by van der Wiel and Wiebes[244] using electron–ion coincidence (see Section 3.4 on electron impact spectroscopy). A study of the Auger spectrum of neon using both photon and electron excitation to produce the initial vacancies shows identical spectra[379] indicating that the Auger processes are independent of the excitation mode. More recently Carlson *et al.* have begun to study Auger spectra of molecules. The differences in the KLL spectra of nitrogen studied using both x-rays and electrons strengthen the evidence provided by other techniques[8, 244–246] for the existence of excited K-shell bound states. In a recent report from the same laboratory Moddeman[380] gives an extensive account of the high resolution Auger and autoionisation electron spectra of N_2, O_2, CO, NO, H_2O, CO_2, CH_4, CH_3F, CH_2F_2, CHF_3, CF_4, SiH_4 and SiF_4.

Siegbahn *et al.*[7, 8, 137, 381] have made extensive studies of autoionisation and Auger electron processes in atoms and molecules at high resolution and high electron energy (3–5 keV). The spectrum of helium shows at least six resonances for the 2snp(^{1}P) doubly excited states autoionising to the 1s(^{2}S) ground state of He^+. Two other recent studies of the autoionisation electron spectra of helium[390, 391] at lower energies and variable scattering angles show interesting features. It is found[390] that the intensity of peaks resulting from optically forbidden transitions from the ground state increases significantly as the impact energy is decreased. The trend is most marked for the triplet state 2s2p(^{3}P). Other states observed are 2s^2(1S), 2s2p(^{3}P), 2p^2(^{1}D), 2s2p(^{1}P), 2s3s(^{1}S), 23sp+ and 24sp+. A study of line shapes[391] at variable scattering angle and electron impact energy shows similar features and it is also of interest to note an apparently significant improvement in signal to noise at large scattering angles (140 degrees). Siegbahn *et al.*[8] also show high quality autoionisation spectra for Ne, Ar, Kr and Xe induced by high energy electron impact. These display a rich structure which is interpreted as being due to transitions of the type (core) nsnp^{6}ml → (core) ns^{2}np^5($^2P_{\frac{3}{2}}$, $^2P_{\frac{1}{2}}$). It is interesting to note that initial transitions to the parity forbidden (core) nsnp^{5}ms states evidently have a high probability even at these high electron impact energies (3–5 keV). Transitions to the nsnp^{5}ms states also appear to have a high relative probability at threshold energies[216]. The Auger spectra of CO, N_2, O_2 and CF_4 have also been reported by Siegbahn *et al.*[8, 381]. For N_2 the results are in good agreement with those of Stalherm *et al.*[247]. The Auger spectrum for CO[8, 381] shows to advantage the superior resolution obtained by the Uppsala group in that vibrational structure is partially resolved on some of the Auger bands involving transitions to ionic states of the valence electrons. An upper limit of 0.18 eV is also obtained for the natural width of the carbon 1s level. More recently an extensive study has been made of Auger processes induced in CH_4, CH_3Br, CH_2Br_2, $CHBr_3$, CBr_4, C_2H_6 and C_6H_6 by 5 keV electron impact[382]. In addition to the energies and intensities of the Auger lines and the chemical shifts, the analysis of the spectra gives the energies and the lifetimes of the doubly ionised molecules. A consideration of line widths and energetics suggest that some of the doubly ionised molecules rapidly dissociate. Energy levels for doubly ionised molecules are determined thus extending the field of experience attainable by mass spectrometry to doubly charged ions with short lifetimes.

It is also possible to study autoionising processes from states excited by ion impact. Rudd *et al.* have carried out a series of experiments with this technique[392] using a variety of hydrogen and rare gas ions to study autoionisation in the rare gases[388, 389, 392–395] and oxygen[396]. Gerber, Morgenstern and Niehaus[397] have studied the energy of electrons ejected in slow ion–atom collisions between Ar and Ar^+ (incident energy (0.2–3 keV). Autoionising of super-excited argon atoms (core) $3s3p^6ns$, np and (core) $3s^2sp^44sns$, np is observed. Edwards, Risley and Geballe[398] have found two autodetaching states of O^-(10.112 and 12.115 eV) by observing the electron energy spectrum of O^- ions excited in collisions with a He target at beam energies of 1–5 keV. In addition the H^-(^1S, ^3P and ^1D) autodetaching states are excited by O^- on H_2 and Ar^- states by O^- on Ar. No structure was observed for O^- on O_2 or Ne. Bydin *et al.*[399] have observed similar processes for the halogen negative ions. A look at this recent literature illustrates the fact that a combination of techniques including ESCA, EIS, Auger spectroscopy (using x-ray and electrons), coincidence experiments and photoabsorption are usually necessary to unravel definitively decomposition processes resulting from inner-shell ionisation and excitation. In this context the continued study of the Auger and autoionisation electron spectra of molecules should provide useful information for the chemist. In particular the methods give significant information on multiply charged ion states which is not obtainable by other techniques.

3.6 PENNING IONISATION ELECTRON SPECTROSCOPY (PIES)

3.6.1 Introduction

The reaction of electronically excited particles with other atoms and molecules is of some importance in high temperature chemistry, discharges, the study of aeronomy and stellar atmospheres. Processes of the general type

$$X + Y^* \rightarrow X^+ + Y + \varepsilon_5$$

are known as Penning Ionisation[400–402]. The measurement of the energy E_5 of the ejected electron gives rise to the technique of Penning Ionisation Electron Spectroscopy (PIES). This method was introduced in 1966 by Cermak[403] using helium metastable atoms to ionise a series of molecular targets. A process closely related to Penning ionisation is that of associative ionisation

$$X + Y^* \rightarrow XY^+ + \varepsilon_5$$

This type of reaction, often known as a Hornbeck–Molnar process[404], has been investigated in rare gas systems by Becker and Lampe[405]. Herman and Cermak[406] have made a study of the relative importance of Penning and associative ionisation.

For the process of Penning ionisation

$$E_5 = E^* - I(X) + \Delta E$$

where E_5 = energy of the ejected electron
$I(X)$ = ionisation energy of X.

The term ΔE arises because, in general, it is to be expected that the relative kinetic energy of the system is different before and after the ionisation process. In practice ΔE can be positive or negative. If ΔE is positive and greater than the relative kinetic energy of the neutral system at infinity (i.e. if $\Delta E = E_5 - [E^* - I(X)] > E_{kin}$, the relative kinetic energy of collision) then associative ionisation will occur if ΔE is negative; either ionisation reaction may then take place. The description of Penning ionisation processes in terms of potential energy diagrams, as well as the development of an understanding of the shifts ΔE, have received extensive treatment by Niehaus and his co-workers[407–410] in a series of recent publications dealing with theoretical and experimental aspects of PIES. The existence of the shifts ΔE as well as the natural broadening of Penning spectra lines makes the technique of limited use for the determination of ionisation potentials when compared with photoelectron spectroscopy. Conversely however the shifts ΔE and the line broadening provide a sensitive probe for the study of interatomic (and molecular) interactions[409] as well as the nature of the Penning process itself. In principles, PIES can provide significant information complementary to the results of reactive scattering with molecular beams.

The general methods and principles of electron energy analysis outlined earlier are equally applicable to PIES. The ionising species Y* have generally been limited to the metastable rare gas atoms which are usually produced by electron bombardment of quasi-molecular beams emerging from a multichannel array. The most common and generally useful metastables have proved to be He 2^3S (19.82 eV) and He 2^1S (20.61 eV) states, although Ne, Ar and Kr $^3P_{0,2}$ states have also been used. Electron bombardment of rare gases results in a mixture of metastables giving overlapping electron spectra (analogous to the use of two wavelengths in PES). In more complex target systems this can lead to undesirable complications in interpretation. It has been demonstrated that a pure He 2^3S beam may be obtained by irradiating the mixed beam with 2^1S–2^1P, (2μ) radiation[408, 411]. Alternatively the relative population of metastable states in the beam may be controlled by varying the electron bombarding energy.

Useful review articles on the production, detection and collisions of electronically excited atoms and molecules have been published by Muschlitz[412, 413]. The total cross-sections for metastable excitation in the rare gases have been studied by Pichanick and Simpson[414], while a variety of other topics relevant to PIES include chemionisation[415, 417], mass spectrometric studies of Penning ionisation[416, 440] and the measurement of Penning ionisation cross-sections by molecular beam[413] and flowing afterglow techniques[418, 419]. The large Penning ionisation cross-sections (typically 10^{-14}–10^{-15} cm^2) lead to large PIES electron currents which are an attractive feature from an experimental standpoint. A short review of some of the early PIES studies has been given by Berry[3].

3.6.2 Recent studies

The pioneering work of Cermak is summarised in a review article[420] where the spectra of some forty compounds are discussed and compared with the results of photoelectron spectroscopy. The apparatus used by Cermak consists basically of a simple electron bombardment source and a Lozier tube for the retarding analysis of electrons. Despite a rather limited resolution

(~0.2 eV) interesting and informative studies were possible and energy shifts corresponding to differences in translational energy were observed. Vibrational structure was evident in the spectra of some diatomic molecules (H_2 and NO) but the limited resolution together with a rising scattered background does not permit a quantitative estimate of the Franck–Condon factors. The collision time for excited thermal neutral particles is several orders of magnitude longer than for photon or electron collisions but nevertheless it has now been shown for some systems that Penning ionisation is apparently governed by the Franck–Condon principle[421, 422, 423]. It is apparent from Cermak's work that the relative probabilities for ionisation to various electronic states is often quite different for Penning and photoionisation. Cermak[424] has also shown how the individual efficiency curves for the production of helium metastable atoms may be determined by the detection of Penning electrons from a given process with variation of the electron bombarding energy. The method offers some advantage over electron scattering in that the curves are not affected by anisotropies in the differential electron scattering cross-section. In more recent work Cermak *et al.* have studied the ionisation of mercury using He*, Ne* and Ar* [425] and also metastable nitrogen molecules[426]. The first three ionisation potentials of mercury were obtained[425] in close agreement with the results of photoelectron spectroscopy, although the state populations are very different. The mechanism of formation of $HeHg^+$ and $ArHg^+$ is also discussed. The study with nitrogen[426] reveals the existence of long-lived excited states of N_2 at 11.8 eV ($E^3\Sigma g^+$). A much higher resolution Penning spectrum of mercury has been obtained by Niehaus *et al.*[407, 427] using helium metastables to observe three ionisation potentials for Hg. While the triplet spectrum is fairly narrow and shows a single peak for each state of Hg^+, the singlet spectra exhibit three broad peaks for each ionic state. Only one of the three singlet peaks is at the normal (800 meV) spacing from the triplet peak while the other two are shifted to lower energies by significant amounts. No definite explanation is offered for the phenomenon, but it is suggested that it may be associated with nuclear rather than electronic motion. An experiment using ^3He may provide further relevant information.

Using an improved retarding potential analyser[407], Niehaus *et al.* have been able to produce much better resolved Penning ionisation electron spectra. The simultaneous use of He 584 Å radiation to produce photoelectrons of known energy has been used to provide an absolute calibration of the electron energy scale and this together with good resolution has enabled the accurate determination of the shifts ΔE. Niehaus *et al.* have made effective use of a quench lamp to obtain 'pure' 2^3S beams[408]. A preliminary study of the energy distributions for the collision pairs X = Ar, Kr, Xe, Hg, C_2H_2 and Y = He, Ne, Ar, Kr showed definite shifts ΔE, and differences in electron distributions caused by He 2^1S and 2^3S on the same target species are attributed to differences in the collision processes. In contrast to the results for Hg already discussed it was found that for the target Ar the triplet distributions are wider and shifted towards higher energies by an amount ($\Delta E \simeq +0.30$ eV) of the order of thermal energies at the experimental temperature. Niehaus *et al.* suggest that this is due to production of some $HeAr^+$ which is in agreement with a strong temperature dependence observed[440] for the ratio $HeAr^+/Ar^+$. In this complementary crossed-beam, mass spectrometer study[440] Hotop and Niehaus also report relative cross-sections for the production of the ions RH^+, RH_2^+ and H_2 by collisions of excited rare gas atoms R^* with H_2, and by using HD the isotope effect has been studied. An interesting additional feature of this work is the study of reactions involving atoms in long-lived high Rydberg states[428]. In the case of argon and krypton, ions are produced only by the high Rydberg states, whereas in the case of helium and neon only the metastable states contribute to a measurable extent. Simple models are proposed to explain the experimental results. For the case of the highly excited Rydberg states the primary process was found to be

$$R^{**} + H_2 \rightarrow R^+ + H_2 + \varepsilon(E = 0)$$

The production of 'zero' energy electrons by this type of process is probably a contributing factor to the rising backgrounds observed in many electron scavenging experiments (see Section 3.4.4).

Although other experiments[421, 422] have indicated that Penning ionisation is essentially a Franck–Condon process the first definitive PIES confirmation has been reported by Hotop and Niehaus[410, 423] for the ionisation of H_2 by $He(2^3S)$. The vibrational quanta for H_2^+ were found to be (in eV) – 0.63(0.43), 0.95(0.88), 1.00(1.00), 0.76(0.89), 0.56(0.70) and 0.65(0.53) in close agreement with the results of PES (values in brackets) obtained at 584 Å in the same spectrometer. For the purpose of such a comparison it is vital that the PES and PIES measurements be made in the same spectrometer since the results will only then be independent of the electron collection efficiency as a function of energy. The small differences in Franck–Condon factors are attributed to either a perturbation of the H–H potential or to a possibly different influence of the electronic part of the transition matrix element describing the Penning ionisation. Nevertheless Hotop and Niehaus consider that the results indicate that Penning ionisation is a good approximation to a vertical process. Using $Ne(^3P_2)$ metastables on H_2, Hotop and Niehaus[410] find good agreement with the results using $He(2^3S)$. An independent study of H_2 using helium metastables in the reviewer's laboratory suggests Franck–Condon factors more closely in agreement with PES values while similar results have also been obtained for the ground state of NO^+ [429]. Further detailed studies by Hotop and Niehaus of Ar, Kr, Xe and Hg [407] in collision with helium metastables provide precise measurements of shifts and peak broadening in the Penning spectra. The shapes and shifts of the distributions are very different for $He(2^1S)$ and $He(2^3S)$ metastables and show a marked change with the relative energy (temperature) of the collision. The magnitude and sign of the shifts vary with size of the target atom and it is reported that the ratio of associative to Penning ionisation can be approximately predicted from the shifts. Following a discussion of the distributions in terms of potential energy curves the results are interpreted on the assumption that the process of Penning ionisation takes place by an electron exchange mechanism

$$X(1) + Y^*(2) \rightarrow X(2) + Y^*(1) + \varepsilon$$

rather than the alternative 'radiative' mechanism

$$X(1) + Y^*(2) \rightarrow X(1) + Y^*(2) + \varepsilon$$

in which the metastable atom is perturbed during the collision, with Penning ionisation being effected by the photon liberated as Y* returns to its ground state. In further studies[408–410] additional evidence would also seem to favour the electron exchange mechanism. A number of theoretical treatments of Penning ionisation are to be found in the literature[430–432], but these involve ideas based on the optical rather than the exchange mechanism. A number of other theoretical treatments of Penning and associative ionisation are to be found in the recent literature[431–438]. Using the quenching techniques to produce He(2^3S) beams, Hotop, Niehaus and Schmeltekopf[408] have measured the singlet to triplet Penning cross-section ratios to be close to unity for various target gases. For the rare gases it is found that the ratio of associative to Penning ionisation increases considerably with decrease in temperature for both He(2^1S) and He(2^3S). Mass spectrometric measurements are consistent with the conclusions drawn from the measured electron energy distributions. A further study[409] of the ionisation of Na, K and Hg by helium metastables leads to the determination of interaction potentials and cross-sections. It is shown that in general the well-depth of the interaction between the metastable and the target particle can be obtained directly from the measured electron distribution. A comparison of Penning and photoelectron spectra for H_2, N_2 and CO [410] shows that in general population of electronic states is quite different for the two processes, while the relative population of vibrational states in each electronic state is very nearly equal. The differences can be attributed to different electronic transition moments for the two processes. Small but definite differences of vibrational populations are explained by a weak distortion of the molecule by the projectile atom during Penning ionisation. In addition, the broadening and shift of Penning peaks is characteristic of the electronic state as well as of the projectile but does not depend on the vibrational quantum number in a given band. The role of vibrational–translational energy conversion in molecules is also discussed and the results indicate that it is particularly effective in the case of H_2. The angular distribution of Penning electrons resulting from collisions of Ar with helium metastables has recently been reported by Hotop and Niehaus[439]. The distribution is assymmetric, favouring large scattering angles (the backward direction) and is also different for He(2^1S) and He(2^3S). The results are interpreted as indicating a considerable alignment of the diatomic at the instant of ionisation.

The refinements introduced by Niehaus and his co-workers have brought a new degree of sophistication to PIES following the pioneering exploratory studies by Cermak. To date there has been no published work using electrostatic deflection analysers for PIES, even though these types of spectrometer have found universal acceptance in other branches of electron spectroscopy. The desirable features of deflection analysers merit their adoption in PIES. In the reviewer's laboratory a 127-degree electrostatic analyser has been used for an extensive series of PIES studies of atoms and molecules using helium metastables[429]. Interesting structure has also been observed[429] in electron spectra resulting from collisions of two metastables in single and mixed rare gas systems.

At low electron energies Penning electron spectra are complicated by a rising background due to Auger electron emission following metastable collisions with the walls of the spectrometer. This undesirable background could be removed by using a modulated target gas beam and phase sensitive detection. With this improvement it should be possible to study many more long-lived electronically excited species of interest to chemists.

References

1. Garstang, R. (1962). *Atomic and Molecular Processes* (Ed. D. R. Bates), 1. (New York: Academic Press)
2. Collin, J. (1968). *Modern Aspects of Mass Spectrometry*, 231. (New York: Plenum Press)
3. Berry, R. S. (1969). *Ann. Rev. Phys. Chem.*, **20**, 357
4. Turner, D. W., Baker, A. D., Baker, C. and Brundle, C. R. (1970). *High Resolution Molecular Photoelectron Spectroscopy*. (New York: J. Wiley)
5. Brundle, C. R. and Robin, M. B. (1971). *Determinations of Organic Structures by Physical Methods Vol. III*, Ed. Nachod/Zuckerman (to be published)
6. Worley, S. D. (1971). *Chem. Rev.*, **71**, 295
7. Siegbahn, K., Nordling, C., Fahlman, A., Norderg, R., Hamrin, K., Hedman, J., Johansson, G., Bergmark, T., Karlson, S., Lindgren, I. and Lindberg, B. (1967). *ESCA: Atomic Molecular and Solid State Structure Studied by Means of Electron Spectroscopy*. (Uppsala: Almquist and Wicksells Boktryckeri AB)
8. Siegbahn, K., Nordling, C., Johansson, G., Hedman, J., Heden, P. F., Hamrin, K., Gelius, U., Bergmark, T., Werme, L. O., Manne, R. and Baer, Y. (1969). *ESCA: Applied to Free Molecules*. (Amsterdam: North Holland Publishing Co.)
9. Hercules, D. M. (1970). *Anal. Chem.*, **42**, 20
10. Yin, L. I., Adler, I. and Lamothe, R. (1969). *Appl. Spectrosc.*, **23**, 41
11. Hollander, J. and Jolly, W. L. (1970). *Accounts Chem. Res.*, **3**, 193
12. Trajmar, S., Rice, J. K. and Kuppermann, A. (1970). *Advan. Chem. Phys.*, **18**, 15
13. Taylor, H. S. (1970). *Advan. Chem. Phys.*, **18**, 91
14. Kuyatt, C. E. (1968). *Methods of Experimental Physics*, **7A**, 1
15. Cermak, V. (1968). *Coll. Czech. Chem. Commun.*, **33**, 2739
16. Rudberg, E. (1930). *Proc. Roy. Soc. (London)*, **A127**, 628
17. Nottingham, W. B. (1939). *Phys. Rev.*, **55**, 203
18. Turner, D. W. and May, D. P. (1966). *J. Chem. Phys.*, **45**, 471
19. Lozier, W. (1930). *Phys. Rev.*, **36**, 1285
20. Al-Joboury, M. I. and Turner, D. W. (1963). *J. Chem. Soc., B*, 5154
21. Frost, D. C., McDowell, C. A. and Vroom, D. A. (1967). *Proc. Roy. Soc.*, **A296**, 566
22. Samson, J. A. R. and Cairns, R. B. (1968). *Phys. Rev.*, **173**, 80
23. Spohr, R. and Puttkamer, E. V. (1967). *Z. Naturforsch*, **22a**, 705
24. Hotop, H. and Niehaus, A. (1969). *Z. Physik*, **228**, 68
25. Fox, R. E., Hickman, W. M., Grove, D. J. and Kjeldaas, T. (1955). *Rev. Sci. Instrum.*, **26**, 1101
26. Hall, R. I., Mazeau, J., Reinhardt, J. and Scherman, C. (1970). *J. Phys. B.*, **3**, 991
27. Golden, D. E. and Zecca, A. (1971). *Rev. Sci. Instrum.*, **42**, 210
28. Gordon, S. M., Haarhoff, P. C. and Krige, G. J. (1969). *Int. J. Mass. Spectrosc. Ion Phys.*, **3**, 13
29. Chantry, P. J. (1969). *Rev. Sci. Instrum.*, **40**, 884
30. Thomas, G. E. and Vogelsburg, F. E. (1971). *Rev. Sci. Instrum.*, **42**, 161
31. Sar-el, H. Z. (1970). *Rev. Sci. Instrum.*, **41**, 561
32. Aksela, S., Karnas, M., Pessa, M. and Suoninen, E. (1970). *Rev. Sci. Instrum.*, **41**, 351
33. Hafner, H., Simpson, J. A. and Kuyatt, C. E. (1968). *Rev. Sci. Instrum.*, **39**, 33
34. Mehlhorn, W. (1960). *Z. Physik*, **160**, 247
35. Purcell, E. M. (1938). *Phys. Rev.*, **54**, 818
36. Simpson, J. A. (1964). *Rev. Sci. Instrum.*, **35**, 1698
37. Kuyatt, C. E. and Simpson, J. A. (1967). *Rev. Sci. Instrum.*, **38**, 103
38. Imhof, R. E. and Read, F. H. (1968). *J. Phys. E. (J. Sci. Instrum.)*, **1**, 859
39. Read, F. H. (1969). *J. Phys. E. (J. Sci. Instrum.)*, **2**, 165
40. Read, F. H. (1969). *J. Phys. E (J. Sci. Instrum.)*, **2**, 679
41. Read, F. H. (1970). *J. Phys. E (J. Sci. Instrum.)*, **3**, 127

42. Heddle, D. W. O. (1969). *J. Phys. E (J. Sci. Instrum.)*, **2,** 1046
43. Heddle, D. W. O. and Kurepa, M. V. (1970). *J. Phys. E (J. Sci. Instrum.)*, **3,** 552
44. Heddle, D. W. O. (1970). *Tables of Focal Properties of Three Element Electrostatic Cylinder Lenses*, J.I.L.A. Report No. 104, University of Colorado.
45. Lassettre, E. N., Skerbele, A., Dillon, M. A. and Ross, K. J. (1968). *J. Chem. Phys.*, **48,** 5066
46. Hughes, A. L. and Rojansky, V. (1929). *Phys. Rev.*, **34,** 284
47. Hughes, A. L. and McMillan, J. H. (1929). *Phys. Rev.*, **34,** 291
48. Clarke, E. M. (1954). *Can. J. Phys.*, **32,** 764
49. Marmet, P. and Kerwin, L. (1960). *Can. J. Phys.*, **38,** 787
50. Turner, D. W. (1968). *Proc. Roy. Soc. (London)*, **A307,** 15
51. Salop, A., Golden, D. E. and Nakano, H. (1969). *Rev. Sci. Instrum.*, **40,** 733
52. Brion, C. E. and Tam, W. C., unpublished work
53. Lloyd, D. R. (1970). *J. Phys. E (J. Sci. Instrum.)*, **3,** 629
54. Harrower, G. A. (1955). *Rev. Sci. Instrum.*, **26,** 850
55. Eland, J. D. H. and Danby, C. J. (1968). *J. Phys. E (J. Sci. Instrum.)*, **1,** 406
56. Roy, D. and Carette, J. D. (1970). *Appl. Phys. Lett.*, **16,** 413
57. Gough, M. P. (1970). *J. Phys. E (J. Sci. Instrum.)*, **3,** 332
58. Boersch, H., Geiger, J. and Stickel, W. (1964). *Z. Physik*, **180,** 415
59. Andersen, W. H. J. and LePoole, J. B. (1970). *J. Phys. E (J. Sci. Instrum.)*, **3,** 121
60. van der Wiel, M. J. (1970). *Physica*, **49,** 411
61. Stamatovic, A. and Schulz, G. J. (1968). *Rev. Sci. Instrum.*, **39,** 1752
62. Stamatovic, A. and Schulz, G. J. (1970). *Rev. Sci. Instrum.*, **41,** 423
63. Ross, K. J. and Garment, R. (1969). *J. Phys. E (J. Sci. Instrum.)*, **2,** 437
64. Samson, J. A. R. (1969). *Rev. Sci. Instrum.*, **40,** 1174
65. Boersch, H., Geiger, J. and Topschowsky, M. (1962). *Phys. Lett.*, **3,** 64
66. Geiger, J. and Topschowsky, M. (1966). *Z. Naturforsch*, **21a,** 626
67. Ehrhardt, H. and Linder, F. (1968). *Phys. Rev. Lett.*, **21,** 419
68. Bendix Corporation, Ann Arbor, Michigan, U.S.A.
69. Mullard Limited, London, England
70. Natalis, P. and Collin, J. (1968). *Chem. Phys. Lett.*, **2,** 414
71. McDowell, C. A. (1963). *Mass Spectrometry*. (New York: McGraw-Hill)
72. Fuchs, V. and Hotop, H. (1969). *Chem. Phys. Lett.*, **4,** 71
73. Stebbings, W. L. and Taylor, J. W. (1971). *Int. J. Mass Spectrom. Ion Phys.*, **6,** 152
74. Branton, G. R., Frost, D. C., Makita, T., McDowell, C. A. and Stenhouse, I. A. (1970). *J. Chem. Phys.*, **52,** 802
75. Samson, J. A. R. and Cairns, R. B. (1968). *Phys. Rev.*, **173,** 80
76. Schoen, R. I. (1964). *J. Chem. Phys.*, **40,** 1830
77. Brundle, C. R., Robin, M. B. and Jones, G. R. (1970). *J. Chem. Phys.*, **52,** 3383
78. Brundle, C. R., Neumann, D., Price, W. C., Evans, D., Potts, A. W. and Streets, D. G. (1970). *J. Chem. Phys.*, **53,** 705
79. Edquist, O., Lindholm, E., Selin, L. E., Sjogren, H. and Asbrink, L. (1970). *Arkiv Fysik*, **40,** 439
80. Vilesov, F. I., Kurbatov, B. L. and Terenin, A. N. (1961). *Sov. Phys. Dokl.*, **6,** 490
81. Dewar, M. J. S. and Worley, S. D. (1969). *J. Chem. Phys.*, **50,** 654
82. Price, W. C. and Turner, D. W. (1970). *Phil. Trans. Roy. Soc. (London)*, **268,** 1
83. Baker, A. D. (1970). *Accounts Chem. Res.*, **34,** 17
84. Lindholm, E. (1969). *Arkiv Fysik*, **40,** 97
85. Lindholm, E. (1969). *Arkiv Fysik*, **40,** 103
86. Lindholm, E. (1969). *Arkiv Fysik*, **40,** 111
87. Lindholm, E. (1969). *Arkiv Fysik*, **40,** 117
88. Lindholm, E. (1969). *Arkiv Fysik*, **40,** 125
89. Lindholm, E. (1969). *Arkiv Fysik*, **40,** 129
90. Lindholm, E. (1970). *Arkiv Fysik*, **40,** 439
91. Cornford, A. B., Frost, D. C., McDowell, C. A., Ragle, J. L. and Stenhouse, I. A. (1970). *Chem. Phys. Lett.*, **5,** 486
92. Hunter, G. and Pritchard, H. O. (1967). *J. Chem. Phys.*, **46,** 2153
93. Asbrink, L. (1970). *Chem. Phys. Lett.*, **7,** 549
94. Brundle, C. R. (1970). *Chem. Phys. Lett.*, **7,** 317
95. Collin, J. E. and Natalis, P. (1969). *Int. J. Mass Spectrom. Ion Phys.*, **2,** 231

96. Edquist, O., Lindholm, E., Selin, L. E. and Asbrink, L. (1970). *Phys. Lett.*, **31A,** 292
97. Edquist, O., Lindholm, E., Selin, L. E. and Asbrink, L. (1970). *Physica Scripta,* **1,** 25
98. Dixon, R. N. and Hull, S. E. (1969). *Chem. Phys. Lett.*, **3,** 367
99. Bahr, J. L., Blake, A. J., Carver, J. H. and Kumer, V. (1969). *J. Quant. Spectrosc. Rad. Trans.*, **9,** 1359
100. Branton, G. R., Frost, D. C., Herring, F. G., McDowell, C. A. and Stenhouse, I. A. (1969). *Chem. Phys. Lett.*, **3,** 581
101. Weiss, M. J. and Lawrence, G. M. (1970). *J. Chem. Phys.*, **53,** 214
102. Weiss, M. J., Lawrence, G. M. and Young, R. A. (1970). *J. Chem. Phys.*, **52,** 2867
103. Natalis, P. and Collin, J. E. (1969). *Int. J. Mass Spectrosc. Ion Phys.*, **2,** 221
104. Brundle, C. R. (1970). *Chem. Phys. Lett.*, **5,** 410
105. Natalis, P. and Collin, J. E. (1968). *Chem. Phys. Lett.*, **2,** 79
106. Brundle, C. R., Neumann, D., Price, W. C., Evans, D., Potts, A. W. and Streets, D. G. (1970). *J. Chem. Phys.*, **53,** 705
107. Edquist, O., Lindholm, E., Selin, L. E., Asbrink, L. Kuyatt, C. E., Mielczarek, S. R., Simpson, J. A. and Fisher-Hjalmer, I., (1970). *Physical Scripta,* **1,** 172
108. Samson, J. A. R. (1968). *Phys. Lett.*, **28A,** 391
109. Pullen, B. P., Carlson, T. A., Moddeman, W. E., Schweitzer, G. K., Bull, W. E. and Grimm, F. A. (1970). *J. Chem. Phys.*, **53,** 768
110. Ragle, J. L., Stenhouse, I. A., Frost, D. C. and McDowell, C. A. (1970). *J. Chem. Phys.*, **53,** 178
111. Brundle, C. R., Robin, M. B. and Basch, H. (1970). *J. Chem. Phys.*, **53,** 2196
112. Hashmall, J. A. and Heilbronner, E. (1970). *Angew. Chem.*, **9,** 305
113. Brailsford, D. F. and Ford, B. (1970). *Mol. Phys.*, **18,** 621
114. Brundle, C. R., Robin, M. B., Basch, H., Pinsky, M. and Bond, A. (1970). *J. Amer. Chem. Soc.*, **92,** 3863
115. Baker, A. D., Betteridge, D., Kemp, N. R. and Kurby, R. E. (1970). *Int. J. Mass Spectrosc. Ion Phys.*, **4,** 90
116. Asbrink, L., Lindholm, E. and Edquist, O. (1970). *Chem. Phys. Lett.*, **5,** 609
117. Momigny, J., Goffart, C. and Natalis, P. (1968). *Bull. Soc. Chem. Belges,* **77,** 533
118. Goffart, C., Momigny, J. and Natalis, P. (1969). *Int. J. Mass. Spectrosc. Ion. Phys.*, **3,** 371
119. Momigny, J. and Lorquet, J. C. (1969). *Int. J. Mass Spectrosc. Ion. Phys.*, **2,** 495
120. Brundle, C. R., Turner, D. W., Robin, M. B. and Basch, H. (1969). *Chem. Phys. Lett.*, **3,** 292
121. Bischof, P., Gleiter, R. and Heilbronner, E. (1970). *Helv. Chim. Acta,* **53,** 1425
122. Praet, M. Th. and Delwiche, J. (1970). *Chem. Phys. Lett.*, **5,** 546
123. Brundle, C. R. and Robin, M. B. (1970). *J. Amer. Chem. Soc.*, **92,** 550
124. Demeo, D. A. and Yencha, A. J. (1970). *J. Chem. Phys.*, **53,** 4536
125. Brundle, C. R., Robin, M. B. and Jones, G. R. (1970). *J. Chem. Phys.*, **52,** 3383
126. Delwiche, J. and Natalis, P. (1970). *Chem. Phys. Lett.*, **5,** 564
127. Delwiche, J., Natalis, P. and Collin, J. E. (1970). *Int. J. Mass. Spectrosc. Ion. Phys.*, **5,** 443
128. Bassett, P. J. and Lloyd, D. R. (1970). *Chem. Phys. Lett.*, **6,** 166
129. Evans, S., Green, J. C., Green, M. L. J., Orchard, A. F. and Turner, D. W. (1970). *Disc. Faraday Soc.*, **47,** 112
130. Braterman, P. S. and Walker, A. P. (1970). *Disc. Faraday Soc.*, **47,** 121
131. Cox, P. A., Evans, S., Hamnett, A. and Orchard, A. F. (1970). *Chem. Phys. Lett.*, **7,** 414
132. Branton, G. R., Brion, C. E., Frost, D. C., Mitchell, K. A. R. and Paddock, N. L. (1970). *J. Chem. Soc. A,* 151
133. Jonathan, N., Morris, A., Smith, D. J. and Ross, K. J. (1970). *Chem. Phys. Lett.*, **7,** 497
134. Jonathan, N., Smith, D. J. and Ross, K. J. (1970). *J. Chem. Phys.*, **53,** 3758
135. Jonathan, N., Smith, D. J. and Ross, K. J. (1971). *Chem. Phys. Lett.*, **9,** 217
136. Cornford, A. B., Frost, D. C., Herring, F. G. and McDowell, C. A. (1971). *J. Chem. Phys.*, **54,** 1872
137. Siegbahn, K. (1969). *Rep. No. UUIP–670.* (University of Uppsala)
138. Yin, L. I., Adler, I. and Lamothe, R. (1969). *Appl. Spectrose,* **23,** 41
139. Helmer, J. C. and Weickert, N. H. (1968). *Appl. Phys. Lett.*, **13,** 266
140. Fadley, C. S., Miner, C. E. and Hollander, J. M. (1969). *Appl. Phys. Lett.*, **15,** 223
141. Fahlman, A., Albridge, R. G., Nordberg, R. and LaCasse, W. M. (1970). *Rev. Sci. Instrum.*, **41,** 596

142. Hedman, J., Heden, P. F., Nordling, C. and Siegbahn, K. (1969). *Phys. Lett.*, **29A,** 178
143. Malmsten, G., Nilsson, O., Thoren, L. and Bergmark, J. E. (1970). *Physica Scripta,* **1,** 37
144. Baer, Y., Heden, P. F., Hedman, J., Klasson, M., Nordling, C. and Siegbahn, K. (1970). *Physica Scripta,* **1,** 55
145. Lindberg, B. J., Hamrin, K., Johansson, G., Gelius, U., Fahlman, A., Nordling, C. and Siegbahn, K. (1970). *Physica Scripta,* **1,** 286
146. Gelius, U., Heden, P. F., Hedman, J., Lindberg, B. J., Manne, R., Nordberg, R., Nordling, C. and Siegbahn, K. (1970). *Physica Scripta,* **2,** 70
147. Thomas, T. D. (1970). *J. Chem. Phys.,* **52,** 1373
148. Thomas, T. D. (1970). *J. Chem. Phys.,* **53,** 1744
149. Davis, D. W., Hollander, J. M., Shirley, D. A. and Thomas, T. D. (1970). *J. Chem. Phys.,* **52,** 3295
150. Fadley, C. S. and Shirley, D. A. (1970). *Phys. Rev. A,* **2,** 1109
151. Schwartz, M. E. (1970). *Chem. Phys. Lett.,* **5,** 50
152. Schwartz, M. E. (1970). *Chem. Phys. Lett.,* **6,** 631
153. Schwartz, M. E. (1970). *Chem. Phys. Lett.,* **7,** 78
154. Basch, H. and Snyder, L. C. (1969). *Chem. Phys. Lett.,* **3,** 333
155. Gelius, U., Ross, B. and Siegbahn, P. (1970). *Chem. Phys. Lett.,* **4,** 471
156. Basch, H. (1970). *Chem. Phys. Lett.,* **5,** 337
157. Bent, H. A. (1971). *J. Chem. Phys.,* **54,** 824
158. Hillier, I. H., Saunders, V. R. and Wood, M. H. (1970). *Chem. Phys. Lett.,* **7,** 323
159. Manne, R. and Aberg, T. (1970). *Chem. Phys. Lett.,* **7,** 282
160. Manne, R. (1970). *Chem. Phys. Lett.,* **5,** 125
161. Coulson, C. A. and Gianturco, F. A. (1970). *Molec. Phys.,* **18,** 607
162. Ha, T. K. and O'Kanski, C. T. (1969). *Chem. Phys. Lett.,* **3,** 603
163. Kramer, L. N. and Klein, M. P. (1971). *Chem. Phys. Lett.,* **8,** 183
164. Chaffee, M. A. (1931). *Phys. Rev.,* **37,** 1233
165. Berkowitz, J. and Ehrhardt, H. (1966). *Phys. Lett.,* **21,** 531
166. Berkowitz, J., Ehrhardt, H. and Tekaat, T. (1967). *Z. Physik,* **200,** 69
167. Hall, L. and Siegel, M. W. (1968). *J. Chem. Phys.,* **48,** 943
168. Auger, P. and Pervin, F. (1927). *J. Phys. Ser. VI,* **8,** 93
169. Schur, G. (1930). *Ann. Phys.,* **4,** 433
170. Tully, J. C., Berry, R. S. and Dalton, B. J. (1968). *Phys. Rev.,* **176,** 95
171. Cooper, J. and Zare, R. N. (1968). *J. Chem. Phys.,* **48,** 942
172. Cooper, J. W. and Manson, S. T. (1969). *Phys. Rev.,* **177,** 157
173. Krause, M. O. (1969). *Phys. Rev.,* **177,** 151
174. Vroom, D. A., Cromeaux, A. R. and McGowan, J. W. (1969). *Chem. Phys. Lett.,* **3,** 476
175. McGowan, J. W., Vroom, D. A. and Cromeaux, A. R. (1969). *J. Chem. Phys.,* **51,** 5626
176. Morgenstern, R., Niehaus, A. and Ruf, M. W. (1970). *Chem. Phys. Lett.,* **4,** 635
177. Samson, J. A. R. (1970). *Phil. Trans. Roy. Soc. London, A,* **268,** 141
178. Harrison, H. (1970). *J. Chem. Phys.,* **52,** 901
179. Sichel, J. M. (1970). *Molec. Phys.,* **18,** 95
180. Buckingham, A. D., Orr, B. J. and Sichel, J. M. (1970). *Phil. Trans. Roy. Soc. London, A,* **268,** 147
181. Manson, S. T. and Cooper, J. W. (1970). *Phys. Rev. A,* **2,** 2170
182. Manson, S. T. and Kennedy, D. J. (1970). *Chem. Phys. Lett.,* **7,** 387
183. Villarejo, D., Herm, R. R. and Inghram, M. G. (1967). *J. Chem. Phys.,* **46,** 4995
184. Villarejo, D. (1968). *J. Chem. Phys.,* **48,** 4014
185. Villarejo, D., Stockbauer, R. and Inghram, M. G. (1969). *J. Chem. Phys.,* **50,** 4599
186. Chupka, W. A. (1968). *J. Chem. Phys.,* **48,** 2337
187. Al-Joboury, M. I. and Turner, D. W. (1967). *J. Chem. Soc. B,* 373
188. Hamrin, K., Johansson, G., Gelius, U., Fahlman, A., Nordling, C. and Siegbahn, K. (1968). *Chem. Phys. Lett.,* **1,** 613
189. Peatman, W. B., Borne, T. B. and Schlag, E. W. (1969). *Chem. Phys. Lett.,* **3,** 492
190. Baer, T., Peatman, W. B. and Schlag, E. W. (1969). *Chem. Phys. Lett.,* **4,** 243
191. Berkowitz, J. and Chupka, W. A. (1969). *J. Chem. Phys.,* **51,** 2341
192. Puttkamer, E. V. (1970). *Z. Naturforsch.,* **25a,** 1062
193. Kerwin, L., Marmet, P. and Carette, J. D. (1969). *Case Studies in Atomic Collision Physics,* 525 (New York: J. Wiley)
194. Simpson, J. A. (1967). *Methods of Experimental Physics.,* **4A,** 84

195. Simpson, J. A. (1967). *Methods of Experimental Physics*, **4A**, 124
196. Moiseiwitsch, B. L. and Smith, S. J. (1968). *Rev. Mod. Phys.*, **40**, 238
197. Rudge, M. R. H. (1968). *Rev. Mod. Phys.*, **40**, 564
198. Kieffer, L. J. (1969). *Atomic Data*, **1**, 19
199. Kieffer, L. J. (1969). *Atomic Data*, **1**, 120
200. McGowan, J. W. (1970). *Science*, **167**, 1083
201. Heddle, D. W. O. and Keesing, R. G. W. (1968). *Advances in Atomic and Molecular Physics*, **4** (New York: Academic Press)
202. Lassettre, E. N. (1969). *Can. J. Chem.*, **47**, 1733
203. Kuppermann, A., Rice, J. K. and Trajmar, S. (1968). *J. Phys. Chem.*, **72**, 3894
204. Taylor, H. S. (1970). *Advan. Chem. Phys.*, **18**, 91
205. Taylor, H. S., Nazaroff, G. V. and Golebiewski, A. (1966). *J. Chem. Phys.*, **45**, 2872
206. Burke, P. G. (1968). *Advances in Atomic and Molecular Physics*, **4.** (New York: Academic Press)
207. Fano, U. (1969). *Atomic Physics*, 209 (New York: Plenum Press)
208. Fano, U. (1969). *Comments Atom. Molec. Phys.*, **1**, 45
209. Bardsley, J. N. and Mandl, F. (1968). *Rep. Prog. Phys.*, **31**, 471
210. Fano, U. and Cooper, J. W. (1968). *Rev. Mod. Phys.*, **40**, 441
211. Bonham, R. A. (1971). *Theory of Electron Impact Spectroscopy of Atoms and Molecules* (Baltimore: University Park Press)
212. Bethe, H. (1930). *Ann. Physik*, **5**, 325
213. Fano, U. (1954). *Phys. Rev.*, **95**, 1198
214. Lassettre, E. N. (1959). *Radiation Res. Supplement*, **1**, 530
215. Compton, R. N., Huebner, R. H., Reinhardt, P. W. and Christophorou, L. G. (1968). *J. Chem. Phys.*, **48**, 901
216. Brion, C. E. and Olsen, L. A. R. (1970). *J. Phys. B*, **3**, 1020
217. Haas, R. (1957). *Z. Physik*, **148**, 177
218. Schulz, G. J. (1959). *Phys. Rev.*, **116**, 1141
219. Schulz, G. J. (1962). *Phys. Rev.*, **125**, 229
220. Schulz, G. J. (1964). *Phys. Rev.*, **135**, 988
221. Skerbele, A., Dillon, M. A. and Lassettre, E. N. (1968). *J. Chem. Phys.*, **49**, 3543
222. Skerbele, A., Dillon, M. A. and Lassettre, E. N. (1968). *J. Chem. Phys.*, **49**, 5042
223. Phelps, A. V. (1968). *Rev. Mod. Phys.*, **40**, 399
224. Geiger, J. and Witmaack, K. (1965). *Z. Physik*, **187**, 433
225. Schulz, G. J. (1962). *Phys. Rev.*, **125**, 229
226. Andrick, D. and Ehrhardt, H. (1966). *Z. Physik.*, **192**, 99
227. Gibson, J. R. and Dolder, K. T. (1969). *J. Phys. B*, **2**, (741)
228. Schulz, G. J. (1959). *Phys. Rev.*, **112**, 150
229. Morrison, J. D. (1963). *J. Chem. Phys.*, **39**, 200
230. Dromey, R. G. and Morrison, J. D. (1970). *Int. J. Mass. Spectrom. Ion Phys.*, **4**, 475
231. Eland, J. H. D., Shepherd, P. J. and Danby, C. J. (1966). *Z. Naturforsch.*, **21a**, 1580
232. Macneill, K. A. G. and Thynne, J. C. J. (1969). *Int. J. Mass Spectrom. Ion Phys.*, **3**, 35
233. Cowperthwaite, R. L. and Myers, H. (1970). *J. Chem. Phys.*, **53**, 1077
234. Winters, R. E., Collins, J. H. and Courchene, W. L. (1966). *J. Chem. Phys.*, **45**, 1931
235. Collins, J. H., Winters, R. E. and Engerholm, G. G. (1968). *J. Chem. Phys.*, **49**, 2469
236. Boersch, H., Geiger, J. and Topschowsky, M. (1969). *Proceedings of the Sixth International Conference on the Physics of Electronic and Atomic Collisions*, 263 (Cambridge, Massachusetts: MIT Press)
237. Geiger, J. and Schmoranzer, H. (1969). *Proceedings of the Sixth International Conference on the Physics of Electronic and Atomic Collisions*, 428 (Cambridge, Massachusetts: MIT Press)
238. Boersch, H., Geiger, J. and Schröder, B. (1969). *Physics of the One and Two Electron Atoms*, 637 (Amsterdam: North Holland)
239. Hartree, D. R. (1928). *Proc. Cambridge Phil. Soc.*, **24**, 426
240. Geiger, J. (1970). *Phys. Lett.*, **33A**, 351
241. Bonham, R. A. and Geiger, J. (1969). *J. Chem. Phys.*, **51**, 5249
242. van der Wiel, M. J., El-Sherbini, Th. M. and Vriens, L. (1969). *Physica*, **42**, 411
243. Ikelaar, P., van der Wiel, M. J. and Tebra, W. (1971). *J. Phys. E*, **4**, 102
244. van der Wiel, M. J. and Wiebes, G. (1971). *Physica*, **53**, 225
245. van der Wiel, M. J., El-Sherbini, Th. M. and Brion, C. E. (1970). *Chem. Phys. Lett.*, **7**, 161

246. Nakamura, M. *et al.* (1969). *Phys. Rev.*, **178,** 80
247. Stalherm, D., Cleff, B., Hillig, H. and Mehlhorn, W. (1969). *Z. Naturforsch.*, **24a,** 1728
248. Burrow, P. D. and Schulz, G. J. (1969). *Phys. Rev.*, **187,** 97
249. Simpson, J. A. (1969). *Mat. Res. and Standards,* **9,** 13
250. Gerber, S. H. (1970). *J. Polymer. Sci., C.* 211
251. Rendina, J. F. and Grojean, R. E. (1971). *Appl. Spectrosc.*, **25,** 24
252. Grojean, R. E. and Rendina, J. F. (1971). *Anal. Chem.*, **43,** 162
253. Lassettre, E. N., Skerbele, A. and Dillon, M. A. (1969). *J. Chem. Phys.*, **50,** 1829
254. Bromberg, J. P. (1969). *J. Chem. Phys.*, **50,** 3906
255. Bromberg, J. P. (1969). *J. Chem. Phys.*, **51,** 4117
256. Bromberg, J. P. (1970). *J. Chem. Phys.*, **52,** 1243
257. Lassettre, E. N., Skerbele, A. and Dillon, M. A. (1970). *J. Chem. Phys.*, **52,** 2797
258. Chamberlain, G. E., Mielczarek, S. R. and Kuyatt, C. E. (1970). *Phys. Rev. A,* **2,** 1905
259. Skerbele, A., Ross, K. J. and Lassettre, E. N. (1969). *J. Chem. Phys.*, **50,** 4486
260. Skerbele, A. and Lassettre, E. N. (1970). *J. Chem. Phys.*, **52,** 2708
261. Skerbele, A. and Lassettre, E. N. (1970). *J. Chem. Phys.*, **53,** 3806
262. Lassettre, E. N. and Skerbele, A. (1971). *J. Chem. Phys.*, **54,** 1597
263. Meyer, V. D. and Lassettre, E. N. (1971). *J. Chem. Phys.*, **54,** 1608
264. Lassettre, E. N. (1970). *J. Chem. Phys.*, **53,** 3801
265. Simpson, J. A. (1969). *Proceedings of the Sixth International Conference on the Physics of Electronic and Atomic Collisions,* 344 (Cambridge, Massachusetts: MIT Press)
266. Krauss, M. and Mielczarek, S. R. (1969). *J. Chem. Phys.*, **51,** 5241
267. Miller, K. J. (1969). *J. Chem. Phys.*, **51,** 5235
268. Hertel, I. V. and Ross, K. J. (1968). *J. Phys. B,* **1,** 697
269. Hertel, I. V. and Ross, K. J. (1968). *J. Phys. B,* **2,** 285
270. Hertel, I. V. and Ross, K. J. (1969). *J. Phys. B,* **2,** 484
271. Hertel, I. V. and Ross, K. J. (1969). *J. Chem. Phys.*, **50,** 536
272. Lassettre, E. N. and Skerbele, A. (1966). *J. Chem. Phys.*, **45,** 1077
273. Doering, J. P. and Williams, A. J. (1967). *J. Chem. Phys.*, **47,** 4180
274. Doering, J. P. and Williams, A. J. (1969). *J. Chem. Phys.*, **51,** 2859
275. Williams, A. J. and Doering, J. P. (1969). *Planet. Space Sci.*, **17,** 1527
276. Doering, J. P. (1969). *J. Chem. Phys.*, **51,** 2866
277. Truhlar, D. G., Rice, J. K., Kuppermann, A., Trajmar, S. and Cartwright, D. C. (1970). *Phys. Rev. A,* **1,** 778
278. Steelhammer, J. C. and Lipsky. S. (1970). *J. Chem. Phys.*, **53,** 4112
279. Steelhammer, J. C. and Lipsky, S. (1970). *J. Chem. Phys.*, **53,** 1445
280. Truhlar, D. G. and Rice, J. K. (1970). *J. Chem. Phys.*, **52,** 4480
281. Trajmar, S., Truhlar, D. C., Rice, J. K. and Kuppermann, A. (1970). *J. Chem. Phys.*, **52,** 4516
282. Trajmar, S., Truhlar, D. C. and Rice, J. K. (1970). *J. Chem. Phys.*, **52,** 4502
283. Brinkmann, R. T. and Trajmar, S. (1970). *Ann. Geophys.*, **26,** 201
284. Trajmar, S., Williams, W. and Kuppermann, A. *J. Chem. Phys.*, to be published
285. Claydon, C. R., Segal, G. A. and Taylor, H. S., to be published
286. Cartwright, D. C. (1970). *Phys. Rev. A,* **2,** 1331
287. Matsuzawa, M. (1969). *J. Chem. Phys.*, **51,** 4705
288. Matsuzawa, M. (1969). *J. Chem. Phys.*. **52,** 5976
289. Konishi, A., Wakiya, K., Yamamoto, M. and Suzuki, H. (1970). *J. Phys. Soc. Japan,* **29,** 526
290. Foo, V. Y., Brion, C. E. and Hasted, J. B. (1971). *Proc. Roy. Soc. (London)* **A322,** 535
291. Ehrhardt, H., Schulz, M., Tekaat, T. and Willmann, K. (1969). *Phys. Rev. Lett.*, **22,** 89
292. Ehrhardt, H., Hesselbacker, K. H. and Willmann, K. (1969). *Proceedings of the Sixth International Conference on the Physics of Electronic and Atomic Collisions,* 217 (Cambridge, Massachusetts: MIT Press)
293. Vriens, L. (1969). *Physica,* **45,** 400
294. Bonsen, T. F. M. and Vriens, L. (1970). *Physica,* **47,** 307
295. Vriens, L. (1970). *Physica,* **47,** 267
296. Gronemeier, K. H. (1970). *Z. Physik,* **232,** 483
297. McGowan, J. W., Williams, J. F. and Curley, E. K. (1969). *Phys. Rev.*, **180,** 132
298. Ehrhardt, H. (1969). *Physics of the One and Two Electron Atoms,* 598 (Amsterdam: North Holland)

299. Golden, D. E. and Zecca, A. (1970). *Phys. Rev. A,* **1,** 241
300. Burke, P. G., Cooper, J. A. and Ormonde, S. (1966). *Phys. Rev. Lett.,* **17,** 345
301. Burrow, P. G. and Schulz, G. J. (1969). *Phys. Rev. Lett.,* **22,** 1271
302. Grissom, J. T., Compton, R. N. and Garrett, W. R. (1969). *Phys. Lett.,* **30A,** 117
303. Grissom, J. T., Garrett, W. R. and Compton, R. N. (1969). *Phys. Rev. Lett.,* **23,** 1011
304. Burrow, P. D. (1970). *Phys. Rev. A,* **2,** 1774
305. Borst, W. L. (1969). *Phys. Rev.,* **181,** 257
306. Hall, R. I., Mazeau, J. and Reinhardt, J. (1970). *Phys. Lett.,* **31A,** 145
307. Weingartshofer, A., Ehrhardt, H., Hermann, V. and Linder, F. (1970). *Phys. Rev. A,* **2,** 294
308. Hall, R. I., Mazeau, J. and Reinhardt, J. (1969). *Phys. Lett.,* **30A,** 427
309. Spence, D. and Schulz, G. J. (1970). *Phys: Rev. A;* **2,** 1802
310. Boness, M. J. W. and Schulz, G. J. (1970). *Phys. Rev. A,* **2,** 2182
311. Read, F. H. (1968). *J. Phys. B,* **1,** 893
312. Read, F. H. (1968). *J. Phys. B,* **1,** 1056
313. Bardsley, J. N. and Read, F. H. (1968). *Chem. Phys. Lett.,* **2,** 333
314. Bensimon, J., Cotte, M., Pedenon, J. F. and Regnaut, C. (1969). *Phys. Lett.,* **30A,** 255
315. Rempt, R. D. (1969). *Phys. Rev. Lett.,* **22,** 1034
316. Brion, C. E. and Olsen, L. A. R. (1970). *J. Chem. Phys.,* **52,** 2163
317. Brongersma, H. H. and Oosterhoff, L. J. (1967). *Chem. Phys. Lett.,* **1,** 169
318. Brongersma, H. H. unpublished work
319. Hasted, J. B. and Awan, A. M. (1969). *J. Phys. B,* **2,** 367
320. Boness, M. J. W., Hasted, J. B. and Larkin, I. W. (1968). *Proc. Roy. Soc. (London)* **A205,** 493
321. Larkin, I. W. and Hasted, J. B. (1970). *Chem. Phys. Lett.,* **5,** 325
322. Andrick, A., Danner, D. and Ehrhardt, H. (1969). *Phys. Lett.,* **29A,** 346
323. Claydon, C. R., Segal, G. A. and Taylor, H. S. (1970). *J. Chem. Phys.,* **52,** 3387
324. Brongersma, H. H., van der Hart, J. A. and Oosterhoff, L. J. (1967). *Fast Reactions and Primary Kinetics,* 211 (New York: Interscience)
325. Brongersma, H. H., Boerboom, A. J. H. and Kistemaker, J. (1969). *Physica,* **44,** 449
326. Brongersma, H. H. and Oosterhoff, L. J. (1969). *Chem. Phys. Lett.,* **3,** 437
327. Knoop, F. W. E., Kistemaker, J. and Oosterhoff, L. J. (1969). *Chem. Phys. Lett.,* **3,** 73
328. Knoop, F. W. E., Brongersma, H. H. and Boerboom, A. J. H. (1970). *Chem. Phys. Lett.,* **5,** 450
329. Curran, R. K. (1963). *J. Chem. Phys.,* **38,** 780
330. Jacobs, G. and Henglein, A. (1964). *Z. Naturforsch.,* **19a,** 906
331. Huebner, R. H., Compton, R. N. and Schweinler, H. C. (1968). *Chem. Phys. Lett.,* **2,** 407
332. Compton, R. N., Stockdale, J. A. and Reinhardt, P. W. (1969). *Phys. Rev.,* **180,** 111
333. Begun, G. M. and Compton, R. N. (1969). *J. Chem. Phys.,* **51,** 2367
334. Brion, C. E. and Eaton, C. R. (1968). *Int. J. Mass. Spectrom. Ion Phys.,* **1,** 102
335. Brion, C. E., Eaton, C R., Olsen, L. A. R. and Thomas, G. E. (1969). *Chem. Phys. Lett.,* **3,** 600
336. Brion, C. E. and Olsen, L. A. R. (1969). *Phys. Rev.,* **187,** 111
337. Brion, C. E. and Thomas, G. E. (1968). *Int. J. Mass. Spectrom. Ion Phys.,* **1**, 25
338. Temkin, A. (1969). *Physics of the One and Two Electron Atoms,* 655 (Amsterdam: North Holland)
339. Madden, R. P., Ederer, D. L. and Codling, K. (1969). *Phys. Rev.,* **177,** 136
340. Hubin-Franskin, M. J. and Collin, J. E. (1970). *Int. J. Mass. Spectrom. Ion Phys.,* **4,** 451
341. Hubin-Franskin, M. J. and Collin, J. E. (1970). *Int. J. Mass Spectrom. Ion Phys.,* **5,** 163
342. Hubin-Franskin, M. J. and Collin, J. E. (1970). *Int. J. Mass Spectrom. Ion Phys.,* **5,** 255
343. Lifshitz, C. and Grajower, R. (1970). *Int. J. Mass Spectrom. Ion Phys.,* **4,** 92
344. Stamatovic, A. and Schulz, G. J. (1969). *Phys. Rev.,* **188,** 213
345. O'Malley, R. M. and Jennings, K. R. (1969). *Int. J. Mass Spectrom. Ion Phys.,* **2,** App. 1
346. Ridge, D. P. and Beauchamp, J. L. (1969). *J. Chem. Phys.,* **51,** 470
347. Imhof, R. E. and Read, F. H. (1969). *Chem. Phys. Lett.,* **3,** 652
348. Lewis, D. and Hamill, W. H. (1969). *J. Chem. Phys.,* **51,** 456
349. Hunter, L. M., Lewis, D. and Hamill, W. H. (1970). *J. Chem. Phys.,* **52,** 1733
350. Lewis, D., Merkel, P. B. and Hamill, W. H. (1970). *J. Chem. Phys.,* **53,** 2750
351. Lewis, D., Merkell, P. B. and Hamill, W. H. (1970). *J. Chem. Phys.,* **53,** 3389

352. Auger, P. (1925). *J. Phys. Radium*, **5**, 205
353. Harris, L. A. (1968). *Anal. Chem.*, **40**, 25A
354. Carlson, T. A. and Krause, M. O. (1966). *Phys. Rev. Lett.*, **17**, 1079
355. Korber, H. and Mehlhorn, W. (1964). *Phys. Lett.*, **13**, 129
356. Mehlhorn, W. (1965). *Phys. Lett.*, **15**, 46
357. Mehlhorn, W. (1965). *Z. Physik*, **187**, 21
358. Korber, H. and Mehlhorn, W. (1966). *Z. Physik*, **191**, 217
359. Mehlhorn, W. and Asaad, W. N. (1966). *Z. Physik*, **191**, 231
360. Mehlhorn, W. (1966). *Phys. Lett.*, **21**, 155
361. Glupe, G. and Mehlhorn, W. (1967). *Phys. Lett.*, **25A**, 274
362. Mehlhorn, W. (1968). *Phys. Lett.*, **26A**, 166
363. Mehlhorn, W. (1968). *Z. Physik*, **208**, 1
364. Mehlhorn, W. and Stalherm, D. (1968). *Z. Physik*, **217**, 294
365. Asaad, W. N. and Mehlhorn, W. (1968). *Z. Physik*, **217**, 304
366. Mehlhorn, W., Stalherm, D. and Verbeek, H. (1968). *Z. Naturforsch.*, **23a**, 387
367. Krause, M. O., Vestal, M. L., Johnston, W. H. and Carlson, T. A. (1964). *Phys. Rev.*, **133**, 385
368. Carlson, T. A. and Krause, M. O. (1965). *Phys. Rev.*, **137**, 1655
369. Carlson, T. A. and Krause, M. O. (1965). *Phys. Rev. Lett.*, **14**, 390
370. Krause, M. O. (1965). *Phys. Lett.*, **19**, 14
371. Carlson, T. A. and Krause, M. O. (1965). *Phys. Rev.*, **140**, 1057
372. Krause, M. O. and Carlson, T. A. (1966). *Phys. Rev.*, **149**, 52
373. Carlson, T. A., Hunt, W. E. and Krause, M. O. (1966). *Phys. Rev.*, **151**, 41
374. Carlson, T. A. and Krause, M. O. (1965). *Bull. Amer. Phys. Soc.*, **10**, 455
375. Carlson, T. A. (1967). *Phys. Rev.*, **156**, 142
376. Krause, M. O., Carlson, T. A. and Dismukes, P. D. (1968). *Phys. Rev.*, **170**, 37
377. Carlson, T. A., Moddeman, W. E., Pullen, B. P. and Krause, M. O. (1970). *Chem. Phys. Lett.*, **5**, 390
378. Carlson, T. A., Moddeman, W. E. and Krause, M. O. (1970). *Phys. Rev. A*, **1**, 1406
379. Krause, M. O., Stevie, F. A., Lewis, L. J., Carlson, T. A. and Moddeman, W. E. (1970). *Phys. Lett.*, **31A**, 81
380. Moddeman, W. E. (1970). *Ph.D. Thesis*, University of Tennessee, Rep. No. ORNL–TM3012
381. Bergmark, T., Spohr, R., Magnusson, N., Werme, L. O., Nordling, C. and Siegbahn, K. (1969). *Rep. UUIP—589*, Uppsala Institute of Physics
382. Spohr, R., Bergmark, T., Magnusson, N., Werme, L. O., Nordling, C. and Siegbahn, K. (1970). *Physica Scripta*, **2**, 31
383. Burhop, E. H. S. (1952). *The Auger Effect and Other Radiationless Transitions* (London: Cambridge University Press)
384. Asaad, W. N. and Burhop, E. H. S. (1958). *Proc. Phys. Soc. (London)*, **71**, 369
385. Lassettre, E. N. and Silverman, S. (1964). *J. Chem. Phys.*, **40**, 1265
386. Simpson, J. A., Mielczarek, S. R. and Cooper, J. (1964). *J. Opt. Soc. Amer.*, **54**, 269
387. Madden, R. P. and Codling, K. (1963). *Phys. Rev. Lett.*, **10**, 516
388. Rudd, M. E. (1964). *Phys. Rev. Lett.*, **13**, 503
389. Rudd, M. E. (1964). *Phys. Rev. Lett.*, **15**, 580
390. Oda, N., Nishimura, F. and Takira, S. (1970). *Phys. Rev. Lett.*, **24**, 42
391. Suzuki, H., Konishi, A., Yamamoto, M. and Wakiya, K. (1970). *J. Phys. Soc. Japan*, **28**, 534
392. Rudd, M. E. (1966). *Rev. Sci. Instrum.*, **37**, 971
393. Rudd, M. E. and Lang, D. V. (1965). *Proceedings of the Fourth International Conference on the Physics of Electronic and Atomic Collisions*, 153 (Hastings on Hudson, N.Y.: Science Bookcrafters)
394. Rudd, M. E., Jorgensen, T. and Volz, D. J. (1966). *Phys. Rev.*, **151**, 28
395. Edwards, A. K. and Rudd, M. E. (1968). *Phys. Rev.*, **170**, 140
396. Rudd, M. E. (1968). *Phys. Rev.*, **169**, 79
397. Gerber, G., Morgernstern, R. and Niehaus, A. (1969). *Phys. Rev. Lett.*, **23**, 511
398. Edwards, A. K., Risley, J. S. and Geballe, R. (1971). *Phys. Rev. A*, **3**, 583
399. Bydin, Y. F. (1967). *JETP Letters*, **6**, 297
400. Penning, F. M. (1927). *Z. Physik*, **46**, 335
401. Penning, F. M. (1929). *Z. Physik*, **57**, 723

402. Kruithof, A. A. and Penning, F. M. (1937). *Physica*, **4,** 430
403. Cermak, V. (1966). *J. Chem. Phys.*, **44,** 3781
404. Hornbeck, J. A. and Molnar, J. P. (1951). *Phys. Rev.*, **84,** 621
405. Becker, P. M. and Lampe, F. W. (1965). *J. Chem. Phys.*, **42,** 3857
406. Herman, Z. and Cermak, V. (1966). *Collect. Czech. Chem. Commun.*, **31,** 649
407. Hotop, H. and Niehaus, A. (1969). *Z. Physik*, **228,** 68
408. Hotop, H., Niehaus, A. and Schmeltekopf, A. L. (1969). *Z. Physik*, **229,** 1
409. Hotop, H. and Niehaus, A. (1970). *Z. Physik*, **238,** 432
410. Hotop, H. and Niehaus, A. (1970). *Int. J. Mass Spectrom. Ion Phys.*, **5,** 415
411. Fry, E. S. and Williams, W. L. (1969). *Rev. Sci. Instrum.*, **40,** 1141
412. Muschlitz, E. E. (1966). *Molecular Beams* (Advances in Chemical Physics, **X,** Ed. J. Ross), 171 (New York: J. Wiley)
413. Muschlitz, E. E. (1968). *Science*, **159,** 399
414. Pichanick, F. M. J. and Simpson, J. A. (1968). *Phys. Rev.*, **168,** 64
415. Berry, R. S. (1970). *Molecular Beams and Reaction Kinetics*, 193 (New York: Academic Press)
416. Jones, E. G. and Harrison, A. G. (1970). *Int. J. Mass Spectrom. Ion Phys.*, **5,** 137
417. Franklin, J. L. (1968). *Advan. Chem. Ser.*, **72,** 1
418. Benton, E. E., Ferguson, E. E., Matson, F. A. and Robertson, W. W. (1962). *Phys. Rev.*, **128,** 206
419. Shaw, M. J., Bolden, R. C., Hemsworth, R. S. and Twiddy, N. D. (1971). *Chem. Phys. Lett.*, **8,** 148
420. Cermak, V. (1968). *Coll. Czech. Chem. Commun.*, **33,** 2739
421. Robertson, W. W. (1966). *J. Chem. Phys.*, **44,** 2456
422. Schmeltekopf, A. L., Fehsenfeld, F. C. and Ferguson, E. E. (1968). *J. Chem. Phys.*, **48,** 2966
423. Hotop, H. and Niehaus, A. (1969). *Chem. Phys. Lett.*, **3,** 687
424. Cermak, V. (1966). *J. Chem. Phys.*, **44,** 3774
425. Cermak, V. and Herman, Z. (1968). *Chem. Phys. Lett.*, **2,** 359
426. Cermak, V. (1970). *Chem. Phys. Lett.*, **4,** 515
427. Fuchs, V. and Niehaus, A. (1968). *Phys. Rev. Lett.*, **21,** 1136
428. Hotop, H. and Niehaus, A. (1967). *J. Chem. Phys.*, **47,** 2506
429. Brion, C. E., McDowell, C. A. and Stewart, W. B., to be published
430. Katsuura, K. (1966). *J. Chem. Phys.*, **42,** 3771
431. Smirnov, B. M. and Firsov, O. B. (1965). *J. Exp. Theoret. Phys.*, **2,** 478
432. Bell, K. L., Dalgarno, A. and Kingston, A. E. (1968). *J. Phys. B*, **1,** 18
433. Bell, K. L. (1970). *J. Phys. B*, **3,** 1308
434. Miller, W. H. (1970). *J. Chem. Phys.*, **52,** 3563
435. Miller, W. H. and Schaeffer, H. F. (1970). *J. Chem. Phys.*, **53,** 1421
436. Matsuzawa, M. and Katsuura, K. (1970). *J. Chem. Phys.*, **52,** 3001
437. Fujii, H., Nakamura, H. and Mori, M. (1970). *J. Phys. Soc., Japan*, **29,** 1030
438. Janev, R. K., Davidovic, D. M. and Tancic, A. R. (1970). *Fizika*, **2,** 165
439. Hotop, H. and Niehaus, A. (1971). *Chem. Phys. Lett.*, **8,** 497
440. Hotop, H. and Niehaus, A. (1968). *Z. Physik*, **215,** 395
441. Rendina, J. F. and Grojean, R. E. (1970). *Appl. Optics*, **9,** 1
442. Callan, E. J. (1961). *Phys. Rev.*, **124,** 793
443. Chen, J. C. Y. (1969). *Theory of Transient Negative Ions* in *Advances in Radiation Chemistry*, Ed. M. Burton and J. M. Magee (New York: J. Wiley)
444. Kessler, K. G. (1969). *Comments At. Molec. Phys.*, **1,** 70
445. Weiss, M. J., Mielczarek, S. R. and Kuyatt, C. E. (1971). *J. Chem. Phys.*, **54,** 1412
446. SSR Instruments Co., Santa Monica, California
447. Brundle, C. R. (1971). *Appl. Spectrosc.*, **25,** 8
448. Hulett, L. D. and Carlson, T. A. (1971). *Appl. Spectrosc.*, **25,** 33

4
Field Ionisation

A. J. B. ROBERTSON
King's College, University of London

4.1 INTRODUCTION

An atom or a molecule may be regarded as a potential well for an electron or electrons, as in Figure 4.1(a). In a very strong electric field (of the order of 10^9 V m^{-1}) the potential well becomes distorted, as shown in Figure 4.1(b). An electron in a level 0 can then pass through the potential barrier by quantum-mechanical tunnelling and escape from the atom or molecule, as shown by the dashed arrow. This process is called field ionisation. Generally the strong field required is produced by concentrating electric lines of force on a metal surface with very high curvature. The atom or molecule is therefore ionised near a metal surface, and the potential diagram can be represented as in Figure 4.2, where the metal surface is considered to be merely a smooth plane. FL represents the energy of the Fermi level in the metal, ϕ is the work function of the metal and I the ionisation energy of the atom or molecule. The line AB represents the increase of the potential energy of an electron as it is moved away from the metal into free space. This linear function is added to the potential-well function of Figure 4.1(a) in a symmetrical manner with respect to the centre of the well.

In Figure 4.2 the level 0 is just at the Fermi level of the metal and so the tunnelling electron can enter a vacant electron state in the metal. However, if the atom or molecule is nearer the metal there is no vacant electron state in the metal into which the electron from level 0 can go. Hence field ionisation cannot occur. According to this model, field ionisation only occurs when the

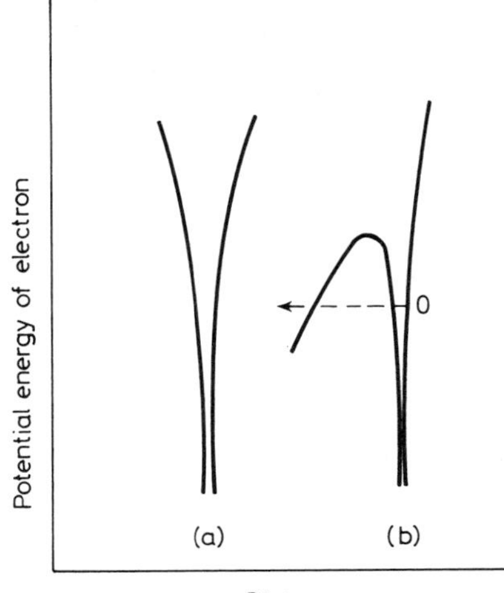

An atom or molecule regarded as a potential well for an electron, (a) in the absence and (b) in the presence of a strong electric field
(From Robertson, A. J. B., *Catalysis of Gas Reactions by Metals*, by courtesy of Logos Press, London)

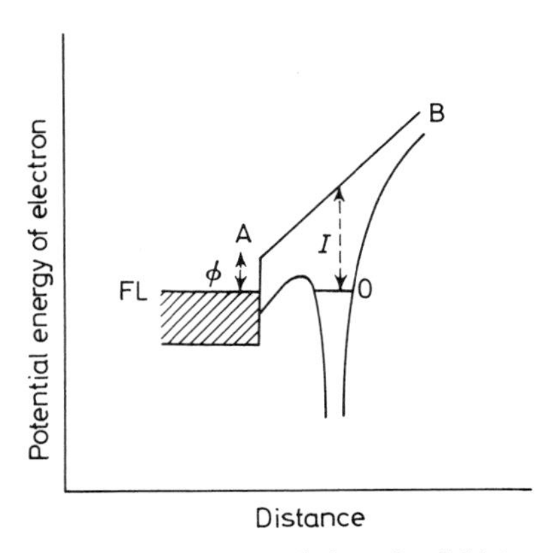

Figure 4.2 Energy relations for field ionisation of an atom or a molecule near a metal surface
(From Robertson, A. J. B., *Catalysis of Gas Reactions by Metals*, by courtesy of Logos Press, London)

distance from the metal surface to the centre of the potential well equals or exceeds a critical distance x_c given by

$$x_c = (I - \phi)/E \tag{4.1}$$

where E is the electric field strength.

Early field ionisation work was carried out in connection with field ion microscopy, with ions formed at a sharp metal tip. The idea of studying the ions formed at a tip under the influence of a strong electric field by mass spectrometry appears to have arisen in three different places[1–5]. Early researches showed that field ionisation mass spectra are generally much simpler than electron impact mass spectra, and in 1955 Gomer and Inghram[3] suggested the use of field ionisation mass spectrometry for the investigation of intermediates in gaseous reactions, and of ions in adsorbed layers with the use of pulsed fields. An adsorbed layer on a metal surface experiences a force when a field is applied just as there is a force on the plates of a charged parallel-plate capacitor. This force may be large enough to pull off adsorbed particles as neutral particles or as ions. This is the process of field desorption.

The simplicity in many cases of field ionisation mass spectra has aroused some interest. An example is provided[6] by a comparison of relative ion intensities in the field ionisation and chemical ionisation mass spectra (see Chapter 5) of eight isomeric decanes. The relative intensity of the molecular ions formed by field ionisation is much greater than that of the $M-1$ ions formed by chemical ionisation. The places of greatest bond fission in field ionisation are quite different from those in chemical ionisation.

4.2 REVIEWS ON FIELD IONISATION MASS SPECTROMETRY

The most comprehensive and systematic review of the entire subject is given in an important book of Beckey[7] whose contributions to the development of the subject have been of outstanding importance. This book discusses the practical production of field ions and the design of ion sources containing lens systems of various kinds. The theory of field ionisation is reviewed and the charge distribution of organic ions in high electric fields is discussed fully. Mechanisms of ion decomposition are discussed. An important contribution is made concerning the formation of condensed layers on a charged metal surface, which can occur when interaction of the dipole moment and the induced dipole moment of a molecule with the electric field leads to surface-layer formation. Beckey also discusses the field ionisation mass spectra of several homologous series of organic substances, and the use of such spectra in structural studies and in quantitative analysis.

A brief and lucid general introduction to the use of field ionisation mass spectrometry in the determination of the structures of organic molecules and in quantitative analyses is provided by an article of Beckey[8]. He points out that field ionisation is a 'soft' method of ionisation imparting little energy to the ion formed, and hence fragmentation is reduced. This article discusses the supply of particles to the ion emitter from the gas phase and by diffusion of adsorbed particles. The use of sharp points, fine wires and sharp edges is reviewed. Beckey also draws attention to the multiple-layer films, composed of a metal base, a thin insulating film, and a perforated covering film, which were used by Gentsch for field ionisation with remarkably low applied voltages, but which do not seem to have been used by other workers. The 'soft' nature of field ionisation can be of value in the structural analysis of carbohydrates by mass spectrometry, reviewed by Gruetzmacher *et al.*[9]. It can also be of value in biochemical applications of high-resolution mass spectrometry, reviewed by Van Lear and McLafferty[14].

A review by Block[10] gives a good general and historical treatment and is a specially good introduction to the use of pulsed voltage techniques and to the study of surface properties and surface reactions. Block has made notable contributions in this last area. Beckey and Comes[11] have reviewed comprehensively techniques of molecular ionisation. Work in Beckey's group up to 1964 has been reviewed[12,13].

The present review mainly considers the period 1966–1970 with emphasis on the more recent work. However, some earlier and some later papers are reviewed. This review aims to be selective rather than exhaustive.

4.3 THEORETICAL CONSIDERATIONS RELATING TO FIELD IONISATION

4.3.1 Simple theory

Figure 4.2 represents a one-dimensional approximation to the problem of field ionisation. The potential energy of an electron in the region between the atom to be ionised and the metal may be approximated by adding to the Coulomb attraction between the ion and the electron a potential due to the applied field, a potential due to the interaction of the electron with its electrical image in the metal, and a potential due to the interaction of the electron with the electrical image of the positive ion in the metal. A formula for the penetration probability of the electron when it, in effect, strikes the barrier can then be deduced. This can be integrated[15]. The frequency with which the electron strikes the barrier may be estimated by calculating, with a Bohr-type theory for an electron revolving round an ion, the number of encounters the electron makes with the barrier. Such calculations account reasonably well for observed field ion currents in many cases.

In this model the electron is pictured as bouncing back and forth in a potential well until at one particular encounter it penetrates the barrier and enters the metal. The concept of transmission time or tunnelling time, sometimes used in connection with barrier penetration, has been discussed[16].

4.3.2 More detailed field ionisation theories

Figure 4.1 represents an atom in a uniform electric field. It can become an ion on losing an electron. A very mathematical paper by Mendelsohn[17] considers an ion in a uniform electric field. Large-Z expansion theory is applied to the non-relativistic ground state of a one-electron ion and this leads to a calculation of ionisation field. This treatment, however, appears to ignore tunnelling. The Schrödinger equation for a one-electron ion with a nucleus of known charge in a uniform field is considered by Mendelsohn and high-order perturbation theory is used.

Time-dependent perturbation theory has been used by Boudreaux and Cutler[18] to compute the probability of field ionisation as a function of the atom–metal separation. The results of these calculations agree better with experiment than those based on one-dimensional barrier penetration probabilities, which predict too wide an ionisation zone. This zone is the region beyond the critical distance in which most of the ionisation occurs. The same authors[19], in a separate three-dimensional quantum-mechanical calculation, view field ionisation as a collision of an atom with an infinitely heavy metal tip in which a rearrangement occurs leaving one of the atomic electrons in the metal. The methods of quantum-mechanical scattering theory are used to calculate the probability of this rearrangement occurring during one encounter of the atom with the metal tip. Fonash and Schrenk[20] have also treated the field ionisation process theoretically as a rearrangement collision. The energy-band structure of the metal is taken into account. They conclude that the topology of the Fermi surface does definitely influence the tunnelling probability. A spherical Fermi surface produces isotropic tunnelling probabilities.

A further theoretical treatment of field ionisation as a three-dimensional rearrangement collision has been given by Sharma and Schrenk[21]. A new feature of their treatment is explicit consideration of the periodic potential variation in the surface region in order to determine its effect on field ionisation tunnelling probabilities at a specific crystallographic plane. The variation of the surface potential along the surface affects the shape of the tail of the electron wave function immediately outside the surface, and the overlap of this wave function with the wave function associated with an incoming atom leads to a variation of the tunnelling probability on a specific plane. A problem which this treatment attempts to elucidate to some extent is that of the observed contrast in field ion micrographs (that is, the photographs produced by the field ion microscope). To calculate the imaging of a tungsten surface by hydrogen with this treatment appears difficult enough, indeed many of the recent abstruse theoretical calculations are concerned with hydrogen atoms, and we are a very long way from exact theoretical discussions of the field ionisation of the large molecules now studied experimentally. A more qualitative treatment than that of Sharma and Schrenk of the idea that field ionisation proceeds preferentially in those regions where the fully occupied orbitals of the image gas atom can overlap with those exposed and only partially occupied orbitals of the surface metal atoms has been given by Knor and Müller[22]. A lucid introductory account to some problems of field ion micrographs has been given by Müller[23].

Another theoretical treatment of field ionisation has been given by Goldenfeld *et al.*[24] who propose an analytical way of finding the probability of the ionic state, depending on field strength, the material of the emitter, and the distance of the atom from the emitter. Their treatment involves a Schrödinger equation for an electron, derived from the atom, in a position beyond the metal in an electric field, and the use of hydrogen-like wave functions.

4.3.3 Structure in field ionisation energy distributions and field-induced resonance states at a surface

The energy spectrum found by mass spectrometry and a retarding energy analyser for H_2^+ and H^+ ions formed by field ionisation of hydrogen on the 110 crystal plane of a tungsten tip was studied by Jason *et al.*[25]. The parent-ion energy distribution showed a main peak with high-energy onset corresponding to a minimum distance from the surface for ionisation. This minimum distance is the critical distance in equation (4.1). In addition, however, an unexpected series of secondary peaks was noted corresponding to lower kinetic energies of the ions, which were therefore formed at greater distances from the surface. Structure was also observed in the energy spectrum of the H^+ ion, whereas the trimer H_3^+ exhibited a simple energy distribution indicating strict surface formation. The original suggestion[25] that these peaks in the energy spectrum might be due to virtual surface states related to the crystal lattice was later seen to be inadequate. An interpretation was proposed[26] involving resonances of the atomic levels with the levels of a one-dimensional triangular well formed by the electric field and the surface. Figure 4.2 shows such a well, bounded on the left by the metal surface and on the right by the energy barrier through which a tunnelling electron emerges. An atom at a distance from the surface exceeding the critical distance gives an electron which enters a level in the metal above the Fermi level, and the existence of discrete energy levels in the triangular well leads to the energy spectrum peaks. The problem has been discussed fully by Jason[27]. The peaks in the energy spectrum for H_2^+ depend critically on the electric field; they are well-resolved with fields of 20–30 GV m^{-1} but are not noticeable with a field of 15 GV m^{-1}. Such fields are larger than those normally employed in field ionisation mass spectrometry. It seems, from the resolution attainable in magnetic-deflection single-focusing mass spectrometers with field ion sources, that the field-induced resonance states are not usually important. If they were, ions would be produced over such a range of energies that considerable loss of mass spectrometer resolution would be noted.

Theoretical arguments have been proposed by Alferieff and Duke[28] which lead to resonances for low-energy ions formed by field ionisation. They use one-dimensional solvable pseudopotential models to calculate field ionisation energy distributions for three cases: first, a step-function clean metal surface; secondly, a step-function metal surface covered with an adsorbed monolayer; and thirdly, a clean metal surface having finite thickness and various shapes. All three cases lead to resonances.

4.3.4 Some miscellaneous topics

4.3.4.1 Field desorption

A discussion of field desorption by Ionov[29] proposes that the adsorbed particles before desorption can be regarded as being in thermal equilibrium with the surface of the adsorbent, and the electric field reduces the heat of adsorption so that desorption occurs.

4.3.4.2 Ionisation of highly excited atoms

Such atoms should ionise much more easily than atoms in their ground states so that ionisation could occur in a weak field and spontaneous ionisation might occur near a metal surface. We note there is an energy gain if this spontaneous ionisation occurs because the ion formed interacts with its electrical image. The ionisation of highly excited noble gas atoms in an electric field in space, and near a metallic surface, has been studied by Kupriyanov[30].

4.3.4.3 Field ionisation in semi-insulating thin films

Reed and Brodie[31] have presented evidence to support the possibility of direct field ionisation of shallow trapping levels in films of materials such as CdS and CdTe. There is supposed to be a direct tunnelling from trapping levels to the conduction band; the process may be described as 'internal field emission'.

Practical field ion emitters often operate because of the existence of semi-conducting projections and one may speculate that an entering electron derived from a molecule which is field ionised could enter a trapping level. In fact, practical ion emitters are far removed from the models used in the more abstruse theories.

4.3.4.4 Field ionisation and the electronic structure of ionised molecules

Lorquet and Hall[32] have calculated by the equivalent orbital method the net positive charge distribution of n-alkane ions (from propane to n-octane) in the presence of an electric field of the order of 10 GV m^{-1}. The long molecular axis can be considered to be parallel to the electric field because of the greater polarisability in the direction of the long axis[33]. If the C—C bond furthest from the charged metal surface producing the field is called α, and the next bond is called β, it is found that the largest fraction of positive charge lies on the β bond which therefore is the one most readily broken. This conclusion was reached first by Beckey[33] in a more qualitative argument.

4.4 FIELD ION EMITTERS

All early work on field ionisation mass spectrometry was carried out with sharp metal tips (usually of tungsten) as the ion emitters. Beckey[34] in 196[illegible] described a field ion source in which the electric field was produced by thin metal wires or sharp metallic edges, and he was the first to use such sources for mass spectrometry. The parallel early work by Robertson *et al.*[35] on the production of positive ions by field ionisation at the surface of a thin wire was carried out with total current measurements and without the use of mass spectrometry. The surface area effective in ion emission from fine wires or metallic edges is much greater than that of an ion-emitting tip, and the wires or edges therefore produce much larger, and more stable, ion currents. In early mass spectrometric work with tips the current carried by the resolved ion beam varied erratically by as much as a factor of two or more, particularly when the ionising electric field was only just great enough to produce ionisation. These current variations were much less with wires and edges as ion emitters; such emitters also gave much better reproducibility of mass spectra.

4.4.1 Sharp tips as ion emitters

Much work with sharp tips has been carried out in connection with field ion microscopy[36]. To increase the current sensitivity, Block (reported by Beckey[34]) used a field ion source consisting of several points arranged in parallel, facing the slit of an electrostatic lens system. Later, Wanless[37] described the use of a tip array consisting of a multiplicity of electrolytically etched tips in parallel, attached to a support by spot welding of their shanks. Arrays used were a single row of tips, a double row of tips, or a number of parallel rows. Wanless combined a tip array with signal averaging of field ion spectra. The object of using a multiple array is to increase the ion current. Tip arrays for use as ion sources are now commercially available.

4.4.1.1 The angular distribution of ions field-emitted from a sharp tip

Beckey *et al.*[38] measured the angular distribution of atoms and molecules field-ionised at platinum tips. For argon, the angular distribution was strongly influenced by layers of adsorbed residual gases on the emitter tip. With argon, the emitted current showed a continuous decay with polar angle, the maximum emission being forward from the tip. In contrast, n-hexane, acetone, water and methanol showed a maximum emission towards the side of the tip, at a polar angle of *c.* 40 degrees from the forward direction. In such cases it was proposed that particles come to the region of ionisation principally from the adsorption layer on the shank of the tip. A similar deduction was made by Jason *et al.*[39], who investigated field ionisation from hydrogen layers adsorbed on tips at 4.2 K, and from the high current obtained they deduced that particle supply to the ionisation region occurs by diffusion of hydrogen from the emitter shank. An alternative view of Nazarenko[40] interprets the field ion current from a tip emitter of parabolic shape with a model in which the neutral particles come to the ionisation zone only from the space around the tip. Experimental measurements of Goldenfeld and Nazarenko[41] on the angular distribution of ions formed on a tip from the polar substances acetone, propionic acid, propionaldehyde, butanone, diethyl ether, ethanol, butanol and water, and the non-polar substances benzene, cyclohexane and pentane, agree with the earlier measurements[38], but are interpreted by them without invoking the diffusion process. However, we note that field-emission microscopy gives much evidence for surface mobility of adsorbed layers on metal tips.

4.4.2 Thin wires as ion emitters

Accounts of the early work on wire emitters have been given[7, 34, 35, 42, 43]. Wires with diameters as small as 2.5 μm have been used. Thin wires are efficient as field ionisers because the wire surface is not perfectly smooth, but contains protrusions where the field is enhanced as compared with a smooth wire[35]. The distribution of these protrusions and its relation to the field enhancement factor has been investigated by Metzinger and Beckey[44]. The total area of the blunter protrusions is much larger than that of the sharper ones. Therefore as the voltage applied to the wire is raised ion emission occurs increasingly on blunter projections, and in consequence the field strength effective in producing field ions is nearly independent of the applied voltage.

A problem in the earlier work with platinum Wollaston wires of 2.5 μm diameter was their fragility. When the wire-cathode voltage exceeded about 15 kV the wire broke. It is desirable to use a micro tensile strength apparatus[45] and select the stronger portions of wire from a given reel. Wollaston wire is of very varying quality. Difficulties arising from the fragility of wires can be avoided by employing three additional stratagems[46]. First, projections are grown at the surface of the wire in very large numbers, with high mechanical strength and with high field enhancement factors. This is the activation process discussed below. Secondly, the mechanical force exerted on the wire by the high electric field is largely compensated by an additional electrode behind the wire which is at the same potential as the cathode in front of the wire. Thirdly, the length of the etched and thin part of the Wollaston wire is reduced to 1.5–2.0 mm. The stress in the wire caused by the electric field is proportional to the square of this length. The wire may be put perpendicular to the length of the cathode slit. The lifetime of a wire when these four stratagems are employed can be several months or more. The resolving power of a thin-wire field ionisation mass spectrometer may decrease with time during analysis of solid substances. This disadvantage can be largely removed by electric heating of the wire[47]. Emitter temperature as a parameter in field ion mass spectrometry has been discussed by Knöppel[48]. The optimum parameters for field ion emitters in mass spectrometers have been studied by Ochterbeck and Beckey[49]. These parameters are emitter activation time, emitter heating, position of the emitter with respect to the slotted cathode plate, and emitter potential. Activated tungsten wires 10 μm in diameter are mechanically very stable and have several advantages over activated blade or foil emitters.

4.4.3 Sharp edges as ion emitters

Commercial razor blades placed 0.2–0.5 mm from a slotted cathode make quite efficient field ion emitters[50]. The ease of ionisation with razor blades of various compounds can be correlated with their ionisation energies[51].

A razor blade is not convenient for ionising substances with ionising energy exceeding *c*. 13 eV. However, etched platinum foils can be used to ionise gases with greater ionisation energy; Cross and Robertson[52] obtained a total current sensitivity of nearly 10^{-3} A torr^{-1} (10^{-5} A N^{-1} m^{2}) for ionisation of nitrogen on a platinum foil originally 8 μm thick which was electrolytically etched to form a sharp edge. However, such edges were found[53] by electron micrograph studies to suffer from lack of durability when used as field ion emitters. Tungsten edges produced by electropolishing were straight and uniformly thinned, and seemed to provide a good compromise between field ionisation efficiency and durability of the edge.

Tips, thin wires and sharp edges as field emitters for mass spectrometry have been compared[46].

4.4.4 Activation of field ion emitters

The field produced by a smooth wire of 2.5 μm diameter with an applied voltage of 10 kV is not great enough for field ionisation of the majority of organic substances. The field, however, can be enhanced by as much as a factor of ten by growing protrusions on the wire. These protrusions grow spontaneously on field anodes when a suitable gas is present and the high field is applied. Similarly, the efficiency of a sharp edge can be enhanced by protrusion growth. Untreated wires and sharp edges have projections on their surfaces which enhance the field, but the number of these is often not enough to make them good ion emitters. The object of conditioning procedures is to produce a vast number of projections on a wire surface or at a sharp edge. Many organic substances have been investigated as conditioning agents[43, 54], and the projections have been studied by electron microscopy and scanning electron microscopy, and by x-ray emission in an electron-microprobe spectrometer[55]. During the earlier work, acetone seemed to be the substance with the highest efficiency for growing protrusions. Later, systematic studies by Migahed concerning the dependence of the activation efficiency on the chemical nature of the activating substance led to the conclusion[56] that the most efficient substances are those having a large permanent dipole moment, a low ionisation energy, and containing a conjugated π-electron system. Benzonitrile was found to be a very good substance for the strong activation of field ion emitters. Nitrobenzene and crotonaldehyde were also efficient activating agents, but less so than benzonitrile. With this last agent, wires of 10 μm diameter can be activated. The projections grown in the presence of benzonitrile are needle-like and seem to consist of polymeric organic semiconducting material.

Derrick and Robertson[57] found that activation of razor blades with 2,2-dimethyl-propanoic acid nitrile (t-butyl cyanide) gave individual protrusions which merged to form a hedge of uniform height, with myriads of projections along its top, in contrast to the individual needle-like projections produced by benzonitrile. They considered the hedge-like form to be the more desirable one. An activated blade prepared with t-butyl cyanide gave total ion current sensitivities (i.e. total current emitted from the blade divided by the pressure in the ion source of the substance being ionised) of the order of 10^{-4} A N^{-1} m^{2}

(10^{-2} A torr^{-1}). The resistivity of the protrusions formed with t-butyl cyanide was estimated from the dimensions of the hedge, determined from electron micrographs, and the voltage drop across the hedge produced by a given current. This voltage drop was found from the displacement of peaks in a mass spectrum. The resistivity of the projections was estimated as about $10^6\ \Omega$ m, showing their semiconducting nature. The presence of some water vapour is beneficial in the formation of the t-butyl cyanide projections[58] and its role needs further study.

Migahed and Beckey[54] report, in a paper containing much valuable information on the production and properties of organic micro-needles on field ion emitters, that benzonitrile has a higher activation efficiency than t-butyl cyanide. Moreover, the heat treatment of the micro-needles led to better results with benzonitrile. Although the total current from t-butyl cyanide activation is comparable to that from benzonitrile activation when a 12 μm tungsten foil is activated, the transmitted current in the mass spectrometer is less with t-butyl cyanide[49]. The benzonitrile needles may be directed more into the forward direction than the t-butyl cyanide projections, so that a larger fraction of the current from the benzonitrile needles reaches the collector.

Practical methods for the activation of field anodes have been fully described by Beckey *et al.*[59]. A study[60] of surface reactions induced by field ionisation of organic molecules has led to a proposed mechanism for the formation of micro-needles from acetone by polymerisation involving radicals formed on the emitter surface in the presence of the strong field.

4.4.5 Field desorption techniques

The large increase of the active emitter area which can be brought about by a suitable activation technique makes it possible to adsorb a sample on to a field anode from solution and subsequently, after evaporation of the solvent, to field desorb the sample in a mass spectrometer. This technique of field desorption mass spectrometry was introduced by Beckey[61] and is specially suitable for thermally unstable substances of low volatility. Beckey obtained a mass spectrum of D-glucose using only *c.* 50 μg of the sugar dissolved in a droplet of water in which the highly activated thin wire anode was immersed. Useful spectra could be obtained for *c.* 15 minutes. Practical details and some experimental procedures for field desorption techniques have been described[62]. A great reduction in thermal decomposition of the sample is possible with the technique.

4.5 FIELD IONISATION SOURCES FOR MASS SPECTROMETERS

A number of detailed descriptions of field ionisation sources have appeared. Beynon *et al.*[63] describe simple modifications which enable field ionisation spectra to be obtained with an AEI MS 7 double-focusing mass spectrometer with a razor blade emitter. Other modifications of commercial mass spectrometers for field ionisation studies have been described[64, 65]. For molecular structural investigations combined electron impact and field ionisation sources are desirable, and several of these have been described[13, 66–69]. Burlingame and his colleagues[68] obtained with a razor blade emitter and a double-focusing mass spectrometer a resolution using field ionisation exceeding 20 000. Chait *et al.*[70] have described a dual field ionisation/electron impact source for a double-focusing mass spectrometer. The main feature of their source is a conveniently replaceable and adjustable razor blade anode. They obtained a maximum sensitivity for field ionisation of 4×10^{-7} A torr^{-1} as measured at the total ion beam monitor, using acetone, and with an object slit width of 0.30 mm. They found optimum sensitivity with the blade parallel to the ion exit slit. The angle of the blade with respect to this slit was very important in determining the sensitivity. These authors note that determination of the elemental composition of a molecule is an accepted keystone in the determination of its structure, and they remark that high-resolution field ion mass spectrometry appears to be superior to classical combustion methods in its accuracy and in its ability to handle sub-microgram samples. A field ion source for the investigation of solid organic substances for structural analysis has been described[71].

4.5.1 Special ion source for measurement of lifetimes of metastable ions

This special ion source has been described by Beckey *et al.*[71]. Ions formed on a thin wire emitter are accelerated to a cathode about 2.5 mm away. At a further distance of about 2 mm is the first of five retarding electrodes, and all of these, or some of them, can be held at an identical but variable retarding potential. The retarding potential is not enough to stop the ions. After the retardation the ions are again accelerated. There is therefore along the region of retardation a 'plateau length *d*' which is of variable length and in which the potential is constant. The residence time of ions on the potential plateau can be varied by altering the retarding potential as well as by altering the plateau length. The ions which dissociate in the plateau region produce fragment ions which have a greater kinetic energy than that of normal metastable ions (see Chapter 8), produced by dissociation after the electric acceleration and before magnetic deflection. In contrast to the normal metastable peaks, the position of the metastable plateau peaks in the mass spectrum is not constant, but varies with the retarding voltage. The metastable plateau peak is displaced to higher masses as the retarding potential is increased. The residence time of the ions on the retarding plateau can be varied from *c.* 10^{-8} to *c.* 10^{-6} s. Measurements of kinetic constants for ionic decay can be made with this source.

4.6 CALCULATION OF ELECTRIC FIELD STRENGTHS AT EMITTER SURFACES AND OF POTENTIAL DISTRIBUTIONS IN ION SOURCES

For a long wire of radius r along the axis of a cylinder of radius R, the electric field E at the surface of the wire is

$$E = V\{r \ln (Rr^{-1})\}^{-1} \tag{4.2}$$

where V is the potential difference between wire and cylinder. This equation assumes a smooth wire surface. The equation may be readily deduced by treating the wire-cylinder arrangement as a cylindrical capacitor.

The field strength at the surface of a wire of circular cross-section running parallel to a plane electrode has been deduced by Gilliland and Viney[72] from a consideration of the potential distribution around a line charge near a plane. For a wire of radius r at a distance l from the plane electrode, and with a potential difference of V between wire and electrode, the field strength E at the point on the wire's circumference which is nearest to the plane electrode is, approximately.

$$E = \frac{V}{\{r \ln(2l/r)\}} \tag{4.3}$$

For the thin wire distant l from a plane with a slit of width $2b$ in it, and with the wire parallel to the slit and vertically above it, Gilliland and Viney[51, 72] find, approximately

$$E = \frac{V}{r \ln\left\{\frac{2(l^2+b^2)^{\frac{1}{2}}}{r}\right\}}\left(1+\frac{rl}{l^2+b^2}\right) \tag{4.4}$$

They used the method of the conformal transformation in which they also considered the electric field at the surface of the sharp edge of a thin blade, the edge being smooth, having radius r, and being distant l from a plate. The sharp edge is parallel with and vertically above a slit of width $2b$ in the plate. For an infinitely thin plate the field at the blade vertex is, approximately[51, 72]

$$E = \frac{2V}{\pi r^{\frac{1}{2}}}\left\{\frac{l+(l^2+b^2)^{\frac{1}{2}}}{l^2+l(l^2+b^2)^{\frac{1}{2}}+b^2}\right)^{\frac{1}{2}} \tag{4.5}$$

which gives

$$E = \frac{2V}{\pi r^{\frac{1}{2}}(l^2+b^2)^{\frac{1}{4}}} \tag{4.6}$$

Brailsford[73] also obtained the result in equation (4.6). Let $z = x+iy$ and $w = u+iv$. The transformation

$$w = ik \cosh z \tag{4.7}$$

where k is a constant, transforms a series of lines parallel to the x axis in the z plane to a series of confocal hyperbolae, symmetric about the v axis, in the w plane. A razor blade edge can be approximated as a hyperbola[74]. Hence equation (4.7) gives results for the edge/flat-plate arrangement. With $c = a+ib$ the further transformation into the c plane given by

$$c = w+\frac{q^2}{4w} \tag{4.8}$$

converts the edge/flat-plate arrangement into an edge/slotted-plate arrangement. In equation (4.8), $2q$ is the slot width. Hence q in equation (4.8) equals b in equation (4.6). We note that in various deductions b has been used in different papers to denote half the slot width and, additionally, an axis in the c plane. Use of equations (4.7) and (4.8) gives equation (4.6). These transformations are also useful in calculating potential distributions in ion sources from which ion trajectories and flight times can be obtained[73].

4.7 KINETICS OF THE UNIMOLECULAR DECOMPOSITIONS OF POSITIVE IONS

Important recent advances in the study of unimolecular ionic dissociation kinetics have been reported by Beckey and his colleagues[75–78]. For the study of very fast ionic dissociation reactions the field ionisation mass spectrometric method is valuable because ions decomposing in space beyond the emitter give product ions which differ in kinetic energy from the same product ions formed at the emitter surface. This happens because the field gradient in front of the emitting anode is very large. The result of decomposition in space beyond the anode is to broaden the peak in the mass spectrum of the product ion formed. From this broadening, kinetics can be deduced.

4.7.1 Peak broadening and ionic dissociation kinetics

To take a definite example, consider a field ionisation source consisting of a sharp edge positioned over a plane slotted cathode, as shown in section in Figure 4.3. The sharp edge may be provided by a razor blade or an etched metal foil. When the razor blade or foil is charged to a sufficiently high potential V_B, field ionisation occurs at the edge. Let this lead to the formation

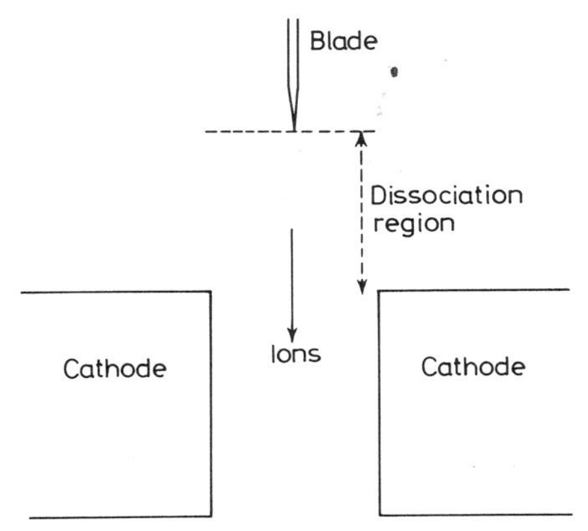

Figure 4.3 An ion source consisting of a sharp edge over a plane slotted cathode (From Derrick, P. J. and Robertson, A. J. B.[58], by courtesy of The Royal Society)

of molecular ions at a potential $V_B - x$ where x is the correction term found necessary by Robertson and Viney[51]. If the edge is free from the protrusions which tend to develop during field ionisation, the term x has the value expected for ionisation at the critical distance, x_c of equation (4.1), from the surface[57].

Consider positively charged molecular ions, each singly charged and of mass M, formed at the potential $V_B - x$ and accelerated towards the cathode. Suppose that an ion which has reached a potential V_1 dissociates to yield a singly charged product ion of mass m and an uncharged particle. The total kinetic energy of the product ion when it reaches a region at zero potential is

$$eV = eV_1 + (em/M)(V_B - x - V_1) \tag{4.9}$$

This equation relates the kinetic energy eV of the product ion to the potential V_1 at which dissociation occurs. Consequently the energy distribution of the product ions, found by mass spectrometric analysis, can be transformed to a distribution according to the potentials at which dissociation occurred. The potential can be related to the spatial position of dissociation, provided the potential distribution between the anode and cathode is known. The knowledge of the potential also allows the time taken for the ion of mass M to reach the place at which it dissociates to be calculated. Hence the dissociation kinetics of the molecular ion are deduced. The method is useful for dissociations occurring in the region between the edge and the cathode, and this means that dissociations occurring within times of 10^{-12}–10^{-9} s after ionisation are accessible to study[79].

The effect of such dissociations is to cause the fragment peak observed in a single-focusing magnetic-deflection instrument to be asymmetrically broadened towards lower masses[12, 80]. The exact analysis of peak shapes has been discussed[58, 76]. Thin wire sources are, of course, as useful as edge sources in such work.

4.7.2 Determination of ionic dissociation kinetics with the retarding potential technique

The special ion source described in Section 4.5.1 has been used to study ionic dissociation kinetics[76]. First, all retarding potentials are kept at zero potential. The ions decomposing in the field-free space are recorded as normal metastables at the apparent mass number $m_n^* = m^2/M$. Then one or several of the retarding electrodes are adjusted to a potential of +2 to +8 kV with respect to the ground state. A new group of 'metastable plateau peaks' then appears in the mass spectrum, with an apparent mass of[76]

$$m_{pl}^* = (m/U_0)\{(U_0 - U_g)(m/M) + U_g\} \tag{4.10}$$

where U_0 is the emitter potential and U_g is the potential in the retarding region. The position of the metastable plateau peak can be varied between two limiting values: $m_{pl}^* = m_n^* = m^2/M$ for $U_g = 0$; and $m_{pl}^* = m$ for $U_g = U_0$. A detailed account of the evaluation of rate constants from the measured intensities of the metastable plateau peaks, and additionally from the normal metastable peaks, has been given[76].

4.7.3 Some results of measurements of ionic dissociation kinetics

From studies of asymmetrical peak broadening, metastable plateau peaks and normal metastable peaks, all applied to a single ionic species undergoing a specified dissociation, unimolecular rate constants for the dissociation over the whole range from *c.* 10^{11} s^{-1} to *c.* 10^{3} s^{-1} can be derived. Beckey and his colleagues[76–78] represent results obtained for a particular specified ionic dissociation reaction by determining an average rate constant $\bar{k}(t)$, which at any time t is related to the number $N(t)$ of undecomposed parent ions by

$$-\mathrm{d}N(t)/\mathrm{d}t = \bar{k}(t)N(t) \tag{4.11}$$

They find generally that any particular ionic dissociation requires for its full kinetic description a continuous distribution of rate constants. The equation given for unimolecular decomposition with such a distribution is[76]

$$\frac{\mathrm{d}N(t)}{\mathrm{d}t} = -\int_{k=0}^{\infty} \frac{\partial N(k,t)}{\partial k} k\mathrm{d}k \equiv -N(t)\bar{k}(t) \tag{4.12}$$

where $N(k,t)$ is the fraction of the parent ions which decompose with the rate constant k at t. The average rate constant $\bar{k}(t)$ is

$$\bar{k}(t) \equiv \int_{k=0}^{\infty} \frac{\partial N(k,t)}{\partial k} k\mathrm{d}k/N(t) \tag{4.13}$$

A dissociation may be described in terms of average rate constants, with the constant $\bar{k}(t)$ having a value which falls as time increases. For many ionic dissociations a doubly logarithmic plot of $\bar{k}(t)$ and t gives a straight line[76, 77]. Values of $\bar{k}(t)$ determined by all three of the experimental methods lie on this line. This is a very significant finding.

We note that a dissociation which may be described in terms of two or more fractions of parent ions, each decomposing with a different rate constant k, which is not a function of time, may equally be described with a continuous

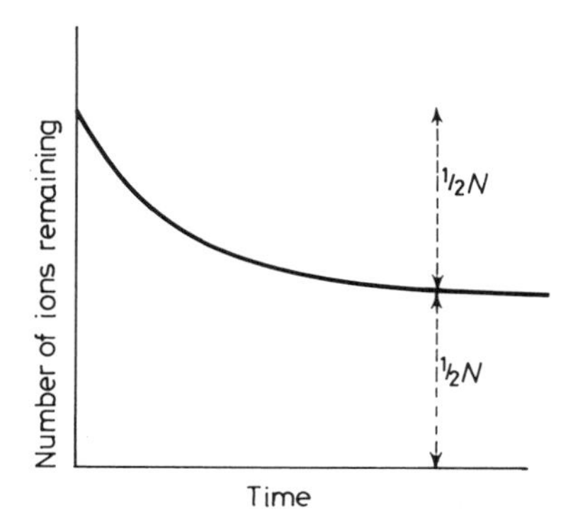

Figure 4.4 Decay curve for an ionic dissociation occurring with two discrete rate constants, one large one and one small one.

distribution of rate constants. To make matters definite, suppose that, out of a large number N of ions, half decompose very rapidly by a unimolecular reaction with a single discrete rate constant, and the other half decompose very slowly, again by a unimolecular reaction with a single discrete rate constant. The first part of the decay curve for the number of ions remaining, as a function of time, is then as shown in Figure 4.4. From such a curve an average rate constant $\bar{k}(t)$ could be found with equation (4.11). It would at first fall continuously with time, eventually attaining an almost constant value when the rapidly decomposing ions have mostly decomposed. The initial value of the average rate constant $\bar{k}(t)$ is half that of the discrete rate constant k which refers only to that group of $\frac{1}{2}N$ molecules decomposing rapidly.

An example of this general situation seems to be provided by the reaction

$$(CH_3)_3CCH_3^{+\cdot} \rightarrow (CH_3)_3C^+ + CH_3$$

for which Derrick and Robertson[58] deduced a single discrete rate constant of 1.0×10^{11} s^{-1} for a group of ions dissociating over the period 13–40 ps. Their results were refined by deconvolution, to allow for instrumental peak broadening. They note, however, that a number or a range of rate constants is needed for a complete description of the dissociation. Tenschert and Beckey[77] give for this dissociation an average rate constant of 5.3×10^{10} s^{-1} after 20 ps. We note that the average constant is less than the discrete constant, as expected from the argument above. However, not all the difference may be explicable in this way. In general terms, however, the agreement between the determinations is good.

From measurements of the lifetimes of ions, the distribution of the internal energy in field-ionised molecules has been deduced[77]. The effect of temperature of the emitter on ionic dissociation kinetics has also been examined[77].

4.8 FRAGMENT IONS IN FIELD IONISATION MASS SPECTRA

The earlier work by Beckey and his colleagues on fragment peaks in field ionisation mass spectra has been reviewed[7]. In these studies, dissociation induced by an electric field has been clarified, and from experiments with paraffins, alcohols, ethers, ketones, amines and olefins a number of deductions of general significance have been made. By 1968, Beckey was able to propose the following rule: 'A sharp normal fragment ion peak in a single-focusing mass spectrometer is due to parent ions decomposing within a time interval of *c.* $10^{-14} - 10^{-12}$ s after field ionisation unless these fragments were already present as neutral free radicals prior to field ionisation. Metastable fragment peaks are due to ions (almost always molecular ions) decomposing within a time interval of several $10^{-8} - 10^{-6}$ s after field ionisation. A broad peak extending over the mass range between the normal fragment peak and the normal metastable peak (which is termed "fast metastable peak") is due to ions decomposing between several 10^{-12} and 10^{-8} s.' (These deductions follow from what has been outlined in Section 4.7.) Beckey concluded that single-step direct bond rupture of organic ions may be observed in a field ion mass spectrometer as fragment peaks or as metastable peaks. In contrast, rearrangement or multi-step processes are usually not observed, or only faintly observed, as fragment peaks on the normal fragment ion mass number, but they appear as metastable peaks. Beckey at that time noted, however, that some fast rearrangement processes might be observed in the future. Such processes were indeed noted later by Chait and Kitson[82], who observed products of rearrangement reactions and multi-step processes appearing as ordinary fragment ions in field ionisation mass spectra. They also drew attention to several papers reporting such fragment ions. Rearrangements of ions when these ions are produced by field ionisation are discussed further in Section 4.10.

That alkyl ions formed from n-octane and n-nonane after field ionisation are produced only by the simplest mechanism of C—C bond rupture was deduced by Goldenfeld and Korostyshevsky[83] from experiments with n-octane-2d_1 and n-nonane-5-^{13}C. Hence field ionisation mass spectra can be used for finding the distribution of isotopes, and may be better for this purpose than electron impact spectra, which often involve processes additional to simple bond rupture. Fragment ions in the field ionisation mass spectra of a number of straight- and branched-chain hydrocarbons and their mono-chloro-substituted derivatives have been investigated by Weiss and Hutchison[84]. Fragmentation was greatly reduced by the presence of a double bond, and increased by the substitution of chlorine for hydrogen.

4.9 METASTABLE PEAKS IN FIELD IONISATION MASS SPECTRA

In a number of recent papers attention has been drawn to the value in structural and analytical investigations of the normal metastable peaks appearing in the field ion mass spectrum at a mass $m_n^* = m^2/M$. These can be relatively more intense in field ionisation spectra compared with electron impact spectra because the time taken for parent ions to reach the field-free region in which dissociation occurs is much less with field ionisation. Equations for calculating flight times of ions in sharp-edge type sources have been derived by Viney[79].

Wanless and Glock[85] noted that simple cleavage rules correlate the abundances of normal metastables observed with 3 octadecane isomers, 6 hexadecane isomers and 18 octane isomers. With long-chain ketones, metastable peaks due to loss of H_2O from the molecular ion vary systematically with the number of degrees of freedom in the molecule[86]. With monoterpenes, metastable peaks are useful in structural studies using both electron impact and field ionisation mass spectra[87].

4.10 REARRANGEMENT PEAKS IN FIELD IONISATION MASS SPECTRA

Rearrangement reactions have been studied by combined electron impact and field ionisation mass spectrometry for the formation of smaller rings from larger ones, for the McLafferty and other hydrogen rearrangement reactions, and for skeletal rearrangement reactions[76]. There has been discussion of the rule (see Section 4.8) that for gas-phase rearrangement reactions the peaks at exactly the normal mass position should be small in field ionisation mass spectra. Beckey[88] has re-formulated the rule to give: 'Rearrangement reactions in the gas phase are slower than the fastest direct bond-fission processes under high-field conditions. Therefore the maxima of such rearrangement peaks in the field ionisation mass spectra are more or less strongly shifted towards smaller mass values.' This formulation is for a single-focusing magnetic-deflection mass spectrometer. With a double-focusing mass spectrometer, Schulze and Richter[89] used the electric sector field as a means for studying gas-phase decompositions of metastable organic ions. They used a 'decoupling technique' to analyse the time requirements and the gas-phase origin of rearrangement processes in molecular ions formed by field ionisation. They established that elimination of RCOOH from aliphatic carboxylic esters can involve very fast hydrogen rearrangement prior to, or concomitant with, C—O bond dissociation: in fact, combination of these two events can occur in 10^{-12}–10^{-11} s after ionisation. The rearrangement product $C_nH_{2n}^{+\cdot}$ can be the largest peak in the spectrum. Several n-propyl alkanoates gave the ion $C_3H_6^{+\cdot}$ as the base peak; this corresponds to the product of a McLafferty rearrangement involving the alkyl rather than the acyl moiety of the molecular ion.

When fragment ions are produced as sharp mass spectral peaks showing no loss of kinetic energy there is the possibility that they are formed by a surface reaction on the emitter[82].

McLafferty-type hydrogen rearrangements and related reactions are often pure statistical processes even in a field ionisation source, and thus fall within the scope of the quasi-equilibrium theory of mass spectra. Levsen and Beckey[90] have studied the influence of the internal energy of organic ions on the rate constants for their reaction from the temperature dependence of field ionisation mass spectra.

4.11 SOME APPLICATIONS OF FIELD IONISATION MASS SPECTROMETRY TO STRUCTURAL INVESTIGATION

This subject is still in its infancy. Brown, Bruschweiler, Pettit and Reichstein[91] in 1970 listed (with references) natural products for which field ionisation mass spectra have been given. These products are 3β-acetoxy-11-oxo-5α-androstane, mono- and di-saccharides, nucleosides, amino acids, peptides, monoterpenes, abcisin II, and somalin. Such spectra have also been given for some pesticides, and for long-chain fatty acids and their esters. Mono- and oligo-saccharides give in electron impact spectra no molecular ion peaks, or only very weak ones, even if the substances are completely methylated. However, field ion spectra can be used for molecular weight determination of simple sugars[92]. A molecular-plus-one peak has been noted with fructose, xylose and glucose, formed by a reaction in the adsorbed state. In these experiments the solid sample was sublimed directly into the ion source. Positive ions of aryl-*O*-glucosides produced by field ionisation undergo cleavage and rearrangement reactions at the glucosidic bond. The influence of substituents in the aglycone has been studied[96], and ρ values have been derived with the Hammett equation. We note, however, that the assumption of a single rate constant for a cleavage or rearrangement reaction of an ion has been shown to be, in general, incorrect (Section 4.7.3). Furthermore, rate correlations of this type should not be based on fragment ion intensities in field ionisation mass spectra when field-induced dissociation or catalytic dissociation produces appreciable fragment ion intensity. Peak-broadening studies are therefore desirable in such work on rate correlations.

The field ionisation mass spectra of a series of nucleosides were found[93] to be extremely simple with all, except guanosine, showing intense molecular ion peaks, and characteristic sugar and base cleavage products as the important fragments. Electron impact spectra, in contrast, were found to be complex. In these experiments a molecular-beam direct-inlet system was used with the sample probe temperature independent of the ion source temperature. The first systematic study by field ionisation of simple peptides was reported by Brown and Pettit[94]. They examined either benzyloxycarbonyl or t-butyloxycarbonyl derivatives of di- to penta-peptide methyl esters containing glycine, alanine, leucine, serine, threonine, proline and tyrosine. 'Sequence peaks' due to simple cleavage and rearrangement peaks were in general more prominent with field ionisation than with electron impact ionisation. A further investigation[95] by field ionisation was made of some cardenolides, or cardiac glycosides, a series of natural products isolated from plants and insects. One series studied was based on the aglycone, digitoxigenin, and another was based on the genin, strophantidin. Important and structurally diagnostic information was revealed by fragment ions in field ionisation mass spectra. These spectra were superior to electron impact spectra in their simpler nature, and because of the greater relative intensity of sequence characteristic peaks.

4.12 ANALYTICAL APPLICATIONS OF FIELD IONISATION MASS SPECTROMETRY

Early work on the possible use of field ionisation mass spectrometers for qualitative and quantitative analysis has been described by Beckey and Wagner[97]. The field ion mass spectrum was found to be useful for obtaining a rapid survey of the composition of a complex mixture of organic substances. Beckey[7] has reviewed later developments.

A possible method for the analysis of paraffin–olefin mixtures has been proposed[85], based on conversion of the olefins to alkyl thiols and the use of metastable peaks in field ionisation spectra. Mead[98] successfully applied field ionisation mass spectrometry to the examination of paraffin waxes and wax fractions in the 300–500 °C boiling range, that is in the molecular weight range 280–700. With a razor blade as emitter in a double-focusing mass spectrometer, adequate sensitivity was obtained, and the spectral reproducibility was of the order of $\pm 5\%$. Very simple spectra, free from fragment ions, were obtained, and from these the qualitative composition and carbon number range of the main hydrocarbon series could be seen at a glance. Series such as di-and tri-cycloparaffins could be readily detected even at low concentrations. Relative sensitivities for the four main hydrocarbon series (normal paraffins, isoparaffins, cycloparaffins and alkylbenzenes) were found to be constant over the molecular weight region studied. The computation involved in quantitative analysis of wax from its field ionisation spectrum was much less than that required for the conventional electron impact spectrum. Analyses of two reference waxes gave results in good agreement with those obtained by conventional mass spectrometric techniques.

Tou[99] noted an intense molecular ion peak in the field ion mass spectra of nine dialkyl phthalates, whereas this molecular ion is either absent or extremely weak in the electron impact spectrum. In the field ion spectra a very strong normal metastable peak was associated with the reaction

$$M^{+\cdot} \rightarrow (M-R+2)^{+} + (R-2)$$

This metastable peak was weak or absent in the electron impact spectrum. The metastable peaks were used for the qualitative analysis of a mixture with field ion mass spectra.

4.13 FIELD IONISATION MASS SPECTROMETRY AND SURFACE PROCESSES

Field ionisation mass spectrometry has been applied to the study of clean metal surfaces and to the study of adsorption phenomena. One experimental method which has been used involves the mass spectrometric analysis of positive ions field-emitted from a sharp metal tip with a clean emitting surface under ultra-high vacuum conditions[100]. Similar experiments have been carried out in the presence of various gases at low pressures, when an adsorbed layer may form. The term 'field evaporation' refers to the evaporation of a metal from its own lattice under the influence of a strong field, and the term 'field desorption' is used when the evaporating species is different from the underlying bulk metal. Field evaporation of metals can be sustained at a maximum rate of *c.* 10 atom layers per second for only about one minute. Therefore the current of metal ions produced when a steady voltage is applied to a metal tip in ultra-high vacuum is very small, and becomes smaller with a constant applied voltage since the tip becomes blunter as field evaporation occurs. Resistance-strip electron multipliers[101] and secondary electron multipliers[102] have been used in mass spectrometers designed to study field evaporation. Pulsed field evaporation mass spectrometry[103–105], in which a sudden voltage pulse of duration 10^{-8} s or less is applied to an emitting tip, is useful in overcoming problems arising from tip blunting, and is especially valuable in the study of adsorption phenomena. Another way of studying adsorption is to obtain kinetic data for the adsorption process from flash filament experiments with a field ion mass spectrometer used as a partial pressure measurement device[106]. The use of field ionisation eliminates problems arising from pyrolytic reactions on the electron-emitting filament in ultra-high-vacuum systems incorporating electron-impact mass spectrometers.

The atom-probe field ion microscope devised by Müller[107] enables mass spectrometric analysis of single field-desorbed atoms to be made. With this elegant technique an atom of interest viewed on a surface by field ion microscopy is selected for examination, and its image is caused to coincide with a small probe hole in the image screen. The selected atom is field desorbed as an ion by application of a voltage pulse to the emitter tip, and if the microscope is aimed properly the ion passes through the probe hole and enters a time-of-flight mass spectrometer (see Chapter 7). The recording system of the mass spectrometer is only operated for a very short time period coinciding with the arrival of the ion, so that problems of electronic noise are overcome. Müller[108] has given an authoritative review of this instrument and of some of the results obtained. Similar instruments constructed by other workers have been described[109,110].

4.13.1 Field evaporation of metals

A brief review of field evaporation theories, with references to earlier work, has been given by Barofsky and Müller[111]. These workers find[101] that at liquid hydrogen temperature, field evaporation of the metal Be, Fe, Cu and Zn gives, respectively, the ions Be^{2+}, Fe^{2+}, Cu^{+} and Cu^{2+} and Zn^{+} and Zn^{2+}. Nickel at liquid nitrogen temperature gives Ni^{2+}. These charges are those predicted from a theory which takes account of the image force and considers direct evaporation of an ion over a field-reduced energy barrier. In the presence of hydrogen all these metals formed hydride ions. Pulsed field evaporation studies[105] of cobalt showed the presence of the metal–gas reaction products CoN_2^{2+} and CoO_2^{2+} formed at 78 K.

The field adsorption and desorption of helium and neon from metal tips has been studied by Müller *et al.*[112], who used the atom probe method. The field desorption of the noble gas ions seems to require simultaneous field evaporation of the tip metal atoms, which sometimes form various metal–noble gas molecular ions. The adsorption of the noble gas atoms is promoted by the field, since an atom of polarisability α in a field E has its potential energy lowered by $\frac{1}{2}\alpha E^2$ by interaction with the field. However, it seems that this lowering of energy is not the only cause of adsorption of imaging gas atoms on the surface of an emitting tip, but in addition a chemically specific shift of electronic charge in the surface atom possibly occurs[113].

4.13.2 Condensed layer formation, molecular interactions and surface reactions

According to the simple model of field ionisation described in Section 4.1, field ionisation does not occur within the critical distance, x_c of equation (4.1), from the surface. Therefore within this distance a condensed layer can form, particularly with polar molecules when the dipole moment interacts with the electric field. Such layer formation can lead to various complications in field ionisation mass spectrometry. A valuable review of surface reactions and molecular interactions studied by field ionisation mass spectrometry has been given by Block[114]. This discusses three mechanisms of ion formation, namely field ionisation, surface ionisation, and the formation of exo-ions which occurs on oxide catalysts. The detection of single ions and the selective use of single-crystal planes in surface studies are other topics reviewed. Block[115] notes that molecular interactions of adsorbed gases interfere with the onset fields of field ions, demonstrating the influence of adsorption and surface structures. One factor involved is the change of work function of the metal produced by adsorption of different substances.

Metzinger and Beckey[116] discovered that the overall field ion current generated by thin wires or by tips changes with the average work function if different organic substances are adsorbed. They discuss the effect fully.

The field ionisation mass spectra of organic molecules often show peaks which originate from reactions with the emitter surface or between adsorbed molecules. The mechanisms of these surface reactions have been studied by Röllgen and Beckey[117] with a field impulse desorption mass spectrometer. Most of the molecules adsorbed on the surface which become ionised are field ionised near their threshold field strength which is attained during the rise of the applied voltage pulse. A result is that mass spectra produced by field impulse desorption differ from spectra produced by the normal method in which a steady voltage is applied to the emitter. The mass lines originating from field dissociation processes are relatively much less intense with pulse desorption. Ions produced by surface reactions, however, are normally enhanced by pulse desorption. The voltage pulse follows after a period of undisturbed reaction and induces reactions in the physically adsorbed or chemisorbed layers on the emitter. The stable products or radicals produced may react during the next impulse-free period. The mechanisms of surface reactions can be studied from the variation of field impulse desorption currents with pulse repetition time, pressure and temperature. With acetone, a protonated molecular ion $(M+1)^+$ is produced mainly by a field-induced proton transfer reaction in the physically adsorbed surface layer. A radical of mass number $(M-1)$ is also produced, and the consecutive reactions of this radical are of primary importance in determining the form of the field ionisation mass spectrum of acetone. Field-induced hydrogen abstraction reactions lead to the formation of radicals which produce polymerisation reactions on the emitter surface. The study of 'field polymerisation' of acetone leads to an elucidation of the phenomenon of growth of micro-needles during the process of activation of emitters with acetone (Section 4.4.4).

Röllgen and Beckey[118], continuing with these researches, found that chemisorbed ions are found on a field ion emitter under high-field conditions. These ions exist within the critical distance x_c from the surface and they are not field desorbed in a certain range of field strengths. Many promotion effects observed in field ionisation mass spectrometry can be interpreted if the chemisorbed ions enhance the ionisation probability of incoming molecules. If there exists resonance between the electronic level of the adsorbed ion and the lowest empty level of the emitter, the ionisation probability is enhanced for a molecule approaching the adsorbed ion (see Section 4.3.3).

With water, reactions in the adsorbed layer on a field ion emitter lead to ions of formula $H_3O^+ \cdot nH_2O$. At room temperature and low fields of $3\ GV\ m^{-1}$, Anway[119] found that n varied from 0 to three or so, but at lower temperatures ions with n of up to nine were noted. Such ions were noted in several earlier researches[7]. Metastable peaks can be produced by the reaction[120]

$$(H_2O)_nH^+ \rightarrow (H_2O)_mH^+ + (H_2O)_{n-m}$$

Anway has discussed processes in the condensed water layer on the emitter in detail, and he has considered some models of this layer.

Block and Moentack[104] used pulse desorption to investigate the electric field dependence of the equilibrium between formic acid and its dimer. Schmidt[121] has used field ionisation mass spectrometry to investigate the interaction of ammonia and of a nitrogen–hydrogen mixture with an iron surface. A general introduction to the use of field ionisation in catalytic studies with metal catalysts has appeared[122].

4.14 SOME MISCELLANEOUS APPLICATIONS OF FIELD IONISATION MASS SPECTROMETRY TO CHEMICAL PROBLEMS

The relatively simple nature of field ionisation mass spectra as compared with electron impact spectra has proved advantageous in a number of researches which have used field ionisation mass spectrometers. The pyrolysis of some polymers and copolymers in the high vacuum system of a field ion mass spectrometer has been investigated by Schüddemage and Hummel and their colleagues[123–127]. Very gentle ionisation at platinum wire emitters produces mainly molecule ions bearing a single positive charge; therefore almost every mass peak in a field-ion mass pyrogram is due to a primary fragment produced by thermal degradation.

The advantages of field ionisation mass spectrometry for the investigation of the distribution of isotopes in organic molecules have been examined by Goldenfeld and Korostyshevsky[128,129]. The decomposition of a number of perchlorates has been investigated by Volk and Schubert[130] who find that field ionisation mass spectrometry gives more information about reactive decomposition products, such as chlorine dioxide, than electron impact mass spectrometry. Brunnée *et al.*[131] have studied the field ionisation mass spectra of some unstable organic compounds with a double-focusing instrument. Several papers by Butzert[132] are concerned with the measurement of free radical concentrations by field ionisation mass spectrometry. Butzert and Beckey[133] used the technique to investigate free radicals formed in heterogeneous decomposition reactions.

Investigations by Damico *et al.*[134] of the field ion spectra of some pesticidal and other biologically significant compounds showed that the structural information which can be derived from field ionisation complements that given by electron impact ionisation. Field ion mass spectra of some alkynes have been reported by Patterson and Seakins[135]. They observed metastable transitions from products of ion–molecule reactions. Nazarenko *et al.*[136] have investigated doubly charged ions observed in the field ion mass spectra of some organic compounds.

During a research on evidence for the existence of protactinium(III) in the solid state, Scherer *et al.*[137] pyrolysed PaI_5 and investigated the decomposition with the help of field ionisation mass spectrometry. They advanced evidence that PaI_3 is formed.

The field ion mass spectrometer was first applied to photochemical studies by Beckey and Groth[138] who studied the photolysis of acetone. Okabe *et al.*[139] photolysed propene, but-1-ene and hydrazine. Additional modes of application of the field ion mass spectrometer to photochemical studies have been reported by Barofsky and Beckey[140]. They used the ion source itself as the reaction chamber, but could not detect photoreactions in the gas phase because of the lack of a sufficiently intense light source. The possibility of using a field desorption mass spectrometer probe to take microsamples from a photochemical system for subsequent field ion mass spectrometric analysis was investigated, and seemed a promising technique.

4.15 NEGATIVE ION EMISSION FROM FIELD-EMITTING METAL SURFACES

Robertson and Williams[141] used a simple mass spectrometer to investigate negative ion emission from thin tungsten and platinum wires field-emitting electrons in various gases at low pressures. Fragment ions were observed from the halogens, CCl_4, SF_6, O_2, C_2H_2 and $C_2(CN)_4$. At least two ion-producing processes occurred, one of which was sputtering. This arose because electrons emitted from the wire produced positive ions by electron impact and these were attracted back to the wire. The other ion-producing process appeared to be a field-induced ionic desorption. Generally no parent molecular negative ions were observed, except with $C_2(CN)_4$ which gave a just measurable yield of molecular negative ions. Projections grew rapidly on the wires when field emission occurred, and they were investigated in detail with a simple cylindrical field emission microscope[142].

References

1. Müller, E. W. (1960). *Advances in Electronics and Electron Physics*, ed. by L. Marton, **13,** 83. (New York: Academic Press)
2. Inghram, M. G. and Gomer, R. (1954). *J. Chem. Phys.*, **22,** 1279
3. Gomer, R. and Inghram, M. G. (1955). *J. Amer. Chem. Soc.*, **77,** 500
4. Inghram, M. G. and Gomer, R. (1955). *Z. Naturforsch.*, **10a,** 863
5. Drechsler, M. (1954). German Patent 954105
6. Beckey, H. D. (1966). *J. Amer. Chem. Soc.*, **88,** 5333
7. Beckey, H. D. (1971). *Field Ionisation Mass Spectrometry*. (Berlin: Akademie-Verlag; Oxford: Pergamon Press)
8. Beckey, H. D. (1969). *Angew. Chem., Int. Edn. Engl.*, **8,** 623
9. Gruetzmacher, H. F., Heyns, K., Muller, D. and Scharman, H. (1967). *Fortschr. Chem. Forsch.*, **5,** 448
10. Block, J. (1968). *Advances in Mass Spectrometry*, ed. by E. Kendrick, **4,** 791. (London: Institute of Petroleum)
11. Beckey, H. D. and Comes, F. J. (1970). *Topics in Organic Mass Spectrometry*, ed. by A. L. Burlingame, 1. (New York: Wiley-Interscience)
12. Beckey, H. D. (1963). *Advances in Mass Spectrometry*, ed by R. M. Elliott, **2,** 1. (Oxford: Pergamon Press)
13. Beckey, H. D., Knöppel, H., Metzinger, G. and Schulze, P. (1966). *Advances in Mass Spectrometry*, ed. by W. L. Mead, **3,** 35. (London: Institute of Petroleum)
14. Van Lear, G. E. and McLafferty, F. W. (1969). *Ann. Rev. Biochem.*, **38,** 289
15. Fiedeldey, H. and Fourie, D. (1960). *Phys. Rev.*, **117,** 924
16. Thornber, K. K., McGill, T. C. and Mead, C. A. (1967). *J. Appl. Phys.*, **38,** 2384
17. Mendelsohn, L. B. (1968). *Phys. Rev.*, **176,** 90
18. Boudreaux, D. S. and Cutler, P. H. (1966). *Surface Sci.*, **5,** 230
19. Boudreaux, D. S. and Cutler, P. H. (1966). *Phys. Rev.*, **149,** 170
20. Fonash, S. J. and Schrenk, G. L. (1969). *Phys. Rev.*, **180,** 649
21. Sharma, S. P. and Schrenk, G. L. (1970). *Phys. Rev.*, **B2,** 598
22. Knor, Z. and Müller, E. W. (1968). *Surface Sci.*, **10,** 21

23. Müller, E. W. (1969). *Quart. Rev. Chem. Soc.*, **23,** 177
24. Goldenfeld, I. V., Korol, E. N. and Pokrovsky, V. A. (1970). *Int. J. Mass Spectrom. Ion Phys.*, **5,** 337
25. Jason, A. J., Burns, R. P. and Inghram, M. G. (1965). *J. Chem. Phys.*, **43,** 3762
26. Jason, A. J., Burns, R. P., Parr, A. C. and Inghram, M. G. (1966). *J. Chem. Phys.*, **44,** 4351
27. Jason, A. J. (1967). *Phys. Rev.*, **156,** 266
28. Alferieff, M. E. and Duke, C. B. (1967). *J. Chem. Phys.*, **46,** 938
29. Ionov, N. I. (1969). *Soviet Phys. Tech. Phys.*, **14,** 542
30. Kupriyanov, S. E. (1967). *Zh. Eksp. Teor. Fiz., Pis'ma Redaktsiyu,* **5,** 245
31. Reed, L. D. and Brodie, D. E. (1968). *Can. J. Phys.*, **46,** 789
32. Lorquet, J. C. and Hall, G. G. (1965). *Mol. Phys.*, **9,** 29
33. Beckey, H. D. (1964). *Les Congrès et Colloques de l'Université de Liège,* **30,** 26. (Liège: l'Université)
34. Beckey, H. D. (1963). *Z. Instrumentenk.*, **71,** 51
35. Robertson, A. J. B., Viney, B. W. and Warrington, M. (1963). *Brit. J. Appl. Phys.*, **14,** 278
36. Müller, E. W. and Tsong, T. T. (1969). *Field Ion Microscopy, Principles and Applications.* (New York: Elsevier)
37. Wanless, G. G. (1968). *Advances in Mass Spectrometry,* ed. by E. Kendrick, **4,** 833
38. Beckey, H. D., Dahmen, J. and Knöppel, H. (1966). *Z. Naturforsch.*, **21a,** 141
39. Jason, A., Halpern, B., Inghram, M. G. and Gomer, R. (1970). *J. Chem. Phys.*, **52,** 2227
40. Nazarenko, V. A. (1970). *Int. J. Mass Spectrom. Ion Phys.*, **5,** 63
41. Goldenfeld, I. V. and Nazarenko, V. A. (1970). *Int. J. Mass Spectrom. Ion Phys.*, **5,** 197
42. Beckey, H. D. (1963). *Z. Anal. Chem.*, **197,** 80
43. Beckey, H. D., Heising, H., Hey, H. and Metzinger, H. G. (1968). *Advances in Mass Spectrometry,* ed. by E. Kendrick, **4,** 817
44. Metzinger, H. G. and Beckey, H. D. (1967). *Z. Physik. Chem. (Frankfurt),* **52,** 27
45. Levsen, K., Heindrichs, A. and Beckey, H. D. (1968). *Messtechnik,* **76,** 9
46. Beckey, H. D., Krone, H. and Roellgen, F. W. (1968). *J. Sci. Instr. (J. Phys. E.),***1,** 118
47. Beckey, H. D. (1968). *Messtechnik,* **76,** 50
48. Knöppel, H. (1970). *Int. J. Mass Spectrom. Ion Phys.*, **4,** 97
49. Ochterbeck, E. and Beckey, H. D. (1971). *Int. J. Mass Spectrom. Ion Phys.*, in the press
50. Robertson, A. J. B. and Viney, B. W. (1966). *Advances in Mass Spectrometry,* ed. by W. L. Mead, **3,** 23
51. Robertson, A. J. B. and Viney, B. W. (1966). *J. Chem. Soc. A,* 1843
52. Cross, C. M. and Robertson, A. J. B. (1966). *J. Sci. Instr.*, **43,** 475
53. Blenkinsop, P. A., Job, B. E., Brailsford, D. F., Cross, C. M. and Robertson, A. J. B. (1968). *Int. J. Mass Spectrom. Ion Phys.*, **1,** 421
54. Migahed, M. D. and Beckey, H. D. (1971). *Int. J. Mass Spectrom. Ion Phys.*, in the press
55. Tou, J. C., Westover, L. B. and Sutton, E. J. (1969). *Int. J. Mass Spectrom. Ion Phys.*, **3,** 277
56. Beckly, H. D., Hilt, E., Maas, A., Migahed, M. D. and Ochterbeck, E. (1969). *Int. J. Mass Spectrom. Ion Phys.*, **3,** 161
57. Derrick, P. J. and Robertson, A. J. B. (1969). *Int. J. Mass Spectrom. Ion Phys.*, **3,** 409
58. Derrick, P. J. and Robertson, A. J. B. (1971). *Proc. Roy. Soc. London. A,* **324,** 491
59. Beckey, H. D., Heindrichs, A., Hilt, E., Migahed, M. D., Schulten, H. R. and Winkler, H. U. (1971). *Messtechnik,* in the press
60. Röllgen, F. W. and Beckey, H. D. (1970). *Surface Sci.*, **23,** 69
61. Beckey, H. D. (1969). *Int. J. Mass Spectrom. Ion Phys.*, **2,** 500
62. Beckey, H. D., Heindrichs, A. and Winkler, H. U. (1970). *Int. J. Mass Spectrom. Ion Phys.*, **3,** Appendix, 9
63. Beynon, J. H., Fontaine, A. E. and Job, B. E. (1966). *Z. Naturforsch.*, **21a,** 776
64. Patterson, B. C. and Poore, K. G. (1967). *J. Sci. Instr.*, **44,** 816
65. Emmerson, J. D. and Job, B. E. (1968). *J. Sci. Instr.*, **1,** 611
66. Korostyshevsky, I. Z. and Goldenfeld, I. V. (1968). *Pribory i Tekhn. Eksperim.*, 146; *Instr. Exptl. Tech. (USSR),* 148
67. Brunnée, C. (1967). *Z. Naturforsch.*, **22b,** 121
68. Schulze, P., Simoneit, B. R. and Burlingame, A. L. (1969). *Int. J. Mass Spectrom. Ion Phys.*, **2,** 183

69. Goldenfeld, I. V., Bondarenko, R. N. and Golovatyi, V. G. (1970). *Pribory i Tekhn. Eksperim.*, 203
70. Chait, E. M., Shannon, T. W., Perry, W. O., Van Lear, G. E. and McLafferty, F. W. (1969). *Int. J. Mass Spectrom. Ion Phys.*, **2,** 141
71. Beckey, H. D., Heindrichs, A. and Tenschert, G. (1967). *Z. Instrumentenk.*, **75,** 195
72. Gilliland, J. M. and Viney, B. W. (1968). *Royal Aircraft Establishment Technical Report* 68271
73. Brailsford, D. F. (1970). *J. Phys. D Appl. Phys.*, **3,** 196
74. Brailsford, D. F. and Robertson, A. J. B. (1968). *Int. J. Mass Spectrom. Ion Phys.*, **1,** 75
75. Beckey, H. D. and Knöppel, H. (1966). *Z. Naturforsch.*, **21a,** 1920
76. Beckey, H. D., Hey, H., Levsen, K. and Tenschert, G. (1969). *Int. J. Mass Spectrom. Ion Phys.*, **2,** 101
77. Tenschert, G. and Beckey, H. D. (1971). *Int. J. Mass Spectrom. Ion Phys.*, in the press
78. Levsen, K. and Beckey, H. D. (1971), in the press
79. Viney, B. W. (1969). *Royal Aircraft Establishment Technical Report 69252*
80. Beckey, H. D. (1959). *Z. Anal. Chem.*, **170,** 359
81. Beckey, H. D. (1968). *Int. J. Mass Spectrom. Ion Phys.*, **1,** 93
82. Chait, E. M. and Kitson, F. G. (1970). *Org. Mass Spectrom.*, **3,** 533
83. Goldenfeld, I. V. and Korostyshevsky, I. Z. (1969). *Int. J. Mass Spectrom. Ion Phys.*, **3,** 404; *Russ. J. Phys. Chem.*, **43,** 1453
84. Weiss, M. J. and Hutchison, D. A. (1968). *J. Chem. Phys.*, **48,** 4386
85. Wanless, G. G. and Glock, G. A. (1967). *Anal. Chem.*, **39,** 2
86. Barber, M., Elliott, R. M. and Kemp, T. R. (1969). *Int. J. Mass Spectrom. Ion Phys.*, **2,** 157
87. Beckey, H. D. and Hey, H. (1968). *Org. Mass Spectrom.*, **1,** 47
88. Beckey, H. D. (1970). *Int. J. Mass Spectrom. Ion Phys.*, **5,** 182
89. Schulze, P. and Richter, W. J. (1971). *Int. J. Mass Spectrom. Ion Phys.*, **6,** 131
90. Levsen, K. and Beckey, H. D. (1971), in the press
91. Brown, P., Bruschweiler, F. R., Pettit, G. R. and Reichstein, T. (1970). *J. Amer. Chem. Soc.*, **92,** 4470
92. Krone, H. and Beckey, H. D. (1969). *Org. Mass Spectrom.*, **2,** 427
93. Brown, P., Pettit, G. R. and Robins, R. K. (1969). *Org. Mass Spectrom.*, **2,** 521
94. Brown, P. and Pettit, G. R. (1970). *Org. Mass Spectrom.*, **3,** 67
95. Brown, P., Bruschweiler, F., Pettit, G. R. and Reichstein, T. (1971). *Org. Mass Spectrom.*, **5,** 573
96. Phillips, G. O., Filby, W. G. and Mead, W. L. (1970). *Chem. Commun.*, 1269
97. Beckey, H. D. and Wagner, G. (1963). *Z. Anal. Chem.*, **197,** 58
98. Mead, W. L. (1968). *Anal. Chem.*, **40,** 743
99. Tou, J. C. (1970). *Anal. Chem.*, **42,** 1381
100. Vanselow, R. and Schmidt, W. A. (1966). *Z. Naturforsch.*, **21a,** 1690
101. Barofsky, D. F. and Müller, E. W. (1968). *Surface Sci.*, **10,** 177
102. Schmidt, W. A. and Frank, O. (1969). *Int. J. Mass Spectrom. Ion Phys.*, **2,** 399
103. Beckey, H. D. and Röllgen, F. W. (1966). *Z. Instrumentenk.*, **74,** 47
104. Block, J. and Moentack, P. L. (1967). *Z. Naturforsch.*, **22a,** 711
105. Barofsky, D. F. (1969). *Int. J. Mass Spectrom. Ion Phys.*, **3,** 156
106. Vanselow, R. and Schmidt, W. A. (1967). *Z. Naturforsch.*, **22a,** 717
107. Müller, E. W., Panitz, J. A. and McLane, S. B. (1968). *Rev. Sci. Instr.*, **39,** 83
108. Müller, E. W. (1970). *Naturwissenschaften*, **57,** 222
109. Turner, P. J. and Southon, M. J. (1969). *Dynamic Mass Spectrometry*, ed. D. Price and J. E. Williams. (London: Heyden and Son)
110. Brenner, S. S. and McKinney, J. T. (1970). *Surface Sci.*, **23,** 88
111. Barofsky, D. F. and Müller, E. W. (1969). *Int. J. Mass Spectrom. Ion Phys.*, **2,** 125
112. Müller, E. W., McLane, S. B. and Panitz, J. A. (1969). *Surface Sci.*, **17,** 430
113. Müller, E. W., Krishnaswamy, S. V. and McLane, S. B. (1970). *Surface Sci.*, **23,** 112
114. Block, J. (1969). *Colloq. Int. Cent. Nat. Rech. Sci.*, **187,** 89
115. Block, J. (1969). *Z. Physik. Chem. (Frankfurt)*, **64,** 199
116. Metzinger, H. G. and Beckey, H. D. (1967). *Z. Naturforsch.*, **22a,** 1020
117. Röllgen, F. W. and Beckey, H. D. (1970). *Surface Sci.*, **23,** 69
118. Röllgen, F. W. and Beckey, H. D. (1971). *Surface Sci.*, **26,** 100
119. Anway, A. R. (1969). *J. Chem. Phys.*, **50,** 2012

120. Goldenfeld, I. V., Nazarenko, V. A. and Pokrovsky, V. A. (1965). *Dokl. Akad. Nauk SSSR,* **161,** 861
121. Schmidt, W. A. (1968). *Angew. Chem., Int. Edn. Engl.,* **7,** 139
122. Robertson, A. J. B. (1970). *Catalysis of Gas Reactions by Metals.* (London: Logos Press)
123. Schüddemage, H. D. R. and Hummel, D. O. (1966). *Koll. Z. Z. Polymere,* **210,** 103
124. Hummel, D. O., Schüddemage, H. D. R. and Pohl, U. (1966). *Koll. Z. Z. Polymere,* **210,** 106
125. Schüddemage, H. D. R. and Hummel, D. O. (1967). *Koll. Z. Z. Polymere,* **220,** 133
126. Ryska, M., Schüddemage, H. D. R. and Hummel, D. O. (1969). *Makromol. Chem.,* **126,** 32
127. Schüddemage, H. D. R. and Hummel, D. O. (1968). *Advances in Mass Spectrometry,* ed. by E. Kendrick, **4,** 857. (London: Institute of Petroleum)
128. Goldenfeld, I. V. and Korostyshevsky, I. Z. (1968). *Dokl. Akad. Nauk SSSR,* **178,** 876
129. Goldenfeld, I. V. and Korostyshevsky, I. Z. (1969). *Isotopenpraxis,* **5,** 151
130. Volk, F. and Schubert, H. (1970). *Explosivstoffe,* **18,** 73
131. Brunnée, C., Kappus, G. and Maurer, K. H. (1967). *Z. Anal. Chem.,* **23,** 17
132. Butzert, H. (1968). *Brennstoff-Chem.,* **49,** 247, 283, 305
133. Butzert, H. and Beckey, H. D. (1968). *Z. Physik. Chem. (Frankfurt),* **62,** 83
134. Damico, J. N., Barron, R. P. and Sphon, J. A. (1969). *Int. J. Mass Spectrom. Ion Phys.,* **2,** 161
135. Patterson, B. C. and Seakins, M. (1967). *Trans. Faraday Soc.,* **63,** 1863
136. Nazarenko, V. A., Goldenfeld, I. V. and Dibrova, P. S. (1969). *Int. J. Mass Spectrom. Ion Phys.,* **2,** 92
137. Scherer, V., Weigel, F. and Van Ghemen, M. (1967). *Inorg. Nucl. Chem. Lett.,* **3,** 589
138. Beckey, H. D. and Groth, W. (1959). *Z. Physik. Chem. (Frankfurt),* **20,** 307
139. Okabe, H., Beckey, H. D. and Groth, W. (1966). *Z. Naturforsch.,* **21a,** 135
140. Barofsky, D. F. and Beckey, H. D. (1970). *Org. Mass Spectrom.,* **3,** 403
141. Robertson, A. J. B. and Williams, P. (1968). *Advances in Mass Spectrometry,* ed. by E. Kendrick, **4,** 847. (London: Institute of Petroleum)
142. Robertson, A. J. B. and Williams, P. (1967). *Twenty-Seventh Annual Conference Physical Electronics,* 144. (Cambridge, Massachusetts: Massachusetts Institute of Technology)

5
Chemical Ionisation Mass Spectrometry

F. H. FIELD
The Rockefeller University, New York

5.1 INTRODUCTION

The technique of chemical ionisation mass spectrometry was invented[1,2] in 1965, and in the years to the present it has undergone an accelerating amount of use. The growth in the past two years or so has been particularly rapid, and it is now clear that the technique is one of significant practical utility and scientific importance. Two reviews of the early aspects of the technique have been published[3,4], but in view of the rapid developments which have occurred it is appropriate to attempt here to describe the present state of the technique.

In chemical ionisation mass spectrometry the ionisation of the substance of interest is effected by ion–molecule reactions rather than by electron, photon, or field ionisation. The initial ionisation in the system is almost always effected by electron impact, although in principle and in a very restricted practice, photon impact can also be used. Thus, we may represent the chemical ionisation process in a general way by the equations:

$$\begin{aligned} R + e &\rightarrow R^+ \\ R^+ + A &\rightarrow A_1^+ + N_1 \\ A_1^+ &\rightarrow A_2^+ + N_2 \\ &\rightarrow A_3^+ + N_3 \\ &\vdots \\ &\rightarrow A_i^+ + N_i \end{aligned} \tag{5.1}$$

where R is the reactant gas, R^+ is a stable reactant ion produced (often by a sequence of ion–molecule reactions) from R, A is the additive molecule under investigation, and the ions A_i^+ are the ions comprising the chemical ionisation mass spectrum of A. A_1^+ is most often the ion produced by adding a proton to A, that is, it is the $(M+1)^+$ ion where M is the molecular weight of A, but occasionally A_1^+ is the $(M-1)^+$ ion formed by hydride ion abstraction from A. The ions $A_2^+, A_3^+, \ldots A_i^+$ represent the various fragment ions formed by the decomposition of A_1^+.

The characteristic feature of the chemical ionisation concept may be looked upon as the utilisation of a specific set of ions (the reactant ions) to effect a specific type of ion–molecule reaction with a wide variety of different compounds. The ions produced by these ion–molecule reactions with a given compound constitute the chemical ionisation mass spectrum of the compound. Because the same reactant ions react with a variety of different compounds, the chemical ionisation mass spectra produced from the different compounds are comparable with each other and reflect differences in structure and chemical reactivities of the compounds or quantitative differences in their concentrations. One of the essential features of the chemical ionisation concept is that it constitutes a practical technique of carrying out qualitative and quantitative analysis and also investigations of ionic properties and reactivities.

An important aspect of the chemical ionisation technique in the fact that with different kinds of reactant ions different kinds of reactions will be involved in the production of the chemical ionisation mass spectra. Thus, for example, if CH_5^+ is used as a reactant, the reaction occurring is proton transfer, and the chemistry involved may be looked upon as an even-electron, gaseous acid–base chemistry. On the other hand, if N_2^+ is used as a reactant ion, electron transfer occurs, and one then has an odd-electron, oxidation–reduction chemistry. It is self-evident that by using different reactant ions the intensity of the chemical ionisation reaction can be varied, that is, reactants with different acid strengths or with different oxidation potentials can be used. We give in Table 5.1 a list of reactants about which the author has at

Table 5.1 Chemical ionisation reactants

Reactant gas	*Major reactant ion*	*Reaction type*	*Reaction intensity*
CH_4	$CH_5^+ + C_2H_5^+$	Acid–base	Strong
C_3H_8	i-$C_3H_7^+$	Acid–base	Moderate
i-C_4H_{10}	t-$C_4H_9^+$	Acid–base	Mild
H_2	H_3^+	Acid–base and electron transfer	Strong
Rare gases (R)	R^+	Electron transfer	Depends on R
N_2	N_2^+	Electron transfer	Strong
H_2O	H_3O^+, $H_5O_2^+$, etc.	Acid–base	Mild
CH_3OH	$CH_3OH_2^+$, $(CH_3OH)_2H^+$	Acid–base	Mild
NH_3	NH_4^+, $N_2H_7^+$, etc.	Acid–base	Mild

least preliminary knowledge. Some properties of the reactants are also given. Only CH_4 and i-C_4H_{10} have to date been used extensively in chemical ionisation studies.

The spectrum of a substance produced by chemical ionisation is usually different from the spectra produced by the other ionisation techniques, and the analytical and scientific utility of the method stems from the unique character of the chemical ionisation spectra. This uniqueness results in turn from the operation of a number of special factors in the chemical ionisation system. First, currently the most widely used reactant gases are those such as methane and i-butane wherein the chemical ionisation is effected by the transfer of massive entities such as protons, hydride ions, or alkyl carbonium ions. For such reactions the chemical ionisation processes are not governed by Franck–Condon considerations, which is in contrast to the important role of the Franck–Condon principle in impact ionisation processes. Secondly, for reactant gases such as methane and i-butane the reactant ions (R^+ in equation (5.1)) are even-electron ions such as CH_5^+ or t-$C_4H_9^+$, and the ions produced by them by reactions with the additive molecules will also be even-electron ions. By comparison, the molecule-ions initially formed by impact ionisation processes are odd-electron ions, and their decomposition chemistry is influenced, and perhaps dominated, by the presence of the odd electron. Clearly, the ions initially formed by even-electron chemical ionisation processes and those formed by impact processes may be expected to decompose by rather different networks of pathways. One would expect the decomposition paths for the odd-electron ions to be more extensive, and this behaviour is observed in fact. Thirdly, the amount of energy contained by the ions initially formed in the chemical ionisation process is generally significantly less than that of ions formed by electron ionisation, and the lower energy of the chemical ionisation process results in smaller amounts of fragmentation. However, the amount of energy involved in chemical ionisation will depend upon the nature of the reactant gas, and for some reactants the energy transfer can be greater than occurs in electron ionisation. To date such high-energy reactants have not been used to any significant degree.

A final factor which operates to make massive-particle-transfer chemical ionisation spectra different from electron ionisation spectra is the relatively localised nature of the ionisation and fragmentation reactions which occur. Depending somewhat on the identity of the compound being measured, the chemical ionisation spectrum results from a random or multiple-point attack of the reactant ion, and if the attack effects some fragmentation, the fragmentation reaction will be of a relatively localised nature. The random attack–localised reaction concept was first advanced[5] to account for the methane chemical ionisation spectra of paraffin hydrocarbons, and the argument advanced was that in, for example, a compound like n-hexadecane that attack of CH_5^+ on the 28 equivalent secondary hydrogen atoms or the 14 equivalent methylene groups would occur at random and that the ensuing reaction would occur at the point of attack. With this hypothesis a convincing rationalisation of the spectra of paraffins and cycloparaffins[6] could be formulated. For compounds containing polar functional groups the attack of the reactant ions is no longer random, but rather it is directed by the ion–dipole forces existing between the functional group and the attacking ion. However, in the case where a molecule contains several functional groups the attack will generally occur on all of the functional groups to one extent or another. As a simple approximation one can think that the attack on the several functional groups occurs at random, although actually the frequency of attacks on the various groups is probably affected by the magnitude of the dipole moment of the group. The fragmentation reactions which occur as a result of the attack of reactant ions will be those appropriate to the groups attacked, and it appears that the spectrum of the molecule is formed by a simple combination of the ions produced by attacks at the several functional groups. This behaviour can be rationalised in terms of the fact that in even-electron ions the charge is more localised than it is, for example, in odd-electron ions because migration of the charge in the even-electron case usually requires a concomitant migration of the atoms in the ion. A contributing factor is the relatively low energy content of the ions in chemical ionisation.

The important practical consequence of this localised, additive decomposition behaviour is that the presence in a polyfunctional molecule of a group or groups which form stable ions will give rise to the formation of the $(M \pm 1)^+$ ions important for molecular weight identification even though the molecule may also contain groups which undergo very facile ionic decompositions. For example, the $(M+1)^+$ ion intensity in 5α-androstane-17β-ol

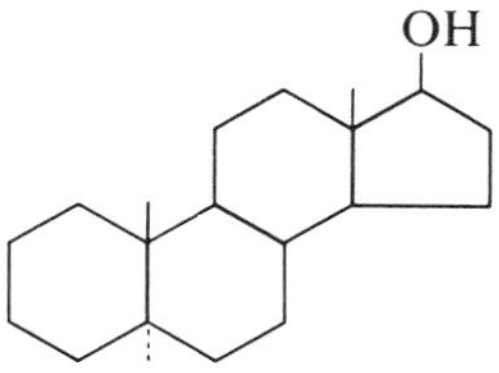

OH

is 3.9% of the total ionisation and that of $(M+1-H_2O)^+$ is 46%[7]. Hydroxyl is easily lost in chemical ionisation because the water formed by protonating hydroxyl is such a good leaving group. The carbonyl group protonates easily to form a stable ionic species, and one finds that the inclusion of a carbonyl group in a hydroxy steroid markedly increases the intensity of $(M+1)^+$. Thus, for 5α-androstane-3-one-17β-ol

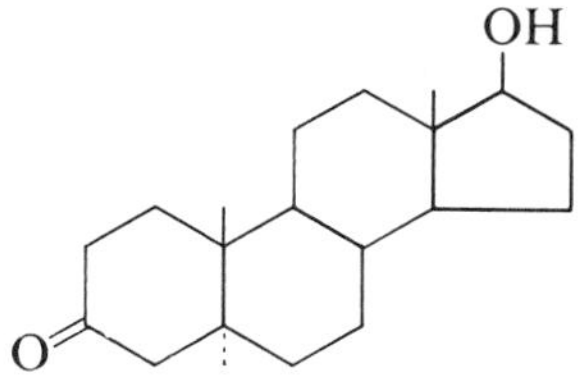

OH

O

the relative intensity of $(M+1)^+$ rises to 50.2%, and the intensity of $(M+1-H_2O^+$ falls to 15.5%[7]. By contrast, it is well known that the electron ionisation spectrum of a polyfuncational molecule can be determined to a large extent by the reactions of one dominant group. As one of many possible examples, we cite the decomposition of *N*,*N*-dimethylphenylalanine[8]

$$\underset{\displaystyle \mathrm{NMe_2}}{\mathrm{PhCH_2\overset{}{C}HCO_2Pr}}$$

for which, in spite of the polyfunctional nature of the molecule, extensive fragmentation dominated and directed by the dimethylamino group occurs.

In summary, the salient characteristics of even-electron chemical ionisation spectra are: (i) the amount of fragmentation occurring is relatively small compared with electron ionisation, (ii) the $(M+1)^+$ and/or the $(M-1)^+$ ion intensities are relatively large, (iii) the ions comprising the spectra tend to be located in the higher mass end of the spectra, and (iv) the fragmentation reactions which occur are significantly different from those observed with electron ionisation. It is to be noted in passing that the $(M \pm 1)^+$ ions are sometimes referred to as quasi-parent ions, but although this term was originally introduced by Munson and Field[1], we now think that it is too indefinite, and we recommend the use of $(M+1)^+$ or $(M-1)^+$, as the case may be.

5.2 EXPERIMENTAL

The overall experimental problem in chemical ionisation mass spectrometry is to produce a set of reactant ions, bring them into contact with the substance to be investigated, and collect the product ions which comprise the chemical ionisation mass spectrum. These procedures can be accomplished, at least in concept, in several ways. One could utilise a tandem mass spectrometer apparatus such as that constructed and utilised for a variety of studies by Futrell, Tiernan and co-workers[9,10]. An ion cyclotron mass spectrometer[11] (see Chapter 6) in principle could be used because the double-resonance ion cyclotron technique enables one to identify the products resulting from ion–molecule reactions of a specific reactant ion. In a recent study[12] a chemical ionisation-type study was made with an ion cyclotron resonance technique, and another study[10] using the tandem method has a distinct chemical ionisation orientation. A third method for making chemical ionisation studies involves a high-pressure mass spectrometic technique, and it has been used in virtually all the chemical ionisation studies reported to date. With it the substance to be investigated is mixed with a large excess of a reactant gas, and the mixture is introduced into the ionisation chamber of the mass spectrometer to produce a pressure which is usually of the order of 0.5–1.5 torr. Ionisation is effected by electron ionisation to produce an ultimate set of ions from the reactant gas; these in turn react with the substance under investigation to produce the chemical ionisation mass spectrum.

In the high-pressure mass spectrometric technique the pressure of the reactant gas and the relative amounts of reactant and additive gases must be such that (i) to a high degree of approximation all of the ionisation effected by the electron ionisation occurs in the reactant gas, (ii) the primary ions of the reactant produced by the electron ionisation react virtually completely with molecules of reactant to produce the stable reactant ions; that is, reactants of primary reactant ions with additive molecules should not occur, and (iii) the stable reactant ions should be formed close enough to the electron beam that they have a maximum opportunity to react with the additive molecules as they drift out of the ionisation chamber. The experience accumulated to date is that these conditions are met for reactant gas pressures in the range 0.5–1.5 torr and ratios of pressures of additive to reactant gas of 10^{-3} or less.

It is desirable to obtain an idea of the number of collisions an ion will experience in the course of its lifetime in the ionisation chamber of the mass spectrometer under chemical ionisation conditions. We write

$$Z_c = kN \tag{5.2}$$

where Z_c = number of collisions per second of the ion with gas molecules, N = number density of gas molecules, and k = rate constant for collisions between ions and molecules. The rate constants for ion–molecule collisions are about 10^{-9} ml mol^{-1} s^{-1} and are independent of ion velocity for relatively low velocities. The number density for a gas at a pressure of 1 torr and a temperature of 150 °C is approximately 2×10^{16}, and thus from (ii) one calculates that an ion undergoes *c.* 2×10^{7} collisions per second. We shall show later that under conventional chemical ionisation conditions and in particular for an ionisation chamber repeller field strength of 12.5 V cm^{-1}, the residence time of an ion in a typical mass spectrometer is *c.* 10^{-5} s. Then the total number of collisions that the ion undergoes during its lifetime in the ionisation chamber is *c.* 200, and reactant ions will have ample opportunity to collide with additive molecules before passing out of the ionisation chamber. At lower repeller voltages, the use of which is quite feasible, an ion can undergo a thousand or more collisions before passing out the ionisation chamber.

The requirements for operating a mass spectrometer in the high-pressure chemical ionisation mode are conceptually very simple: the flow out of the ionisation chamber must be maintained at a low value, and the pumping capacity of the machine must be sufficiently high to maintain the requisite low pressure along the path of the ion beam from the ionisation chamber to the ion collector. Differential pumping on the source and the analyser volumes is utilised. Several groups working in the chemical ionisation field have published more or less detailed descriptions of the mass spectrometers used in their investigations. The original apparatus used by Munson and Field[1] was the Esso Chemical Physics Mass Spectrometer described earlier by Field[13]. Futrell and Wojcik[14], Michnowicz and Munson[15], Arsenault, Althaus and Divekar[16], and McCloskey and co-workers[17] have given descriptions of modifications of CEC 21-110 mass spectrometers for chemical ionisation use. Fales *et al.*[18] describe modifications of an AEI MS-9 mass spectrometer for chemical ionisation service. The descriptions by Futrell and Wojcik and by Fales and co-workers are the most elaborate. The chemical ionisation process has been patented[2], and exclusive United States manufacturing and sales rights for the process are held by Scientific Research Instruments Corporation, Baltimore, Md.

The Esso Chemical Physics Mass Spectrometer in this laboratory has been extensively used for chemical ionisation and other high-pressure work, and its source and pumping design is fairly representative of current practice. The ion source is constructed to be as gas tight as possible, and ideally one wishes that the only openings in the source are the holes for electron entrance and ion exit. In the Esso machine both of these openings have the dimensions 3×0.05 mm, which corresponds to a total aperture area of 0.30 mm^2. In work done to date using this machine, the pumping has been accomplished by 300 l s^{-1} pumps on the source and analyser regions of the mass spectrometer. For a source pressure of *c.* 1 torr the pressure in the ionisation chamber envelope is *c.* 5×10^{-4} torr and that in the analyser chamber is *c.* 8×10^{-6} torr. The ion accelerating voltage of the machine is 3000 V and satisfactory operation is obtained. Scattering of the ion beam is negligible; peak shape and resolution are unchanged from their low-pressure values; and the overall operation of the instrument is not affected in an adverse way by operation in the chemical ionisation mode. The machine has been operated with source pressures up to *c*. 2 torr before arcs occur. The upper limit on the allowable pressure varies with ion acceleration voltage and the identity of the reactant gas, and it doubtless will also vary from one mass spectrometer to another. With our instrument using methane or isobutane as reactant gas, arcs occur when the pressure in the source envelope becomes of the order of several μm. Quite recently we have replaced the original source envelope pump with one with a rated speed of 1200 l s^{-1}, and using it we have obtained satisfactory performance of the equipment with source pressures of the order of 4–5 torr.

Several other instrumental modifications desirable for chemical ionisation operation should be mentioned. Perhaps most important, one can calculate that a beam of 70 V electrons traversing a path of 1 cm through i-butane at a pressure of 1 torr experiences an intensity diminution of many orders of magnitude (allowing one ionising collision per electron), and thus the production of significant numbers of collectable ions in an ionisation chamber operated under chemical ionisation conditions requires that the length of the ionisation chamber be as short as possible and the energy of the ionising electrons be significantly higher than is the case in electron ionisation mass spectrometry. Current practice is to operate at electron energies in the range 200–600 V. Because the electron entrance hole is necessarily small in area and because many of the electrons do not penetrate all the way through the ionisation chamber, the trap current cannot be used to control the filament, and instead the control is maintained on the emitted electron current. The role played by the repeller electrode and consequently the utility of this electrode in chemical ionisation mass spectrometry seems to vary from one group of workers to another. Fales *et al.*[18] report that the ion current is maximised when the ion repeller voltage is zero with respect to the ionisation chamber. Michnowicz[7] reports that maximum sensitivity is achieved at zero repeller voltage when he has a clean source, but after a few weeks of operation of the mass spectrometer the intensity passes through a broad maximum at a repeller voltage of *c*. 3 V. By contrast, in the Esso Chemical Physics Mass Spectrometer no ions are extracted from the source at zero repeller voltage, and the ion intensity at first increases and then reaches a plateau as the repeller voltage is increased to 10–15 V. In our opinion, no real understanding exists of the reasons for these differences in repeller behaviour.

Futrell and Wojcik[14] report achieving a resolution of 56 700 at 10% valley in their CEC 21-110 mass spectrometer operated in the chemical ionisation mode, and Fales and co-workers[18] report operating their AEI MS-9 in the chemical ionisation mode at a resolution of 10 000. All the evidence available indicates that the resolving power of high-resolution instruments is not degraded by operating in this mode. In addition, Fales and co-workers point out that the dependence of resolution on the width of the first adjustable slit of their mass spectrometer is much reduced compared with that encountered in low-pressure operation, and this effect presumably results from the narrow width of the source slit used in the chemical ionisation mode. It should be kept in mind that the energy distribution of the ions emerging from the high-pressure chemical ionisation source may well be quite narrow, possibly not greater than the thermal kinetic energy distribution, and if this is in fact true it will make the attainment of high resolution a simpler task with chemical ionisation than with low-pressure mass spectrometry. Finally, there is reason to think that the need for high resolution capabilities may be less in chemical ionisation mass spectrometry than in electron ionisation mass spectrometry. The amount of fragmentation is less in the chemical ionisation mode; the fragmentation processes are simpler with less possibilities for rearrangement and consequently the number of different atomic compositions which might reasonably have to be considered as possibilities for a given ion and distinguished by high-resolution mass measurements will be considerably less.

Gases at pressure of the order of 1 torr undergo electrical breakdown relatively easily and in so doing become conductors. Since the ion sources of deflection-type mass spectrometers normally operate at potentials ranging up to 8–10 kV, care must be exercised to prevent arcs from occurring through the reactant gas entering the ionisation chamber. This gas must be prevented from coming in contact with grounded metal, for in the absence of elaborate precautions (see below) grounding establishes a conduction path through which the arcs propagate. The arcs render the machine temporarily inoperable and may cause equipment damage. In the absence of adequate shielding there is also some danger of injury to personnel. Our experience has been that the use of an all-glass gas-introduction system solves the problem with a minimum of difficulty. If metal is used in the inlet it must be floated at source potential, or alternatively, if it is grounded, the rather elaborate anti-arcing potential dropping resistance network described by Futrell and Wojcik[14] must be used.

We now consider the important question of sensitivity of the chemical ionisation technique and particularly its sensitivity in comparison to that of conventional electron ionisation. We give the answer at the outset. The evidence accumulating indicates that chemical ionisation sensitivity is at least equal to and probably greater than the sensitivity of electron ionisation. Because of the different source configurations and conditions in the two methods, accurate quantitative information about optimum sensitivities of each cannot be obtained with any of the instruments utilised to date. The most significant information on comparative sensitivities currently available is that of Fales *et al.*[18] obtained with their AEI MS-9 mass spectrometer. They made measurements with two compounds, di-(2-ethylhexyl) phthalate and strychnine. The compound investigated was evaporated from the direct-introduction probe, and the spectra were measured alternatively in the chemical ionisation and electron ionisation modes. The total ion currents were obtained by summing the intensities of ions between $m/e = 800$ and 120, and in this way the contribution from the reagent gas ions in the chemical ionisation mode is not included. In the case of di-(2-ethylhexyl)phthalate the chemical ionisation total ion current is 1.43 times greater than the electron ionisation current, and for strychnine the chemical ionisation current is 4.58 times greater than the electron ionisation current. Fales and co-workers observe that in fairness to the electron ionisation method it should be remembered that although the electron ionisation source they used met the manufacturer's specifications for sensitivity, it is not optimised for electron ionisation use in that the electron entrance aperture and ion exit apertures were smaller than is normal in the electron ionisation mode.

Consequently, it is desirable to calculate expected sensitivities in the two modes of operation taking into account aperture size differences. We assume the pressure of the material under investigation (for simplicity referred to as 'additive') is the same in both modes, namely, 1×10^{-5} torr, a reasonable but arbitrary value. We assume methane to be the reactant in the chemical ionisation mode, and we take its pressure in the ionisation chamber to be 1 torr. In our mass spectrometer the electron entrance slit dimensions for chemical ionisation are 3×0.05 mm, whereas a typical electron entrance slit for electron ionisation would have the dimensions 3×1 mm. Thus the slit area for electron ionisation is 20 times larger than that for chemical ionisation. Our ion exit slit for chemical ionisation has the dimensions 3×0.05 mm, whereas for electron ionisation the value is 3×0.25 mm. Thus the ratio of areas of this slit is 5. We should mention that an ion beam width of 0.25 mm in our machine produces a resolution of *c.* 500. Higher resolution would require correspondingly narrower beams.

For electron ionisation we calculate the ratio of the ion current (I_a) to electron current (I_e) from the relationship

$$\begin{aligned} I_a/I_e &= NQL \\ &= (2 \times 10^{11})(2 \times 10^{-15})(0.3) \\ &= 1.2 \times 10^{-4} \end{aligned} \qquad (5.3a)$$

In this expression N is the particle density in the source, and the value of 2×10^{11} ml^{-1} corresponds to the assumed pressure of 1×10^{-5} torr. Q is the ionisation cross-section, and the assumed value of 20 Å^2 is a reasonable electron ionisation cross-section for a moderate-sized molecule. L is the path length over which collectable ions are formed, namely, the ion exit slit length of 0.3 cm. To facilitate the comparison with the chemical ionisation case, we shall assume that I_e is the total electron current incident upon the electron entrance electrode, and it might be thought of as constituting unit incident electron current.

In the chemical ionisation case the electron are completely stopped, and we must utilise radiation chemistry-like calculations to determine the number of positive ions formed. The energy required to produce an ion pair in the reactant gas may conservatively be taken to be 30 eV, and if the energy of the ionising electrons is 600 V, 20 reactant ions will be produced per electron passing into the ionisation chamber. However, because the electron entrance hole is smaller in the chemical ionisation case than in the electron ionisation case, a smaller fraction of the electrons incident upon the electron entrance electrode will actually pass into the ionisation chamber, and the electron currents in the ionisation chamber in the two cases will be proportional to the areas of the electron entrance holes. We saw above that the chemical ionisation area is 1/20 that for electron ionisation, and thus the electron current in the ionisation chamber in the chemical ionisation case will only be $0.05\,I_e$, letting I_e, as above, represent the incident electron current. We let I_r represent the current of reactant ions, and since 20 reactant ions are produced per electron in the ionisation chamber, we conclude that $I_r = I_e$. The current of additive ions, in which we are ultimately interested, is calculated from the expression

$$\begin{aligned} I_a/I_r &= Nk\tau \\ &= (2\times 10^{11})(1\times 10^{-9})(1\times 10^{-5}) \\ &= 2\times 10^{-3} \end{aligned} \tag{5.3b}$$

In this expression N is the particle density of the additive molecules, and the value of 2×10^{11} corresponds to the assumed pressure of 1×10^{-5} torr. k is the rate constant for a bimolecular ion–molecule reaction, for which a typical value is 10^{-9} ml mol^{-1} s^{-1} and τ is the residence time of the reactant ions in the ionisation chamber. A conservative estimate of a value for this quantity is 10 μs, and the resulting value of I_a/I_r is 2×10^{-3}. Since we showed above that $I_r = I_e$, we obtain the result that in the chemical ionisation case $I_a/I_e = 2\times 10^{-3}$, which is to be compared with the value of 1.2×10^{-4} for the same quantity in the electron ionisation case. Thus, the additive ion current in the ionisation chamber per unit electron current incident upon the electron entrance electrode is *c.* 17 times larger in the chemical ionisation case than in the electron ionisation case.

The ion exit slit area is smaller in the chemical ionisation source, and thus the efficiency of extraction of the ions from the source will probably not be as large in the chemical ionisation case. The most conservative position which can be adopted is that the extraction efficiences will be in the same ratio as the ion exit slit areas, which we assumed above to have the value of 5. Using this factor, we conclude that the additive ion current emerging from the chemical ionisation chamber per unit incident electron current is *c.* 3 times larger than that found for electron ionisation. For certain instruments and instrumental conditions the chemical ionisation intensity loss at the ion exit slit may not be proportional to the areas, and we look upon this calculation as constituting the worst case for diminution of chemical ionisation sensitivity because of ion beam extraction problems.

Two other factors affecting chemical ionisation sensitivity need be considered. First, the most meaningful sensitivity criterion for a mass spectrometer is the ion current per unit mass of material present in the ionisation chamber. Because the ionisation chamber in the chemical ionisation mode has, as we have seen, smaller electron entrance and ion exit holes, and is otherwise gas tight, the volatilisation of a given mass of a substance of interest in the chemical ionisation source should produce a higher pressure for a longer time than would be the case in the more open electron ionisation source. This increase in pressure translates itself directly into increased sensitivity. Secondly, since the amount of fragmentation in chemical ionisation is smaller than in electron ionisation, the ionisation produced is concentrated in a smaller number of ions, which also leads to a significantly enhanced sensitivity.

Finally, we point out that there is reason to think that the chemical ionisation sensitivity can be increased over that calculated. We assumed an ion residence time of 10 μs, but it is quite feasible to operate at lower repeller voltage so as to increase this quantity by a factor of 5 or more. We pointed out earlier that using larger pumps on the machine permit operation at a source pressure of 4–5 torr and this, coupled with a corresponding increase in the electron energy, would increase the amount of ionisation occurring in the source. Furthermore, since the ion residence time is proportional to the ratio of field strength to pressure, the increased pressure would also increase the ion residence time. We cannot estimate the possible increase in sensitivity resulting from such a mode of operation, but it might well be of significant magnitude.

The agreement between our calculated sensitivities and those found experimentally by Fales *et al.*[18] is gratifying. A calculation similar to that given here but using slightly different assumptions has been given by Futrell and Wojcik[14]. The results of their calculation are very much the same as those obtained here.

5.3 LISTING AND CLASSIFICATION OF REACTIONS OBSERVED IN EVEN-ELECTRON CHEMICAL IONISATION MASS SPECTROMETRY

The even-electron spectra of an appreciable number of compounds and compound types have been determined, and while many more compounds and compound types remain to be investigated, we think it appropriate here to classify and summarise the kinds of reactions presently known to be involved in the formation of even-electron chemical ionisation spectra. More specifically, we will consider primarily methane and i-butane spectra, which we will consider to be characteristic of all even-electron spectra. The difference between the spectra obtained with methane and i-butane is almost completely one of the extent of fragmentation occurring, with i-butane producing the smaller amount. The evidence upon which the reactions to be given rests is not as elaborate as that available for many electron ionisation mechanisms. For example, very little deuterium tracer mechanistic studies have been made. However, the reactions occurring in chemical ionisation (for the rest of this discussion it is to be assumed that even-electron chemical ionisation is being considered) are for the most part simple and self-evident from the masses of the ions involved. Some information is available from the presence of metastable ions (see Chapter 8). Finally, and perhaps most important, it was evident from the earliest investigations that the reactions occurring in chemical ionisation mass spectrometry followed most of the rules developed for condensed-phase ionic chemistry, which is not surprising since the ions involved in the condensed-phase reactions are, like the gas-phase ions, even-electron ions. Making the assumption that the condensed-phase rules are generally applicable, it has been relatively easy to rationalise most of the reactions occurring in the gas phase. Also, one does not usually find in chemical ionisation spectra multiple fragmentation of bonds within a given ion and sequential decompositions of ions. Thus, the overall task of identifying the reactions occurring is appreciably simpler then is the case in electron ionisation.

We classify the reaction types occurring in chemical ionisation as follows.

5.3.1 Addition reactions

Reactant ions can transfer charged entities to additive molecules to produce addition ions. The entity transferred may comprise part or all of the reactant ions. The most important entity transferred is the proton to form the $(M+1)^+$ ion. Thus,

$$A + CH_5^+ \rightarrow AH^+ + CH_4 \tag{5.4}$$

The AH^+ ion in equation (5.4) can decompose to give fragment ions, but they will be considered in later categories, and we count as addition ions only those with a lifetime long enough to permit their collection in the mass spectrometer. Other addition ions observed in methane chemical ionisation are $(M+29)^+$ and $(M+41)^+$, where the $m/e = 29$ and 41 entities added to M are the $C_2H_5^+$ and $C_3H_5^+$ ions found in the methane plasma. The $(M+1)^+$ ion is found with all other reactant ions which can be considered as Brønsted acids, but the extent to which the proton will be transferred will depend upon the strength of the acid. Thus, t-$C_4H_9^+$ from i-C_4H_{10} is too weak an acid to protonate water, and no H_3O^+ ion is formed by i-butane chemical ionisation. In propane, an $(M+43)^+$ ion is formed by the addition of $C_3H_7^+$ ion to the additive, and in i-butane one finds $(M+39)^+$ and $(M+57)^+$ from the addition of $C_3H_3^+$ and $C_4H_9^+$, respectively. In water one observes $(M+19)^+$ from the addition of H_3O^+, and doubtless analogous addition ions will be observed as more reactant gases are utilised. The intensities of the association ions such as $(M+29)^+$ in methane and $(M+57)^+$ in i-butane depend inversely upon the temperature of the ionisation chamber, and from our qualitative observations they seem also to depend upon the base strength of the additive molecules, although the evidence on this latter point is quite sketchy. Nonetheless, we think we have observed that various strong bases yield relatively smaller amounts of ions such as $(M+29)^+$, and we think the reason is that the greater basicity promotes transfer of a proton from the reactant ion rather than simple association with the reactant ion. The $(M+1)^+$ ion is without doubt the most significant ion produced by chemical ionisation mass spectrometry, since for most additives it gives information about molecular weight, and it is the ion from which most fragmentation paths start.

The association ions such as $(M+29)^+$ and $(M+41)^+$ occur at m/e values higher than the molecular weight of the additive, and while we first thought this might constitute something of a problem in the interpretation of spectra, in practice it turns out that it is a desirable characteristic in the sense that it enables one to identify with greater certainty the molecular weight of the additive. The relative intensities of the $(M+1)^+$, $(M+29)^+$, and $(M+41)^+$ ions in methane chemical ionisation spectra usually constitute a pattern of sufficient degree of constancy that it can be used as a diagnostic for the molecular weight of the additive. Roughly speaking, the intensity of $(M+29)^+$ is 10% of the intensity of $(M+1)^+$, and that of $(M+41)^+$ is somewhat smaller than that of $(M+29)^+$. In our laboratory we consider the appearance of three ions with these relative intensities separated by 28 and 12 mass units respectively to constitute proof that the intense peak is the $(M+1)^+$ ion, and this gives immediately the molecular weight of the additive. The absence of ions 28 and 40 mass units higher than an ion considered as a possibility for the $(M+1)^+$ ion generates a strong suspicion that the ion in question is not $(M+1)^+$, but it is not taken as complete proof. As was mentioned above, if the additive is a very strong base, the $(M+29)^+$ and especially the $(M+41)^+$ ions will be quite weak.

Another type of addition reaction that should be mentioned is that giving rise to the formation of $(2M+1)^+$ ions. Such ions are formed by the reaction

$$AH^+ + A \rightarrow A_2H^+ \qquad (4.5)$$

and are observed when the additive molecule A has moderate or strong basić properties, e.g. alcohols, esters, amines. The formation of A_2H^+ is bimolecular in A, and thus the intensity of A_2H^+ will vary with the square of the pressure of A in the ionisation chamber. The tendency toward formation of AH_2^+ in chemical ionisation will be significantly greater than the tendency for the occurrence of ion–molecule reactions in electron ionisation because the residence time of AH^+ in the ionisation chamber under the high-pressure chemical ionisation conditions will be greater by an order of magnitude or more than residence times in electron ionisation. Thus, the additive pressures at which these protonated dimers are observed are significantly lower than the pressures at which ion–molecule reactions occur in electron ionisation. As with additive ions such as $(M+29)^+$, etc., when one is forewarned and knows what to expect, the occurrence of these $(2M+1)^+$ ions serve as another diagnostic for the identification of the $(M+1)^+$ ion. The intensities of $(2M+1)^+$ ions are strongly dependent upon ionisation chamber temperature, and at very high temperatures such as 200–300 °C their intensities are small or zero.

It is appropriate at this point to digress momentarily from our consideration of even-electron chemical ionisation phenomena and point out that in odd-electron chemical ionisation reactions as, for example, with nitrogen or rare gases as reactants, one often finds significant amounts of AH^+ ions formed even though no proton transfer from the reactant ions can occur. These ions are doubtless formed by ion–molecule reactions of the additive in the same way that $(M+1)^+$ ions of weak intensity are observed in some electron ionisation spectra. However, because the residence times in the chemical ionisation experiments are greater than in the electron ionisation experiments, the $(M+1)^+$ intensities in chemical ionisation can be much larger than those found in electron ionisation. These intensities will depend upon the square of the pressure of additive molecules and, indeed, it is very easy to obtain nitrogen chemical ionisation spectra which partake of many of the characteristics of methane chemical ionisation spectra. A true nitrogen chemical ionisation spectrum is obtained only by careful control of the additive pressure.

5.3.2 Abstraction reactions

For compounds of low basicity, reaction, when it occurs, generally involves abstraction of a negative entity, which may sometimes be looked upon as a dissociative proton transfer. For example, in normal paraffin hydrocarbons the most intense ion in the spectrum is the $(M-1)^+$ ion formed by abstraction of H^- from the molecule[5]. In fluorocarbons intense peaks are found at $(M-19)^+$, which corresponds to the abstraction of F^- from the molecule[19]. For example, one finds

$$\text{n-}C_6F_{14} + CH_5^+ \rightarrow C_6F_{13}^+ + HF + CH_4 \quad (5.6)$$
$$m/e = 319$$

and the $m/e = 319$ ion comprises 20–30% of the total ionisation of the fluorocarbon. For very weakly basic molecules, especially those with very strong bonds, little or no chemical ionisation occurs. Thus, for hexafluorobenzene very little ionisation occurs, and that which does is mostly at M^+. For O_2 the basicity is so low that no ionisation at all occurs with methane reactant. Abstraction reactions will occur with molecules containing basic groups, but they are usually overshadowed by other reactions such as proton transfer. However, if a functional group in the molecule can stabilise a charge, certain abstraction reactions are enhanced. Thus, in tetramethylethylenediamine the $(M-1)^+$ intensity is 30.2% of the amine total ionisation[20]. The reaction producing the $(M-1)^+$ ion may be written as

$$(CH_3)_2NCH_2CH_2N(CH_3)_2 + C_2H_5^+ \rightarrow$$
$$(CH_3)_2N^+{=}CHCH_2N(CH_3)_2 + C_2H_6 \quad (5.7)$$

and the hydride ion loss is promoted by the opportunity of referring the charge to the adjacent nitrogen atom.

As an extension of the concept, we will include under abstraction reactions electrophilic attack on C—C bonds because the overall reaction can be looked upon as a kind of alkide ion abstraction reaction. For example, in the spectra or normal paraffin hydrocarbons alkyl ions are observed at all carbon numbers up to and including $(M-1)^+$, and it has been suggested[1, 5] that these ions (except $(M-1)^+$) are formed by alkide ion abstraction from the paraffin molecule. Thus,

$$\text{n-}C_{16}H_{34} + CH_5^+ \rightarrow C_{10}H_{21}^+ + C_6H_{14} + CH_4 \quad (5.8)$$
$$m/e = 141$$

The electrophilic attack can occur at random along the carbon chain, which accounts for the approximately equal intensities of the alkyl ions (other than $(M-1)^+$ ions) formed and also gives rise to the concept of random attack of the reactant ions. In complete analogy to the situation encountered with hydride ion abstraction reactions, the presence of a charge-stabilising functional group enhances the electrophilic attack at certain bonds. In n-tributylamine[1] the intensity at $m/e = 142$ corresponding to loss of propyl is 19.9% of the amine ionisation. The reaction is

$$(\text{n-}C_4H_9)_3N + CH_5^+ \rightarrow (\text{n-}C_4H_9)_2N^+{=}CH_2 + C_3H_8 + CH_4 \quad (5.9)$$
$$m/e = 142$$

and the large intensity observed is the result of the stabilising influence of the nitrogen atom.

5.3.3 Molecule-ion formation

With methane chemical ionisation one finds with some compounds a small amount of odd-electron ions formed along with the dominant even-electron ions. The reaction occurring is electron transfer to appropriate components of the methane plasma. Ethylene molecule-ion ($C_2H_4^+$) comprises 2–3% of the methane plasma at 1 torr and the recombination energy of $C_2H_4^+$ is given by Lindholm[21] as somewhat less than 10.5 eV. This is a relatively high recombination energy, and many additive molecules will react with $C_2H_4^+$ in an electron transfer reaction to produce additive molecule-ions, which we designate as M^+. Since the relative amount of ethylene molecule-ion in the methane plasma is small, the M^+ intensities produced by it will also be small, typically of the order of 1–2% or less of the additive ionisation. However, a major constituent of the methane plasma is $C_2H_5^+$, and the recombination energy of this ion will be somewhat less than the ionisation potential of C_2H_5, which is 8.25 eV [22]. Consequently, additive molecules with ionisation potentials less than about 8.2 eV can undergo electron transfer reactions with $C_2H_5^+$, and because the $C_2H_5^+$ concentration is high, M^+ ions with moderate intensity can be produced. The ionisation potentials of tertiary amines are less than 8 V [22], and thus we find that the M^+ intensity in n-tributylamine[1] is 7.7% of the amine ionisation.

5.3.4 Functional group removal

In a molecule which consists of a relatively non-basic portion such as an alkyl or naphthene group and a more basic, polar group, the attack of the reactant ions will occur primarily at the functional group. Protonation and/or ion addition reactions can occur, and this attack of the reactant ions can result in cleavage of the functional group to one degree or another depending upon the identities of the functional group and the reactant ion. Examples of reactions involving functional group removal are

$$\text{n-}C_{10}H_{21}OH + CH_5^+ \rightarrow C_{10}H_{21}^+ + H_2O + CH_4 \qquad (5.10)$$

Ref. 1 $m/e = 141$

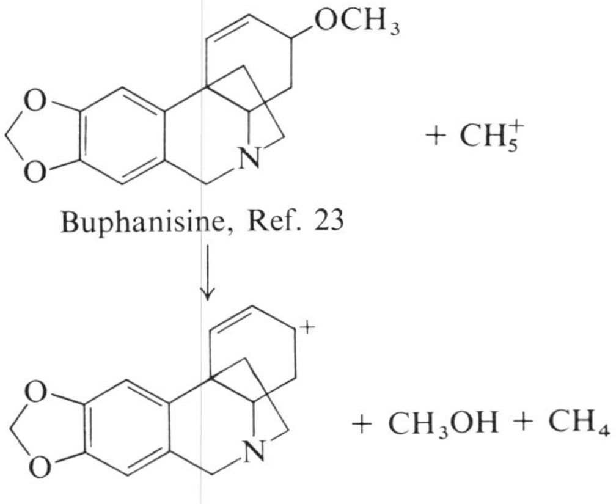

OCH₃

O O N + CH_5^+

Buphanisine, Ref. 23

↓

+

O O N + CH_3OH + CH_4 (5.11)

$$\text{Cl—CH}_2\text{CH—CH}_2\text{(O)} + \text{CH}_5^+ \rightarrow \overset{+}{\text{C}}\text{H}_2\text{—CH—CH}_2\text{(O)} + \text{HCl} + \text{CH}_4 \quad (5.12)$$

Ref. 1

As can be seen from these examples, we consider the functional group removal reaction to encompass primarily the removal of relatively small functional groups as the corresponding neutral molecules. The distinction between this reaction type and the abstraction reaction is admittedly rather arbitrary, but in our opinion useful.

In some cases one observes the formation of ions smaller than that produced initially by removal of the functional group, and it is reasonable to believe that these smaller ions result from further decomposition of the first ion formed. For example, in the methane chemical ionisation spectrum of n-decanol, alkyl and alkenyl ions containing less than 10 carbon atoms are observed, and these are probably formed by decomposition of the decyl ion produced in reaction (5.10).

The relative tendencies of the various functional groups to undergo these reactions has not been systematically studied, but from general experience with chemical ionisation spectra one knows that —OH is lost as H_2O quite easily, —NH_2 is lost as NH_3 with difficulty, halogens are lost as HX and acetoxy is lost as CH_3COOH relatively easily, etc. However, no experimental information is available concerning the order of the tendency for different groups to leave the protonated molecule, that is, the order of the different groups as leaving groups. As an aid to our thinking it will be useful to keep in mind that for the reactions with which we are concerned, namely,

$$\text{R—X} + \text{CH}_5^+ \rightarrow \text{R—XH}^+ + \text{CH}_4 \quad (5.13a)$$

$$\text{R—XH}^+ \rightarrow \text{R}^+ + \text{XH} \quad (5.13b)$$

the decomposition reaction (5.13b) involves the separation of R^+ from the molecule XH. The reverse of this reaction is obviously the addition of R^+ to XH, and if R were H^+ the reverse reaction would be the reaction upon which the proton affinity of XH is based. We suggest that the tendencies for decomposition of a series of R—XH^+ ions where R remains constant and X varies will be inversely related to the proton affinities of the XH molecules; that is, when the proton affinity of XH is low, the decomposition tendency of R—XH^+ according to (5.13b) will be high, and conversely. Using this criterion, we suggest the following order of leaving groups for a few commonly encountered groups. The groups are arranged in increasing order of the tendency of the protonated group to leave the $(M+1)^+$ ion,

$$\text{NH}_2 < \text{CH}_3\text{S} \simeq \text{C}_6\text{H}_5 \simeq \text{CH}_3\text{O} \simeq \text{OC(:O)H} < \text{CN} \simeq \text{SH} < \text{OH} < \text{I} \simeq \text{Cl} \simeq \text{Br}$$

The proton affinities upon which this order is based are those of Haney and Franklin[24].

The effect of the identity of R in R—XH^+ is as one would expect from conventional concepts of the relationship between the structures and energies of ions. For example, if R represents alkyl groups, the decomposition of R—XH^+ will increase in the order R = primary < R = secondary < R = tertiary. Thus, for the alcohols[25], 84% of the ions in ethanol are $(M+1)^+$

ions, whereas for hexan-2-ol the corresponding figure is 2 %, and for 2-methylheptan-2-ol no perceptible amount of $(M+1)^+$ is formed.

5.3.5 Heterolytic cleavage

This name is used to designate heterolytic bond rupture α to a protonated functional group. For example,

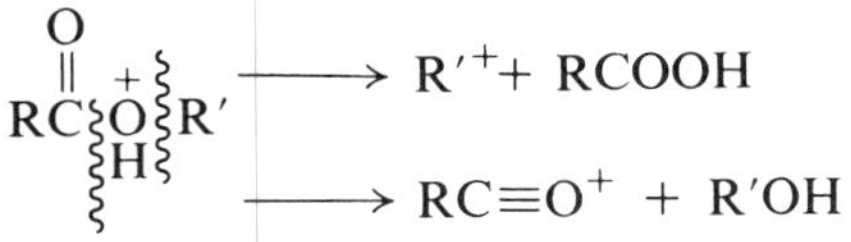

$$RC(=O)\overset{+}{O}(H)R' \longrightarrow R'^{+} + RCOOH \quad (5.14a)$$

$$\longrightarrow RC{\equiv}O^{+} + R'OH \quad (5.14b)$$

Ref. 26

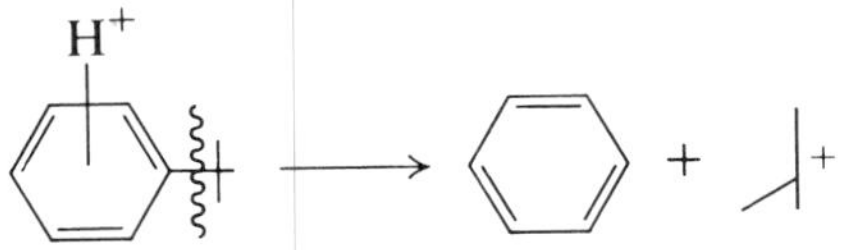

$$C_6H_5C(CH_3)_3 \cdot H^+ \longrightarrow C_6H_6 + (CH_3)_3C^+ \quad (5.15)$$

Ref. 27

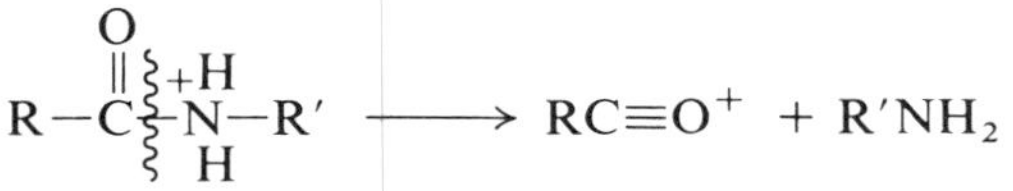

$$R{-}C(=O){-}\overset{+}{N}H_2{-}R' \longrightarrow RC{\equiv}O^{+} + R'NH_2 \quad (5.16)$$

Peptides, Ref. 28

The extent to which the cleavage reaction occurs depends, as usual, upon the energies of the fragments formed and the acid strength of the protonating agent. Thus, in reaction (5.15), because of the low energy of the t-butyl ion, this ion comprises about 80 % of the methane chemical ionisation of t-butyl benzene. On the other hand, alkyl ion formation occurs to a significantly smaller extent for n-alkyl benzenes. The distinction between heterolytic cleavage and functional group removal (Section 5.3.4) is very small and is introduced primarily for classification convenience.

5.3.6 β-Fission

Conventional β-fission reactions of carbonium ions are frequently encountered in chemical ionisation. Examples are shown in equations (5.17)

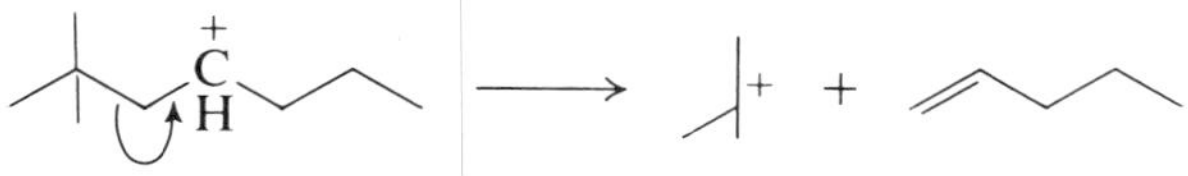

$$(CH_3)_3C{-}CH_2{-}\overset{+}{C}H{-}CH_2CH_2CH_3 \longrightarrow (CH_3)_3C^{+} + CH_2{=}CH{-}CH_2CH_2CH_3 \quad (5.17)$$

Ref. 5

and (5.18). The extent to which the β-fission reaction occurs depends in predictable ways upon the energies of the ions formed. Thus, in branched paraffin hydrocarbons containing one or more quaternary carbon atoms

HO OH

+ (5.18)

O O

Ref. 29

the probability of β-fission processes producing tertiary carbonium ions is so large that $(M-1)^+$ ion intensities are much lower than those found in straight-chain paraffins.

5.3.7 Hydrogen transfer reactions

We consider in this section reactions of the type often referred to in electron ionisation mass spectrometry as McLafferty rearrangements. We point out a significant difference between the expected behaviour with respect to hydrogen transfer reactions in chemical ionisation as compared with electron ionisation. In chemical ionisation involving ionisation by proton attachment to a functional group, the number of electrons of the functional group available to participate in bonding is diminished by the ionisation. By contrast, in electron ionisation the number of electrons available or energetically promoted sufficiently to participate in bonding is increased by ionisation at the functional group.

The smaller availability of electrons for bonding in the chemical ionisation case diminishes or eliminates the possibility that the functional group can participate in hydrogen transfer reactions. An excellent illustration is given by ketones. No systematic study of ketones has been published, but we have measured several spectra in our laboratory, and we find, for example, that in octan-2-one little or no McLafferty-type rearrangement occurs. Indeed, little fragmentation of any type occurs even using methane as reactant gas, as is evidenced by the fact that the sum of the $(M+1)^+$, $(M+29)^+$ and $(M+41)^+$ ions comprise *c.* 75% of the total octanone ionisation. By contrast, in electron ionisation the rearrangement peak at $m/e = 58$ is the second most intense in the spectrum. Thus, for hydrogen transfer rearrangement reactions to occur to a significant extent in chemical ionisation it is necessary to have two unsaturated centres in the molecule as in an ester or an unsaturated ketone, etc. Several slightly different types of hydrogen transfer reactions are observed and thus we shall establish subcategories.

5.3.7.1 Hydrogen transfer–olefin reactions

Many carboxylic acid esters exhibit intense peaks at m/e values corresponding to the formation of the protonated carboxylic acids. It has been suggested[26] that the reaction occurring is reaction (5.19). No direct evidence exists that the protonation is indeed upon the ether oxygen and nothing is known about

$$R{-}C(=O)\cdots\overset{+}{O}(H)\cdots CH_2{-}CH(R'){-}H \longrightarrow R{-}C(={}^{+}OH){-}OH + R'CH{=}CH_2 \qquad (5.19)$$

the degree of site specificity of the proton transferred to form the protonated carboxylic acid.

5.3.7.2 *Alkyl group exchange in esters*

This reaction is analogous to that in Section 5.3.7.1 differing only in that the ionisation is effected by attack of an alkyl ion from the plasma of the reactant gas. An illustration of the reaction is given in reaction (5.20) where the attack-

$$R{-}C(=O)\cdots\overset{+}{O}(C_2H_5)\cdots CH_2{-}CH(R'){-}H \longrightarrow R{-}C(={}^{+}OH){-}OC_2H_5 + R'CH{=}CH_2 \qquad (5.20)$$

ing ion is taken to be $C_2H_5^+$ from the methane plasma. An analogous reaction results from the attack of $C_3H_5^+$ from the methane plasma. The intensities of the ions produced by reactions such as (5.20) are generally significantly smaller than those of the reactions (5.19) promoted by proton transfer.

5.3.7.3 *Ketene loss reactions*

Esters which contain an additional basic centre and which for energetic reasons cannot undergo hydrogen transfer reactions (Section 5.3.7.1) can transfer a proton to form protonated aldehydes with the concomitant loss of ketene-type molecules. Thus, for vinyl propionate[26] reaction (5.21) takes

$$O{=}C(CH(H)CH_3)\cdots\overset{+}{O}(H){-}CH{=}CH_2 \longrightarrow \underset{m/e\,=\,45}{CH_3CH{=}\overset{+}{O}H} + CH_3CH{=}C{=}O \qquad (5.21)$$

place. The $m/e = 45$ ion comprises 12.3% of the total ionisation of vinyl propionate, and in an analogous reaction in phenyl propionate the intensity of the ion produced by the loss of methyl ketene is 40% of the ester ionisation. Thus, the reaction can occur with a high probability.

It has also been reported[30] that a nitrogen analogue of ketene, H—N=C=O, can be eliminated to form a protonated alcohol.

5.3.7.4 *Carboxylic acid displacement by t-* $C_4H_9^+$

This reaction occurs in esters which contain a centre of unsaturation in addition to the ester group and it involves the initial addition of t-butyl ion to the second centre of unsaturation. Thus, it has been found[31] that t-butyl ion displaces acetic acid from methoxymethyl acetate, and the reaction is shown in equation (5.22). The $m/e = 101$ ion is the $(M-3)^+$ ion, which

$$CH_3-C(=O)-O-CH_2-\overset{+}{O}(-CH_3)-C(CH_3)_2-CH_2-H \longrightarrow \underset{m/e\,=\,101}{\overset{+}{O}(CH_3)(CH_2)(C(CH_3)_2)\text{ ring: } CH_2-\overset{+}{O}-CH_3,\ CH_2-C(CH_3)_2} + CH_3COOH \qquad (5.22)$$

appears to be found frequently in the i-butane chemical ionisation mass spectra of acetate esters. The loss of 3 a.m.u. involved in forming this ion is the difference in the mass of the t-butyl ion added and the acetic acid molecule lost. For formate esters the ion observed is $(M+11)^+$; for propionate esters it is $(M-17)^+$, etc. Reaction (5.22) is written as if it involves the migration of pairs of electrons, but since the charged centre is not directly involved in the transformations represented in equation (5.22), it could as well be written with single electron migrations. No direct evidence on the point is available.

5.3.7.5 *Loss of elements of* HCHO+CO

An unusual hydrogen transfer reaction which has been observed in two compounds can be illustrated by the production of $m/e = 63$ from methylthiomethyl acetate[32] (equation (5.23)). No evidence exists as to the exact

$$H_2C(H)-C(=O)-O-CH_2-\overset{+}{S}H-CH_3 \longrightarrow \underset{m/e\,=\,63}{CH_3-\overset{+}{S}H-CH_3} + \begin{cases} H_2CO + CO \\ \text{or} \\ H_2C\text{—}C(=O)\text{—}O \text{ (ring)} \end{cases} \qquad (5.23)$$

identity of the neutral products formed, and of the possibilities given in equation (5.23), the formation of the α-lactone seems rather unlikely because of the high strain-energy in this small-ring compound. As was the case in reaction (5.22), representing reaction (5.23) in terms of electron pair migrations is done arbitrarily and without experimental evidence. Although reaction (5.23) *a priori* does not seem to be a very probable reaction, in fact, in the i-butane chemical ionisation spectrum of methylthiomethyl acetate the $m/e = 63$ ion comprises *c.* 20% of the total ionisation of the ester.

5.3.7.6 *Hydrogen transfer with loss of elements of N-acylimmines + CO*

This is a reaction type closely related to the formation of α-lactones or aldehydes + CO (Section 5.3.7.5). In this case the reaction occurs in a protonated peptide[28], and the reaction may be illustrated by

$$\mathrm{R{-}\overset{O}{\overset{\|}{C}}{-}N(H){-}CH(R'){-}C{=}O\,({-}\overset{+}{N}H_2R'')} \longrightarrow \mathrm{NH_3R''} + \begin{cases} \mathrm{R{-}\overset{O}{\overset{\|}{C}}{-}N{-}C{=}O \text{ (ring with } C(H)(R')\text{)}} & \alpha\text{-lactam} \\ \mathrm{R{-}\overset{O}{\overset{\|}{C}}{-}N{=}CHR' + CO} & N\text{-acylimmine} \end{cases} \quad (5.24)$$

As was the case in reaction (5.23), no evidence exists as to whether the neutral products in equation (5.24) are in the form of the α-lactam or in the form of *N*-acylimmine + CO, but, as before, the strain in the former compound would seem to argue against its formation. Reaction (5.24) is an important one in the chemical ionisation sequencing of peptides in that it constitutes the basis for obtaining the *C*-terminus sequence of the amino acids.

5.3.8 Backside-assisted decompositions

This type of reaction requires the presence of at least two basic centres in the molecule and involves the decomposition of the protonated molecule by

$$\mathrm{(H_3C)_2N{-}CH_2CH_2{-}\overset{+}{N}H(CH_3){-}CH_2CH_2{-}N(CH_3)_2} \longrightarrow \mathrm{(H_3C)_2N{-}CH_2CH_2{-}NH(CH_3)} + \mathrm{\triangleright\overset{+}{N}(CH_3)_2} \quad (5.25)$$

virtue of an internal nucleophilic attack by the unprotonated basic centre. Thus, with alkyl polyamines reaction (5.25) occurs[20]. In certain amino acids a similar reaction occurs (reaction (5.26))[33].

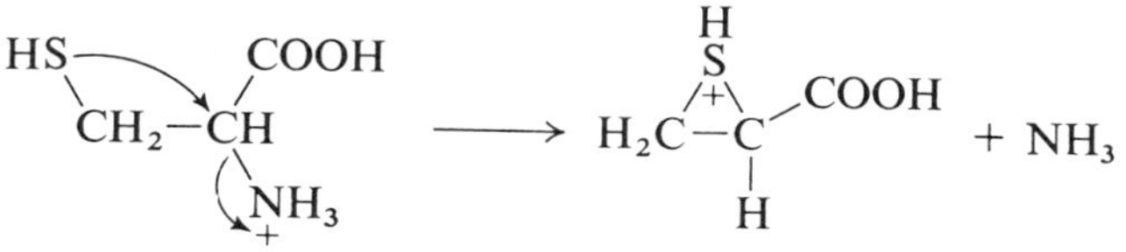

$$HSCH_2CH(COOH)\overset{+}{N}H_3 \longrightarrow H_2C\!-\!\overset{+}{S}H\!-\!CH(COOH) + NH_3 \qquad (5.26)$$

In general, loss of ammonia from amino acids does not occur unless assisted by backside attack as illustrated in equation (5.26).

5.3.9 Protonated acid anhydride formation

Ortho-diesters exhibit the usual ester decomposition reactions, but in addition the nearness of two carbalkoxy groups leads to the formation of anhydride-like ions[34]. For example, one finds:

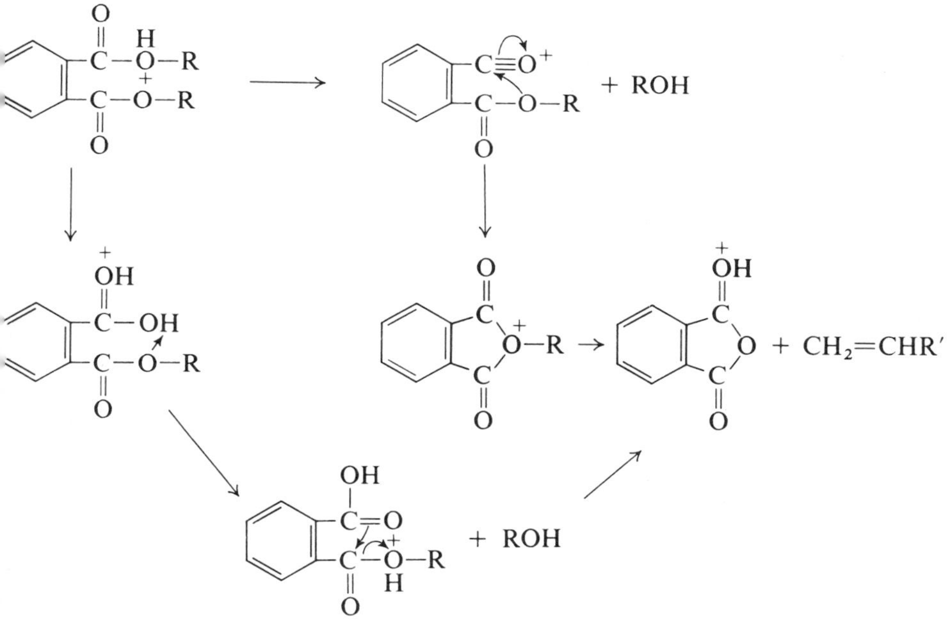

+ ROH

+ $CH_2{=}CHR'$

+ ROH

The occurrence of these dual reaction pathways is suggested by the observation of appropriate metastable ions.

5.4 MASS SPECTRA OF DIFFERENT COMPOUND TYPES

5.4.1 Alkanes

The chemical ionisation mass spectra of alkane hydrocarbons using methane as reactant gas has been investigated by Field, Munson and Becker[5]. The paraffin hydrocarbons have played an important role in the development of contemporary organic mass spectrometry in spite of their inertness as neutral compounds and their extensive fragmentation in electron ionisation. The early post-World War 2 development of electron ionisation mass spectrometry was dominated by studies of hydrocarbons because of the interest of petroleum companies and the discovery of chemical ionisation really resulted from an investigation of paraffin hydrocarbons and the realisation that the chemical ionisation spectra of these compounds differed drastically from the electron ionisation spectra. In normal alkanes the $(M-1)^+$ ion (formed by abstraction reactions (Section 5.3.2)) dominates the spectrum and fragment alkyl ions with about equal intensities (much smaller than that of $(M-1)^+$) comprise the remainder of the spectrum. By contrast, as it is well known, in electron ionisation the largest intensities in paraffins are observed in the C_3–C_5 region and the intensities of ions with higher m/e values exhibit a steady decrease in intensity. The intensity of the $(M-1)^+$ ion in chemical ionisation is independent of the length of the normal alkane and the form of the spectrum of n-tetratetracontane (n-$C_{44}H_{90}$) is very similar to that of n-octane. The analytical utility of these features of the chemical ionisation spectra is self-evident.

In branched paraffins the $(M-1)^+$ intensity tends to be lower than in normal paraffins, but the decrease in intensity with increasing branching is not as sharp as the corresponding decrease in the intensities of the parent ions produced by electron ionisation. Single branching has little effect on the chemical ionisation $(M-1)^+$ intensity, and multiple single branching produces at most a moderate $(M-1)^+$ ion intensity decrease. Indeed, the only structural feature which profoundly affects the $(M-1)^+$ intensity is a quaternary carbon, for this feature permits the ready occurrence of β-fission reactions (see Section 5.3.6). A quantitative rationalisation of the diminution of $(M-1)^+$ intensities with increasing branching utilising the random attack–localised reaction concept of the chemical ionisation of paraffins has been made[5].

5.4.2 Cycloparaffins

The methane chemical ionisation mass spectra of cycloparaffins[6] differ meaningfully from those of alkanes, although it appears that the same general type of reaction (Section 5.3.2) is involved in the production of the ions. The addition of a proton to a cycloalkane produces an $(M+1)^+$ ion with empirical formula $C_nH_{2n+1}^+$, and extensive fragmentation of this ion to smaller ions with the same empirical formula also occurs. The initial chemical ionisation attack also involves hydride-ion abstraction to produce an $(M-1)^+$ ion with empirical formula $C_nH_{2n-1}^+$. Fragmentation of this ion occurs to produce smaller ions with the same empirical formula. Thus, the chemical ionisation spectra of cycloparaffins consist essentially of two series of ions, namely, the alkyl series, $C_nH_{2n+1}^+$, and the alkenyl series, $C_nH_{2n-1}^+$. The $(M-1)^+$ ions are the most intense with relative intensities between 27 and 74 % of the total ionisation, depending on the structure of the cycloalkanes. These intensities are, on the whole, appreciably higher than $(M-1)^+$ intensities found in alkanes. The spectrum of cyclo-C_6D_{12} using methane as reactant has been determined[6]. No isotopically mixed analogues of $C_6D_{11}^+$ are formed, which shows that no H–D exchange occurs between cyclo-C_6D_{12} and $CH_5^+ + C_2H_5^+$ in the chemical ionisation process or between $C_6D_{11}^+$ and CH_4 in the subsequent travels of the ions out of the ionisation chamber. Furthermore, it may be concluded from the spectrum that the $C_6D_{11}^+$ ions formed must have the cyclohexyl structure rather than an alkenyl ion structure. Small intensities of $C_3H_7^+$ ions are observed in the spectrum of protonated cyclohexane, and in cyclo-C_6D_{12} the analogous ions exhibit isotopic mixing, which indicates strongly that extensive rearrangements and hydrogen migration occur during the lifetime of the hexyl ions which break down to produce the fragment propyl ions.

5.4.3 Alkenes and alkynes

The chemical ionisation mass spectra of a number of alkenes and alkynes have been investigated using methane as reactant[35]. The spectra of mono-olefins resemble the cycloparaffins discussed in the previous section in that two series of ions are produced, namely, alkyl ($C_nH_{2n+1}^+$) and alkenyl ($C_nH_{2n-1}^+$). However, the amount of fragmentation producing relatively low m/e fragment ions is greater in the olefins than in the cycloparaffins. The alkyl ions are produced by proton transfer to the double bond (Section 5.3.1), and the alkenyl ions are formed by hydride ion abstraction (Section 5.3.2). Allylic hydrogens are abstracted most readily, and $(M-1)^+$ intensities are high in compounds with a large number of allylic hydrogens. For example, the $(M-1)^+$ intensity in 2,3-dimethylbut-2-ene is 34%, which is about a factor of 3 larger than the $(M-1)^+$ intensity for but-1-ene. For polyolefins, cyclic olefins and acetylenes two series of ions are produced and these result from initial protonation or initial hydride ion abstraction. The presence of multiple double bonds or of an olefinic ring structure in a molecule increases the intensity of the $(M+1)^+$ ion appreciably above that observed in acylic mono-olefins. Thus, the $(M+1)^+$ ions are the most intense in the spectra of 1,3,5-heptatriene with a relative intensity value of 55%. By contrast, the presence of a triple bond in the molecule does not significantly enhance the intensity of the $(M+1)^+$ ion as compared with values for mono-olefins.

5.4.4 Aromatics

In aromatic compounds the π-electrons provide a basicity in the molecules which markedly affects the chemical ionisation mass spectra. As with aromatic hydrocarbons in the condensed phase, attack by the electrophilic reactant ions is facilitated, and both addition complexes between the attacking ions and the aromatic substrate (Section 5.3.1) and displacement reactions on the aromatic nucleus (illustrated in Section 5.3.5) are to be observed[27]. Aromatic rings are too stable to be fragmented in chemical ionisation, and thus in benzene no ions smaller than M^+ are observed. The most intense ion is

$(M+1)^+$, which is assumed to have the cyclic benzenium ion structure. When alkyl groups are substituted on the ring, fragmentation processes are observed. Substitution of a methyl group on the ring offers the possibility of the formation of benzyl ion by an abstraction reaction (Section 5.3.2), and in toluene the intensity of $(M-1)^+$ is about 3% of the toluene ionisation. The intensity of $(M-1)^+$ depends on the number of hydrogen atoms which can be removed and it rises to a value of 24% in pentamethylbenzene. Compounds containing a tertiary alkyl group substituted on the ring undergo a facile alkyl ion displacement reaction (illustrated in reaction (5.15)). This reaction is identical with the condensed-phase dealkylation process. For aromatic molecules containing a side chain with at least three carbon atoms and available hydrogens, a hydrogen transfer reaction (Section 5.3.7.1) occurs, and the reaction involves loss of olefin to form a lower *m/e* ion with formula $C_nH_{2n-5}^+$.

5.4.5 C_7H_8 isomers

Investigations of the methane chemical ionisation spectra of toluene, cycloheptatriene, and norbornadiene have been made[36], and of particular interest are the results with toluene-α-d_3 and cycloheptatriene-7-*d*. It is concluded from these studies that the formation of $C_7H_7^+$ from both toluene and cycloheptatriene does not involve randomisation of the hydrogen and deuterium atoms in the molecules, a result which is in sharp contrast to the randomisation found in electron ionisation. However, in methane chemical ionisation in these compounds, when ionisation is followed by C—C fragmentation, mixing of hydrogen occurs. The behaviour with respect to isotope mixing observed with the C_7H_8 isomers parallels exactly that observed with deuteriated cyclohexane.

5.4.6 Chemical ionisation of benzene by rare gas reactants

A study has been made of the spectra of benzene produced by different gases[37], and this is an illustration of the initial production of odd-electron ions by the chemical ionisation technique. Xe, Kr, Ar, and Ne were the rare gases used and they have recombination energies ranging from 12.1 to 21.6 eV [21]. Thus, ionisation of the benzene should occur readily in all cases. As an illustration of the ease of ionisation, for mixtures containing 1.0 torr of rare gas and about 10^{-3} torr of benzene, while the initial ionisation by the electrons produces mainly rare gas ions, predominantly hydrocarbon ions are collected.

The degree and type of fragmentation produced in the electron transfer reaction varies with the recombination energy of the rare gas used. Almost no fragmentation results from the reaction of Xe, for the relative intensity of $C_6H_6^+ = 94\%$. Using Kr, the ions produced still include 78% $C_6H_6^+$, but ionisation with Ar gives mainly the fragments $C_4H_4^+$ (30%), $C_6H_5^+$ (16%), and $C_3H_3^+$ (14%). With Ne as reactant the parent ion intensity is only *c.* 1%, and extensive fragmentation including the production of two-carbon atom fragments occurs. The electron ionisation spectrum of benzene has certain similarities and certain differences from the chemical ionisation spectra generated by Kr, Ar and Ne. A rather broad range of energy is transferred to the benzene molecules by electron ionisation, whereas the energy transferred by the chemical ionisation process falls within narrow limits. Thus, the electron ionisation spectrum exhibits features which are found separately in the spectra produced by several different charge-transfer agents having a range of recombination energies.

5.4.7 Esters

Esters comprised the first non-hydrocarbon compound type for which a systematic chemical ionisation investigation was made[26], and because a number of important chemical ionisation reactions were discovered in the process, it constituted a felicitous choice. As a typical example, the spectrum of n-heptyl propionate is given in Table 5.2. The largest addition-type ion

Table 5.2 Relative ion intensities for n-heptyl propionate
CH_4 reactant; M.W. = 170; $P_{(CH_4)}$ = 1.0 torr
(From Munson and Field[26], by courtesy of the American Chemical Society)

m/e	*Ion*	*Intensity/%*
213	$(M+C_3H_5)^+$	0.5
201	$(M+C_2H_5)^+$	0.1
173	$(M+1)^+$	5.0
171	$(M-1)^+$	0.6
117		0.6
115	Alkyl exchange	3.1
103	Alkyl exchange	9.0
99	$C_7H_{15}^+$	4.0
98	$C_7H_{14}^+$	1.5
97	$C_7H_{13}^+$	8.8
75	$C_2H_5C(OH)_2^+$	39.0
57	$C_2H_5C{\equiv}O^+$	16.0

(Section 5.3.1) is the $(M+1)^+$ ion at m/e = 173. Alkyl exchange ions (Section 5.3.7.2) are observed at m/e = 115 (allyl ion attack) and m/e = 103 (ethyl ion attack). Heptyl ion (m/e = 99) and propionyl ion (m/e = 57) produced by heterolytic cleavage (Section 5.3.5) are formed with moderate intensity and also one observes the $C_7H_{13}^+$ ion (m/e = 97), an alkenyl ion produced by loss of H_2 from heptyl ion. The most intense ion in the spectrum is that of protonated propionic acid (m/e = 75) produced by a hydrogen transfer reaction (Section 5.3.7.1).

The chemical ionisation spectra of the alkyl esters of propionic acid contain a relatively small number of ions, and almost all of the ions found in the spectra give meaningful information concerning the identity and structure of the ester producing the spectrum.

5.4.8 Alcohols

The i-butane chemical ionisation mass spectra of a number of saturated monohydroxylic alcohols have been determined[25] to establish the general spectral patterns of this class of compound. The following ion types comprise large fractions of the spectra of alcohols: the alkyl ion formed by protonation of the hydroxyl followed by the loss of water (Section 5.3.4), $(M-1)^+$ formed by hydride abstraction, the protonated molecule $(M+1)^+$, and the association complexes of the molecule with the $m/e = 39$ and $m/e = 57$ ions of the i-butane plasma. When methane is used as reactant gas, further fragmentation of the alkyl ion (R^+) occurs and this fragmentation results in the presence of alkyl and alkenyl ions with lower numbers of carbon atoms in the spectra.

Variations occur in the mass spectra with the structures of the alcohols. For the smallest alcohols, methanol to the propanols, the protonated molecules dominate the spectra, and these protonated molecules react with another molecule even at quite low alcohol pressures to produce protonated dimers. For alcohols containing five or more carbon atoms the R^+ ion intensity dominates the spectra. In tertiary carbinols the R^+ ions comprise almost all of the observed spectra (99–95% of the alcohol ionisation), and this dominance of R^+ formation is the consequence of the fact that low-energy tertiary carbonium ions can be formed by a simple bond cleavage.

$(M-1)^+$ ions are generally found for all alcohols except the tertiary alcohols. These ions, along with $(M+1)^+$ ions, are of value because they give information concerning the molecular weight of the alcohol.

The chemical ionisation spectra of alcohols are such that they can be of considerable utility in the analysis of alcohols. The electron ionisation spectra of aliphatic alchols involve extensive fragmentation and, except for the smallest alcohols, ions in the molecular weight regions are small or non-existent. Oxonium ions are formed in large abundance; thus the electron ionisation spectra emphasise oxygen-containing fragments of the molecules. On the other hand, i-butane chemical ionisation of the larger alcohols emphasises the hydrocarbon fragments R^+. One may immediately conclude that for alcohols electron ionisation mass spectrometry and chemical ionisation mass spectrometry complement each other nicely. In addition, however, for primary and some secondary alcohols the chemical ionisation $(M-1)^+$ ion provides evidence for the molecular weight of the molecules.

5.4.9 Substituted benzophenones

The chemical ionisation spectra of a number of benzophenones have been determined with different reagent gases by Michnowicz and Munson[15]. Two main types of ions are observed: (i) addition ions (predominantly $(M+1)^+$ ions) produced by addition reactions (Section 5.3.1) and (ii) substituted and unsubstituted benzoyl ions produced by functional-group removal (Section 5.3.4) where the functional group removed is either the phenyl moiety or the substituted phenyl moiety of the substituted benzophenone molecule. The reagent gases utilised were methane, ethane, propane, and i-butane and the amount of fragmentation observed decreased with this order of reagent gases. It was also found that the amount of fragmentation occurring with a given reagent gas increased significantly as the ionisation chamber repeller voltage was increased in the range 0–75 V. This behaviour is attributed quite reasonably to the higher kinetic energy content of the reactant ions at the higher repeller voltages. The authors point out that this phenomenon may have some analytical utility.

The relative amounts of substituted and unsubstituted benzoyl ions formed varies with the identity of the substituent on the benzophenone molecule, but the variation is not that which might be expected from *a priori* considerations. Thus one would think that the presence of an electron-releasing substituent would enhance the relative amount of substituted benzoyl ion produced. This is indeed observed in the electron ionisation spectra of substituted benzophenones[38], but the chemical ionisation results with propane as reactant gas are reversed, that is, the relative intensity of the unsubstituted benzophenone ion is enhanced by the presence of an electron-releasing substituent. Michnowicz and Munson suggest that the difference in behaviour observed with the two types of ionisation results from the fact that in the electron ionisation case the substituent affects the energy of the fragment ion formed, but in the chemical ionisation case it influences the portion of the benzophenone molecule on which the initial protonation occurs. Thus, in electron ionisation the presence of an electron-releasing substituent lowers the energy of the substituted benzoyl ion with a resultant enhancement of intensity, but in chemical ionisation the substituent increases the amount of initial protonation on the substituted benzene ring of the molecule. Since fragmentation of this $(M+1)^+$ ion produces only unsubstituted benzoyl ions, the presence of the substituent enhances the relative intensity of the unsubstituted fragment ion. This is an illuminating example of how different factors may operate in electron and chemical ionisation.

5.4.10 Alkaloids

Fales, Lloyd and Milne[23] have investigated the methane chemical ionisation spectra of 29 alkaloids. The $(M+1)^+$ ion is never absent in the chemical ionisation spectra and it is invariably more abundant than M^+ ion in the corresponding electron ionisation spectrum. The chemical ionisation identification of aliphatic hydroxyl and methoxyl groups is always possible and skeletal information from rearrangement reactions is generally absent. In a number of cases significant information about molecular structure is available from the fragmentation reactions occurring in chemical ionisation; for example, the authors state that the chemical ionisation mass spectrum of ephedrine

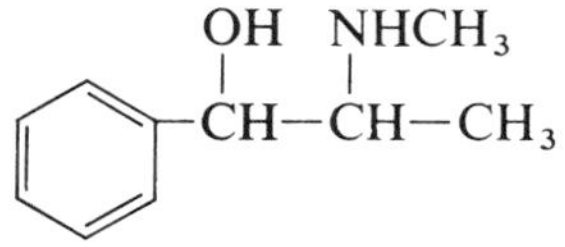

$$C_6H_5-\underset{\displaystyle OH}{CH}-\underset{\displaystyle NHCH_3}{CH}-CH_3$$

M.W. = 165

could stand alone as a complete structure proof. The $(M+1)^+$ ion at $m/e = 166$ permits identification of a molecular formula; loss of water to give the ion at $m/e = 148$ or of methylamine to give that at $m/e = 135$ reveals the presence of the two functional groups, —OH and —NHMe, and the ions at $m/e = 58$ ($MeCH{=}NH^+Me$) and $m/e = 107$ ($PhCH{=}OH^+$) can be re-assembled to give the correct structure. Chemical ionisation does not fragment complicated ring systems, and information obtainable in electron ionisation about this feature of the molecular structure is not available. On the other hand, because of the localised-attack concept, sensitive bonds in molecules are not always fragmented and thus the catastrophic fragmentations sometimes observed in electron ionisation occur to a much smaller extent in chemical ionisation.

5.4.11 Amino acids

The methane chemical ionisation mass spectra of 21 amino acids have been measured by Milne, Axenrod and Fales[33]. The $(M+1)^+$ ion is observed in all cases, and it is an intense ion except in those cases (glutamine glutamic acid and ornithine) where cyclisation with loss of water or ammonia might be expected to be particularly easy. Even in these cases the $(M+1)^+$ intensities are 5–9% of the base peak. Valine, $Me_2CHCH(NH_2)CO_2H$, gives a chemical ionisation spectrum typical of that of aliphatic amino acids, and its spectrum consists of three ions: $(M+1)$, relative intensity (R.I.) = 58; $(M+1-H_2O)^+$, R.I. = 5; and $(M+1-COOH_2)^+$, R.I. = 100. The two fragment ions are produced by the functional group removal reaction (Section 5.3.4), where the groups removed are OH and COOH. In the case of hydroxy aliphatic amino acids, the loss of either OH or COOH is followed by some loss of another water molecule. Thus, threonine ($MeCHOHCH(NH_2)CO_2H$) exhibits the ions $(M+1-H_2O-COOH_2)^+$, R.I. = 23 and $(M+1-2H_2O)^+$, R.I. = 9 in addition to the ion types listed above. For substituted and aromatic amino acids other decomposition reactions are observed, the most important of which is probably loss of the amine group. This necessarily involves a backside-assisted decomposition (Section 5.3.8) as illustrated in reaction (5.25). Milne, Axenrod and Fales point out that in electron ionisation mass spectra the amine fragment ($RCH{=}NH_2^+$) is generally a dominant ion in the spectrum and the molecule-ion intensities are very low except for methionine, phenylalanine and tryptophan. In chemical ionisation, on the other hand, attention is focused on ions such as $(M+1-H_2O)^+$, $(M+1-NH_3)^+$, and $(M+1-COOH_2)^+$. In the case of an unknown compound, the chemical ionisation spectrum would be highly diagnostic of an amino acid, while the electron ionisation spectrum might be of additional aid in the determination of the structure of the amino acid side chain.

5.4.12 Amino acid phenylthiohydantoin derivatives

It is appropriate at this point to discuss another aspect of the mass spectrometry of the amino acids, namely, the chemical ionisation spectra of phenylthiohydantoin (PTH) derivatives, which are formed during the Edman degradation of proteins[39]. In the Edman degradation method for obtaining the amino acid sequence of proteins, phenyl isothiocyanate is added to the protein or peptide, and it reacts with the *N*-terminal amino acid to produce the phenylthiohydantoin derivative of the amino acid. Identification of this compound corresponds to the identification of the *N*-terminal amino acid, and repetitive utilisations of the reaction permits the identification of the amino acid sequence in the protein or peptide. The reactions are shown in equation (5.28). Methyl isothiocyanate can also be used to make methyl-

$$C_6H_5{-}N{=}C{=}S + NH_2{-}\underset{H}{\overset{R}{C}}{-}\overset{O}{\overset{\|}{C}}{-}\underset{H}{N}{-}\sim$$

$$\searrow$$

$$C_6H_5{-}\overset{H}{N}{-}\overset{S}{\overset{\|}{C}}{-}\underset{H}{N}{-}\underset{H}{\overset{R}{C}}{-}\overset{O}{\overset{\|}{C}}{-}\underset{H}{N}{-}\sim$$

$$\downarrow \qquad (5.28)$$

$$\text{Phenylthiohydantoin ring: } C_6H_5{-}N{-}C({=}S){-}N(H){-}C(H)(R){-}C({=}O){-}(N) \quad + \; NH_2{-}\sim$$

Phenylthiohydantoin derivative

thiohydantoin derivatives. The Edman method is the standard method for protein sequencing and thus the identification of the thiohydantoin derivatives is a matter of much practical importance.

Fales and co-workers[39] have determined the i-butane chemical ionisation spectra of the thiohydantoin derivatives of 20 amino acids. The heterocyclic ring of the phenylthiohydantoin system is a relatively basic unit and this coupled with the mildness of t-$C_4H_9^+$ as a Brønsted acid causes the $(M+1)^+$ ion to be the major ion for the PTH derivatives of 17 of the 20 acids investigated. For the other three amino acid derivatives (arginine, lysine and *S*-carboxymethylcysteine), ions suitable for the identification of the compounds exist. In practical sequencing work it is necessary to know quantitatively the amount of PTH derivative formed and in the technique described by Fales and co-workers quantitative measurements are obtained by a form of the isotopic dilution technique by use of deuteriated PTH derivatives.

Information concerning the relative sensitivity of chemical and electron ionisation mass spectrometry is available from this investigation. The PTH derivative of leucine was introduced into the direct introduction probe of the mass spectrometer and the electron ionisation and chemical ionisation spectra were alternatively determined as the sample evaporated from the probe. The signal strength observed in each mode was measured and it was found that the sensitivity of the chemical ionisation method was roughly 100 times that of the electron ionisation technique. It is also of interest that the authors mention that the chemical ionisation sensitivity is probably 10–100 times greater than the gas chromatographic method, which is often times used as a means of identifying PTH derivatives. On the other hand, the gas chromatographic method has an advantage in that the elution patterns from earlier steps provide a continuous monitor of the presence of overlapping sequences due to protein impurities and non-specific cleavage.

5.4.13 Peptides

To date, two groups have contributed information concerning the application of chemical ionisation mass spectrometry to the very important problem of peptide and protein sequencing. Mass spectrometry has been utilised since 1959 to obtain information about the sequences of amino acids in proteins[40]. The molecule-ions formed from peptides cleave at points (usually at the amide bonds) to form sets of fragment ions from which information concerning the amino acid sequence can be deduced. A serious problem in the mass spectrometric sequencing method is the fact that peptides are large polar molecules, and their volatility in their natural state is so small as virtually to preclude the occurrence of sufficient volatilisation to permit mass spectrometric measurements. The problem can be ameliorated by preparing suitable derivatives of the peptides[40]. Esterification of the carboxylic acid group and acylation of the free amine group eliminates the formation of zwitterions in the peptides, and consequently the volatility is much increased. Permethylation of the amide nitrogens eliminates hydrogens which can participate in hydrogen-bonding interactions and this results in further enhancement of volatility. Another method of volatility enhancement, described a number of years ago by Biemann[41], involves reduction of the carbonyl groups in the peptide with $LiAlH_4$ to produce polyamino alcohols in which both zwitterion and hydrogen bond interactions are virtually eliminated.

In electron ionisation peptide sequencing, one encounters the problem that extensive fragmentation of the peptides occurs in the ionisation process, and consequently, the spectra produced are complicated both in terms of the numbers of ions formed and by the widespread occurrence of re-arrangement processes which interfere with and obscure the sequence-determining fragmentation processes unless high-resolution mass measurements are made. The fragmentation lowers the intensities of ions in the high-mass end of the spectra, often quite drastically, and thus the sensitivity of the method is significantly lower than one would like.

In methane chemical ionisation mass spectrometry of esterified acylated peptides, protonation seems to occur primarily on the nitrogen atoms and the $(M+1)^+$ ions thus formed fragment to form acyl ions by heterolytic cleavage (Section 5.3.5) illustrated by reaction (5.16)[28]. This is analogous to the most important reaction occurring in electron ionisation, namely,

$$\mathrm{R{-}\overset{\overset{\cdot O^{+}}{\|}}{C}{-}NH{-}R'} \longrightarrow \mathrm{R{-}C{\equiv}O^{+} + R'N^{\cdot}H} \qquad (5.29)$$

The acyl ions produced in reaction (5.29) and reaction (5.16) permit the deduction of the amino acid sequence starting at the amine end of the molecule and thus are referred to as *N*-terminal sequence-determining ions. In addition to these ions, one observes in chemical but not in electron ionisation the formation of ammonium-type ions produced by hydrogen transfer (reaction (5.23), Section 5.3.7.6). The reaction as written requires the presence of hydrogen atoms on amide nitrogens and one would expect that the reaction would not occur for permethylated peptides. Kiryushkin and co-workers[28] state that indeed such ions are not observed in permethylated peptides, but Gray, Wojcik and Futrell[42] observe the formation of minor amounts of this type of ion. They suggest that its formation involves the transfer of a hydrogen from the α carbon of the peptidyl fragment to the amino nitrogen. These ammonium ions also contain information concerning the amino acid sequence and they are referred to as *C*-terminal sequence-determining ions since they may be thought of as originating at the carboxyl end of the peptide molecule.

Kiryushkin and co-workers[28] point out that the presence of two sequence-determining series of ions, i.e., the *N*- and *C*-terminal series, in the chemical ionisation spectrum of a peptide reduces significantly the amount of uncertainty and indeterminacy in deducing the amino acid sequence from the observed spectrum. If one attempts to determine the sequence from just the *N*-terminal sequence-determining ions at low resolving power (with no information as to the empirical formulae of the ions), the correct sequence will usually be found, but only as one of a number of alternative sequences. However, when the *C*-terminal sequence-determining ions are also available, many of the alternative possibilities are eliminated and usually only one sequence compatible with both the *N*- and *C*-terminal ions remains. This constitutes a considerable advantage of the chemical ionisation method.

Table 5.3 Chemical ionisation mass spectrum of a peptide
(From Kiryushkin *et al.*[28], by courtesy of Heyden & Son Ltd.)

Peptide	*Sequence-determining ions: mass (abundance)*	
	N-*terminal*	C-*terminal*
Prop-trp-leu-val-pro-leu-ala-OMe	243(21), 356(14) 455(7), 552(8), 665(1), 768(11).	104(13), 217(20), 314(100), 413(70), 526(53).

Kiryushkin and co-workers[28] give a tabulation of the *N*- and *C*-terminal sequence-determining ions in 12 small peptides. In these compounds the sequence-determining ions of both types are always present and they often, if not always, have an appreciable intensity. We give in Table 5.3 as a typical example, the ions and intensities found by Kiryushkin and co-workers for one peptide.

Gray, Wojcik and Futrell[42] have determined the methane chemical ionisation spectra of several acetylated permethylated peptides. The amino acid sequence is obtained from the *N*-terminal sequence-determining ions and the spectra produced are simple and easy to interpret. The authors emphasise that because of the smaller amount of fragmentation occurring in chemical ionisation and because of the relatively small intensity differences observed between high and low-mass ions, the chemical ionisation method can provide sequence information with much smaller amounts of sample than are needed for electron ionisation measurements. For example, for acetylated and permethylated penta-alanine, 200 nmol of peptide derivatives were required to give an electron ionisation spectrum containing discernable intensities of the highest-mass sequence-determining fragments. On the other hand with chemical ionisation, adequate spectra were obtained with as little as 2 nmol. Similar results were obtained for the other peptides studied, and it seems to be established that the sensitivity of the chemical ionisation method is significantly higher than that of electron ionisation. Because of this greater sensitivity the chemical ionisation method will require lower vapour pressures than electron ionisation mass spectrometry, and spectra should be obtainable at lower temperatures. A corollary of this is that larger peptides should reach an adequate vapour pressure without decomposing.

5.4.14 Polytertiary alkylamines

The chemical ionisation mass spectra of five polytertiaryamines and of trimethylamine have been determined by use of nitrogen, methane, and i-butane as reactant gases[20]. The amounts of fragmentation observed decrease in the order of reactants nitrogen > methane > i-butane. The spectra with nitrogen were very similar to those obtained by electron ionisation. The intensities of the $(M+1)^+$ ions formed with methane and i-butane as reactant gases always comprised a significant fraction of the amine ionisation and, of course, the intensities were always greater with i-butane than with methane. With i-butane the $(M+1)^+$ intensities ranged from 93% of the amine ionisation for trimethylamine to 64% of the amine ionisation for n-hexamethyltriethylenetetramine,

$$(H_3C)_2N-CH_2CH_2-N(CH_3)-CH_2CH_2-N(CH_3)-CH_2CH_2-N(CH_3)_2$$

M.W. = 230

Thus the intensity of $(M+1)^+$ for polyamines is relatively insensitive to the length of the chain and this behaviour is similar to that observed and discussed previously for peptides and paraffin hydrocarbons. It seems fair to conclude that lack of dependence of the amount of fragmentation on chain length is a characteristic of even-electron chemical ionisation behaviour and this is in contrast to the opposite behaviour observed with electron ionisation.

A moderate amount of fragmentation occurs in methane chemical ionisation and the dominant reactions are hydride ion abstraction and electrophilic attack on C—C bonds (Section 5.3.2) and backside-assisted decompositions (Section 5.3.8) illustrated by reaction (5.24).

5.4.15 Esters of di- and tri-carboxylic acids

The i-butane chemical ionisation mass spectra of a series of esters of di- and tri-carboxylic acids have been reported[34]. These compounds, which are widely used as plasticisers, are difficult to identify by electron ionisation mass spectrometry because their molecule ions are of weak or zero abundance and the chief decomposition paths involve loss of the side chains which distinguish one compound from another. In the case of phthalate esters, this decomposition produces a very characteristic ion at $m/e = 149$,

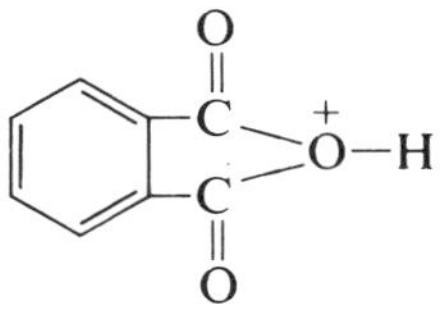

and this ion is taken as being representative of phthalates as a class. In i-butane chemical ionisation, on the other hand, the $(M+1)^+$ ions are observed in all of the 13 compounds investigated and in all but a few cases it constitutes the most intense peak in the spectrum. The spectra are such as to permit distinguishing between the several different esters.

The decompositions which are observed to occur involve the combination of the usual reactions found in esters (Sections 5.3.5 and 5.3.6) coupled with protonated acid anhydride formation (Section 5.3.9, reaction (5.26)). Significant differences are observed between the amount of decomposition occurring with *o*-, *m*- and *p*-diesters and between maleate and fumarate diesters. The differences are exactly as one would predict from the structures of the esters.

5.4.16 Barbiturates

The methane chemical ionisation spectra of eight barbiturates have been determined[43]. Extensive decomposition of the molecule-ion occurs in electron ionisation of this class of compound, and this makes the problem of distinguishing one barbiturate from another a difficult or impossible task. By contrast, the $(M+1)^+$ ions obtained in chemical ionisation are all of high intensity and, indeed, are the most intense peaks in the spectra. Of the eight compounds studied, all but two could be identified solely on the basis of the m/e value of the $(M+1)^+$ ion. It has been shown that the chemical ionisation method is of practical value in the clinical identification of barbiturates in suspected overdose cases, for the injection into the chemical ionisation mass spectrometer of a chloroform extract of the stomach washings from an overdose patient led to a rapid and straightforward identification of the drug ingested.

5.4.17 Dangerous drugs

In a work related to the barbiturate investigation described above, Milne, Fales and Axenrod[30] have investigated the possibility of using chemical ionisation mass spectroscopy as a clinical method for the identification of commonly used drugs. The i-butane chemical ionisation mass spectra of 48 commonly used drugs have been measured and the majority of these give only an $(M+1)^+$ ion, which can be used for simple and unequivocal identification of the compound.

5.4.18 Photodimers of cyclic $\alpha\beta$-unsaturated ketones

It has been shown[29] that information concerning the structures of photodimers of cyclic $\alpha\beta$-unsaturated ketones can be obtained from the chemical ionisation of the dimers with methane. A number of isomeric dimers is formed in photolysis and these can be divided into two groups designated as head-to-head (h–h), and head-to-tail (h–t), which differ in that the former is a 1,4-diketone, while the latter is a 1,5-diketone.

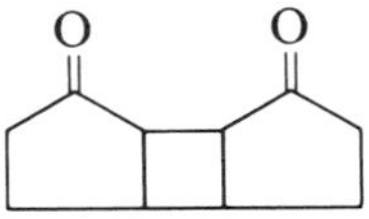

(h–h)

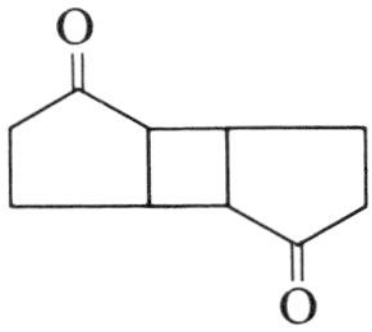

(h–t)

The mass spectra consist mainly of two ions, namely, $(M+1)^+$ ion and the $(M/2+1)^+$ fragment ion. The amount of formation of the fragment ion in the head-to-head isomer is markedly less than that in the head-to-tail isomer and this constitutes a means of distinguishing the two isomers. Fourteen different photodimers from variously substituted cyclopentenones and cyclohexenones have been investigated, and this difference in behaviour between the isomers appears without exception. It is suggested that the decomposition occurs by a sequence of two β-fission reactions (Section 5.3.6) and in the head-to-head dimer the first of these results in an ionic structure with the positive charge next to the carbonyl group, i.e.

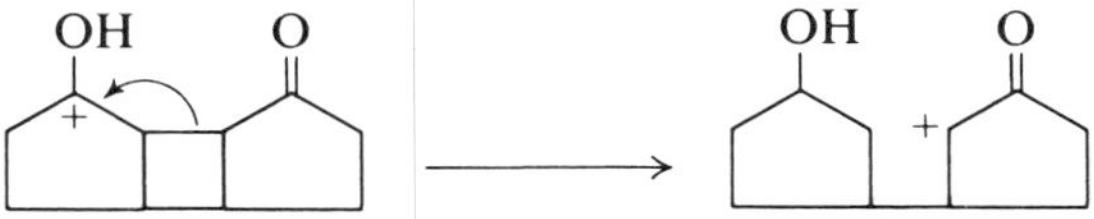

(5.30)

This is a relatively high energy structure, and thus the fragmentation process occurs to a smaller extent than that in the head-to-tail isomer, for which the β-fission process does not place the charge next to the carbonyl group. Clearly, small differences in ionic energies have a profound effect on the spectra produced and it is one of the desirable characteristics of chemical ionisation mass spectrometry that spectra are significantly affected by energy differences with magnitudes comparable to those of importance in condensed-phase phenomena.

5.4.19 Miscellaneous

One of the earliest uses of chemical ionisation mass spectroscopy as a tool in bio-organic research was the determination of the structure of the antibiotic, botryodiplodin[16],

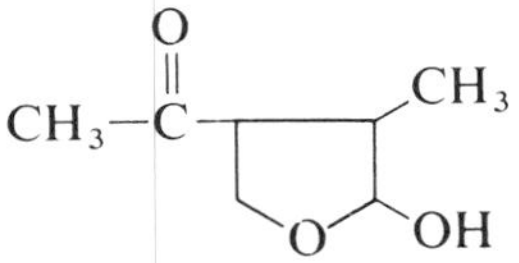

The molecular weight of the compound was determined as was its elemental composition from a determination of its exact mass. The determination of elemental compositions of certain fragment ions also proved to be of use in determining the structure of the molecule.

The methane chemical ionisation spectra of two nucleosides, namely, thymidine and adenosine have been determined[44]. Principal ions observed are the $(M+1)^+$ ions and the protonated free base. Fewer structural details are generally available from chemical ionisation spectra as compared with their electron impact counterparts, although the $(M+1)^+$ ion enhancement in the former case offers a major advantage.

An early preliminary investigation of the chemical ionisation spectra of a number of complex molecules was carried out by Fales, Milne and Vestal[45]. The compounds studied at 2-deoxy-D-ribose, codeine, crinamine, haemanthamine, *O*-methylpellotine, alanylvaline and cholestanone. Finally, at the time of writing only one inorganic compound seems to have been investigated by chemical ionisation, namely borazine[46]. Methane, ethane and n-butane are used as reagent gases, and the reactions observed are primarily addition reactions (Section 5.3.1) and H^- abstraction reactions (Section 5.3.2). The abstraction reactions occur to the greatest extent with methane as reactant and become progressively less with ethane and n-butane.

5.5 PHYSICAL CHEMICAL STUDIES

The chemical ionisation studies so far described have been almost completely concerned with the determination of the mass spectra produced from different types of compounds and with the deduction of the ionic chemistry involved in the production of the ions observed. The work has been largely qualitative and descriptive and a primary goal has been the development of a mass spectrometric technique of practical analytical utility. However, recently a series of chemical ionisation investigations of a more physical, quantitative nature have been undertaken and these have involved the determination of rate constants, activation energies and frequency factors (entropies of activation) for gaseous ionic reactions. In addition, it appears to be possible to obtained information concerning gaseous ionic equilibria including the determination of equilibrium constants, and it may be possible to obtain information concerning the relative energies of gaseous ions.

A method has been devised to determine from experimental quantities the rate constants for the unimolecular decompositions of gaseous ions occurring under chemical ionisation conditions in the mass spectrometer, and from the temperature variations of these rate constants activation energies and entropies may be deduced. Most of the measurements made are quite new, and consequently caution has been exercised in drawing conclusions concerning the exact physical and chemical significance of the results obtained. However, it appears that a physical chemistry of the gaseous ionic systems involved in chemical ionisation is developing. It is to be hoped that comparisons of the physical chemistry of these systems with that of condensed-phase systems will lead to an increase in understanding of both. Thus it seems quite possible that chemical ionisation mass spectrometry will have a utility related to the information it provides about the physical chemistry of various ionic systems which will be of importance comparable with its analytical importance.

It is desirable at this point to call attention to a characteristic aspect of the chemical ionisation process. When chemical ionisation is effected at a pressure of *c.* 1 torr in the mass spectrometer ionisation chamber, the ions comprising the mass spectra are produced by any of the several reactions listed in Section 5.3 and after formation the ions undergo collisions with molecules of the reactant gas before passing out of the ionisation chamber. Thus, unlike the conditions which obtain in conventional ionisation (low-pressure electron and photon ionisation), the ions produced by chemical ionisation are not isolated, for, indeed, as was shown in the Section 5.2, an ion may experience a thousand or more collisions in the ionisation chamber. In summary, under chemical ionisation conditions the ions are not formed and collected in the absence of interactions with surroundings and the interaction which does occur will have a significant effect on the mass spectra and upon the temperature dependence of the mass spectra.

5.5.1 Mathematical formulation of chemical ionisation kinetics

The first formulation of the rate expression for the unimolecular decomposition of ions under chemical ionisation conditions was made for the i-butane chemical ionisation of benzyl acetate[47], and we shall use an ester (ROAc) in our exposition although our formulation will be different in detail from the earlier one.

The t-butyl reactant ions are formed from the i-butane in the near vicinity of the electron beam and they drift to the ion exit slit of the ionisation chamber under the influence of the repeller field. Protonated ester molecules (taken as a typical example) are formed along the path of the t-butyl ions by proton-transfer reactions and after formation the protonated ester ions drift toward the ion exit slit under the influence of the repeller field. In the process of doing this they undergo decomposition reactions which may be looked upon as a set of parallel, competing first-order reactions, that is,

$$\mathrm{ROAcH^+} \xrightarrow{k_i} \begin{array}{l} \mathrm{P_1^+ + F_1} \\ \mathrm{P_2^+ + F_2} \\ \vdots \\ \mathrm{P_i^+ + F_i} \end{array} \tag{5.31}$$

where P_i = ith product ion. The generalised equation for the formation of the protonated ester ion is

$$\mathrm{ROAc + t\text{-}C_4H_9^+} \xrightarrow{k_f} \mathrm{ROAcH^+ + C_4H_8} \tag{5.32}$$

and we may treat this system as involving the consecutive reactions (5.31) and (5.32) occurring in elements of volume drifting between the electron beam and the ion exit slit under the influence of the repeller field.

The rate equations applicable to the system are

$$\frac{d(ROAcH^+)}{dt} = k_f(t\text{-}C_4H_9^+)(ROAc) - \sum_i k_i(ROAcH^+) \tag{5.33}$$

where Σk_i is the sum of the rate constants for all the decomposition reactions undergone by $ROAcH^+$. Equation (5.32) is pseudo-first order, and so

$$(t\text{-}C_4H_9^+) = (t\text{-}C_4H_9^+)_0 \exp(-k_f(ROAc)t) \tag{5.34}$$

Substituting in (5.33) we obtain

$$\frac{d(ROAcH^+)}{dt} = k_f(ROAc)(t\text{-}C_4H_9^+)_0 \exp(-k_f(ROAc)t) - \sum_i k_i(ROAcH^+) \tag{5.35}$$

Equation (5.35) is the equation for consecutive reactions and its solution is

$$(ROAcH^+) = \frac{k_f(ROAc)(t\text{-}C_4H_9^+)_0}{k_f(ROAc) - \sum_i k_i}\left[\exp(-\sum_i k_i t) - \exp(-k_f(ROAc)t)\right] \tag{5.36}$$

where $(ROAcH^+)$ is the concentration of $(ROAcH^+)$ in the element of volume at time t.

The initial concentration of $ROAcH^+$, that is the $ROAcH^+$ formed by equation (5.32) is

$$(ROAcH^+)_0 = (t\text{-}C_4H_9^+)_0[1 - \exp(-k_f(ROAc)t)] \tag{5.37}$$

Dividing (5.36) by (5.37) gives

$$\frac{(ROAcH^+)}{(ROAcH^+)_0} = \left(\frac{k_f(ROAc)}{k_f(ROAc) - \sum_i k_i}\right) \frac{\exp(-\sum_i k_i t) - \exp(-k_f(ROAc)t}{1 - \exp(-k_f(ROAc)t)} \tag{5.38}$$

The mass spectrometer ion currents are proportional to ion concentrations when $t = \tau$, the time required for the $t\text{-}C_4H_9^+$ ion to drift out of the ionisation chamber. Thus

$$\frac{I_{ROAcH^+}}{I^0_{ROAcH^+}} = \frac{(ROAcH^+)}{(ROAcH^+)_0} \tag{5.39}$$

We here ignore the small error made by differences in drift velocities of $t\text{-}C_4H_9^+$ and $ROAcH^+$.

Equation (5.38) is awkward to handle and at low concentrations of ROAc a simplification can be made. Thus by expanding the quantity $\exp(-k_f(ROAc)\tau)$ to the linear term, we obtain

$$\frac{I_{ROAcH^+}}{I^0_{ROAcH^+}} = \frac{\exp(-\sum_i k_i\tau) - 1 + k_f(ROAc)\tau}{k_f(ROAc)\tau - \sum_i k_i\tau} \tag{5.40}$$

Dividing out the right-hand side of (5.40) gives

$$\frac{I_{ROAcH^+}}{I^0_{ROAcH^+}} = \frac{1 - \exp(\sum_i k_i\tau)}{\sum_i k_i\tau}\left(1 + \frac{k_f(ROAc)}{\sum_i k_i} + \left(\frac{k_f(ROAc)}{\sum_i k_i}\right)^2 + \ldots +\right) \tag{5.41}$$

For sufficiently small values of (ROAc) all terms higher than the first can be ignored, and we get

$$\frac{I_{\mathrm{ROAcH^+}}}{I^0_{\mathrm{ROAcH^+}}} = \frac{1-\exp(-\sum_i k_i\tau)}{\sum_i k_i\tau} \tag{5.42}$$

which has been obtained previously[47] by a different argument.

The quantity $\sum_i k_i\tau$ is obtained from the experimental values of $I_{\mathrm{ROAcH^+}}/I^0_{\mathrm{ROAcH^+}}$ by solving equation (5.42) using an iterative technique. The term $\sum_i k_i$ is obtained if one knows τ (see below). In turn the rate constants for the formation of the product ions P_i (equation (5.31)) are obtained from

$$k_i = (I_{P_i^+}/\sum_i I_{P_i^+})\sum_i k_i \tag{5.43}$$

Since the pressure in the ionisation chamber in chemical ionisation studies is sufficiently high that the ions undergo many collisions in passing out of the ionisation chamber, ion mobility considerations must be invoked to calculate the residence times τ. A discussion of these considerations is beyond the scope of this chapter but a satisfactory treatment is given in the monograph of McDaniel[48]. As a matter of interest the expression used for a mobility of various substituted benzyl acetate esters is

$$K = 1.81\times 10^3 T[1+(16/M^+)]^{\frac{1}{2}} \tag{5.44}$$

where K = ion mobility in cm s^{-1}/stat volt cm^{-1}, T = absolute temperature and M = molecular weight of ion. This equation applies to the drift of an ion of $m/e = M^+$ in CH_4 at 1.00 torr. For ion drift in i-C_4H_{10} at 0.70 torr

$$K = 8.70\times 10^2 T[1+(58/M^+)]^{\frac{1}{2}} \tag{5.45}$$

The residence time in the mass spectrometer ionisation chamber is given by

$$\tau = \frac{d}{KF} \tag{5.46}$$

where τ = residence time, d = drift distance and F = field strength (stat volts cm^{-1}). A typical residence time in the Esso Chemical Physics Mass Spectrometer is 12.6 μs for protonated benzyl acetate in i-butane at a repeller voltage of 5.0 V (practical), d = 0.20 cm, and T = 373 K.

5.5.2 Temperature variations of chemical ionisation kinetics

The study of the kinetics of chemical ionisation reactions was initially stimulated by the experimental observation that under some conditions chemical ionisation spectra were very temperature dependent, and upon kinetic analysis the result was obtained that the rate constants for unimolecular reactions such as (5.31) obey the Arrhenius equation. This result is unprecedented in mass spectrometer kinetics, and consequently it points up in a striking way the difference between chemical ionisation and conventional low-pressure mass spectrometry.

It is useful to consider this matter in terms of the quasi-equilibrium theory (see Chapter 1) of mass spectra[49, 50]. This theory yields an expression for the rate constant of an ionic decomposition reaction occurring in the mass spectrometer ionisation chamber which is the product of a frequency factor and an energy factor. In its simplest form this is

$$k(E) = \nu[(E-\varepsilon)/E]^{S-1} \tag{5.47}$$

where ν is the frequency factor, E is the total energy content of the molecule, ε is the activation energy for the reaction of interest and S is the effective number of oscillators in the reacting ion. The experimental rate constant is obtained by integrating $k(E)$ over the energy distribution function appropriate to the system under consideration. In the case of low-pressure electron or photon impact the energy distribution in the isolated ions existing in these systems is determined by the functions which characterise the transfer of energy from the impacting electron or photon beam. Several examples of such energy distribution functions are given by Rosenstock and Krauss[50] and we need not discuss them further than to say that they are not Boltzmann distributions. In chemical ionisation processes, however, since collisions do occur, one can expect the energy distribution in the reacting ions either to be or to approximate to a Boltzmann distribution. One can immediately conclude that this difference in the energy distribution functions will constitute further reason for differences in the spectra produced by the chemical ionisation technique.

More important, however, it has been shown by Magee[51] that if the quasi-equilibrium expression for the rate constant as a function of energy is integrated over the Boltzmann distribution, the conventional absolute reaction rate theory expression for the rate constant is obtained:

$$k = \frac{kT}{h}\frac{F^{\ddagger}}{F}\mathrm{e}^{-\varepsilon/kT} \tag{5.48}$$

In this equation F represents the partition function and the other terms have the usual significance. This result is in accord with one's intuitive feeling that if sufficient collisions occur in the ionisation chamber, the reactions occurring will be thermally activated and the rate expression of conventional kinetics will apply. Unlike the situation in low-pressure electron impact mass spectrometry, temperature is a meaningful variable in that it is a parameter characterising the energy distribution of the system. When the energy distribution is thus defined, one can reasonably expect to use conventional techniques (Arrhenius plots) to obtain information about energies and entropies of activation for ionic processes occurring in the mass spectrometer. We shall see that this expectation is realised in fact.

5.5.3 Kinetic results for benzyl acetate and t-amyl acetate

The i-butane chemical ionisation spectra of benzyl acetate and t-amyl acetate have been investigated at a number of temperatures and the rate constants for the decomposition of the protonated esters to benzyl and t-amyl ions, respectively, have been obtained[47] at the several temperatures from equations (5.42) and (5.43). The rate constants obey the Arrhenius relationship, and activation energies and frequency factors obtained from the Arrhenius plots are given in Table 5.4.

It is postulated that the sequence of events involved in the decomposition of esters is that a proton is transferred from t-$C_4H_9^+$ to the ester in a slightly exothermic process, but sufficient collisions occur in the ionisation chamber

Table 5.4 Kinetic quantities for the formation of $C_7H_7^+$ and t-$C_5H_{11}^+$
(From Field[47], by courtesy of the American Chemical Society)

Ion	*Activation energy/* kcal mol^{-1}	*Log A*	k_{373}/s^{-1}
$C_7H_7^+$	12.3 ± 0.9	11.2	1.0×10^4
t-$C_5H_{11}^+$	12.4 ± 0.9	12.4	11×10^4

to remove the exothermicity and establish a Boltzmann distribution of energies in the protonated ester ions. The decomposition of the protonated ester ions then occurs from the Boltzmann distribution. The experimentally determined activation energy is the energy difference between the ground state of the protonated ester and the energy required for the formation of the carbonium ion and acetic acid. The frequency factors in the rate equations will have their usual kinetic significance and will constitute measures of the entropies of activation of the ionic dissociation process.

If this analysis is correct, or substantially correct, the activation energies for comparable reactions provide measures of the relative energies of product ions produced by the thermal decompositions and the frequency factors provide information about entropies of activation. It is clear that reactions such as those producing benzyl ion, (equations (5.31) and (5.32) with R = C_7H_7) are analogous to $A_{AL}1$ acid-catalysed solvolysis reactions in solution and the significance of the kinetic quantities in the gas-phase chemical ionisation process is analogous with that in condensed-phase solvolysis. In the gas phase solvent interactions are absent and the prospect is that much new and valuable information will be obtained from this technique.

The kinetic quantities given in Table 5.4 are unexpected. The fact that the rate constant for the formation of t-amyl ion is greater than that of the formation of benzyl ion is in keeping with condensed-phase solvolysis results, but in gas phase this is completely the consequence of a larger frequency factor since the activation energies for the two processes are identical. There is no condensed phase solvolytic precedent for this observation, although it has been pointed out[47] that in the gas phase the energies of the charge centres in benzyl and t-amyl ions are approximately equal. A rationale for the lower frequency factor in the benzyl acetate case has been given[47]. In the ground state of the protonated benzyl acetate ion free rotation can occur around the phenyl–methylene bond, but in the transition state developing p-orbitals on the methylene interact with the p-orbitals on the phenyl group and convert this free rotation to a torsional vibration.

5.5.4 Substituted benzyl acetates

The effect of substituents on the energies and reactivities of aromatic compounds is perhaps as extensively investigated and as well understood as any aspect of physical organic chemistry, and consequently investigations of substituted benzyl acetates were undertaken[52] to help provide an understanding of the factors involved in chemical ionisation temperature effects. The compounds studied were benzyl acetates substituted in the *para*-position by methoxy, fluoro, methyl, chloro and nitro groups. The ease with which the protonated molecules decompose to produce benzyl ions depends markedly upon the identity of the substituent, and the variation observed is in the direction expected. Thus the methoxy ester ion decomposes to *p*-methoxybenzyl ion very rapidly even at the lowest temperature attainable in the mass spectrometer (*c.* 40 °C), whereas the nitro compound did not decompose to nitrobenzyl ion at all even at the highest temperature attainable (*c.* 250 °C). Temperature studies were made for all the compounds, and the kinetic quantities obtained are given in Table 5.5. Significant trends with the identity of the substituent occur for all of the quantities tabulated.

Table 5.5 Kinetic quantities for *p*-substituted benzyl acetates
(From Field[52], by courtesy of the American Chemical Society)

Substituent	E_a/kcal mol^{-1}	Log *A*	*Log* k_{300}
CH_3O	4.3 ± 0.7	9.0 ± 0.5	5.9
F	9.0 ± 0.7	9.8 ± 0.5	3.3
CH_3	10.3 ± 1.0	12.1 ± 0.7	4.6
H	12.3 ± 0.9	11.2 ± 0.7	2.3
Cl	12.5 ± 0.9	11.8 ± 0.6	2.7
NO_2			< -4.5*

*Calculated from failure to observe formation of $NO_2 \cdot C_6H_4 \cdot CH_2^+$ at 254 °C.

A Hammett plot (log K_X/K_H vs. σ^+ for the substituents X) was constructed for the substituted benzyl acetates and the correlation was good for all the substituents except nitro, for which the observed decomposition rate constant was much too low. However, the overall correlation was such as to indicate that the chemical ionisation behaviour is determined by the same factors as are involved in condensed-phase ionic chemistry. Thus these results constitute evidence that the chemical ionisation results will be meaningfully related to condensed-phase phenomena.

5.5.5 Methoxymethyl formate and acetate

The acid-catalysed hydrolysis of simple partial acylals has been studied in the condensed phase[53], and a study of the behaviour of methoxymethyl formate and methoxymethyl acetate in chemical ionisation mass spectrometry was undertaken[31] to provide a comparison of the gaseous- and solution-phase chemistry of these simple partial acylals. Methoxymethyl formate and methoxymethyl acetate undergo hydrolysis in aqueous acid by an $A_{AL}1$ mechanism.

$$\mathrm{R{-}\overset{\overset{\displaystyle O}{||}}{C}OCH_2OCH_3 \overset{H^+}{\rightleftharpoons} R{-}\overset{\overset{\displaystyle {}^{+}OH}{||}}{C}{-}OCH_2OCH_3 \xrightarrow[\text{Slow}]{-RCOOH} {}^{+}CH_2OCH_3 \xrightarrow[\text{Fast}]{H_2O} HCHO + CH_3OH + H^+} \quad (5.49)$$

In aqueous hydrochloric acid the formate hydrolyses about three times faster than acetate and it was hoped that in chemical ionisation mass spectrometry the same pattern would be observed, namely, that under the same conditions the formate would be more extensively ionised to the methoxymethyl cation. The i-butane chemical ionisation spectrum of methoxymethyl acetate is of interest in two ways: (i) although the compound is an ester and one would expect the formation of $CH_3OCH_2^+$ ion by reactions (5.31) and (5.32), in fact no $CH_3OCH_2^+$ ion is produced even at the highest temperature investigated (*c.* 250 °C); and (ii) a moderately intense $(M-3)^+$ ion is observed and this led to the first recognition of the reaction involving carboxylic displacement by t-$C_4H_9^+$ (see Section 5.3.7.4, reaction (5.22)). Analogous behaviour is observed with methoxymethyl formate.

Since, as we have seen, benzyl acetate and t-amyl acetate easily form benzyl and t-amyl ions under these conditions, one must conclude that some property of the methoxymethyl ion (possibly a relatively high energy) decreases its ease of formation. Consequently, it is necessary to use a stronger gaseous acid to investigate the relative ease of formation of methoxymethyl cation from methoxymethyl formate and acetate. Methane chemical ionisation spectra of the compounds were obtained at several temperatures and copious amounts of methoxymethyl cation were formed from both esters by means of the reaction

$$\mathrm{RCOOCH_2OCH_3 \xrightarrow{CH_5^+} RCOOCH_2OCH_3{\cdot}H^+ \rightarrow \underset{m/e\,=\,45}{CH_3OCH_2^+} + RCOOH} \quad (5.50)$$

Rate constants were calculated, and Arrhenius plots for the rate constants at several temperatures were constructed. The kinetic parameters for the formation of methoxymethyl ion from the formate and acetate esters are

Table 5.6 Kinetic parameters for formation of $CH_3OCH_2^+$
(From Weeks and Field[31], by courtesy of the American Chemical Society)

Compound	E_a/kcal mol^{-1}	A/s^{-1}	k_{300}/s^{-1}
$HCOOCH_2OCH_3$	4.8	4.5×10^9	1.4×10^6
$CH_3COOCH_2OCH_3$	4.1	6.5×10^8	7.5×10^5

given in Table 5.6. k_{300} is the rate constant at 300 K, and the rate constants for the two compounds differ by a factor of two with methoxymethyl formate having the higher value. In solution the methoxymethyl formate hydrolysis is faster by a factor of 3, so the relative magnitudes of the rate constants in the gas phase and in solution agree very well. The values for the activation energies and the frequency factors obtained for the two compounds are not susceptible to simple explanations.

5.5.6 Methylthiomethyl acetate and propionate

Investigations of these compounds have been made[32] to determine whether carboxylic acid displacement by t-$C_4H_9^+$ (Section 5.3.7.4) would occur with another kind of nucleophilic centre (sulphur) in the alkyl portion of the ester. Indeed, the reactions are found to occur readily in the i-butane chemical ionisation of the compounds, producing $m/e = 117$ ions, which comprise $(M-3)^+$ for the acetate and $(M-17)^+$ for the propionate ester.

Whilst most of the features of the i-butane chemical ionisation spectra of the sulphur compounds are similar to those of their oxygen analogues, an important difference exists, namely, methylthiomethyl ion is produced in high abundance,

$$\begin{array}{l} RCOOCH_2SCH_3 + t\text{-}C_4H_9^+ \longrightarrow RCOOCH_2SCH_3{\cdot}H^+ + C_4H_8 \\ \qquad\qquad\qquad\qquad\qquad\qquad\qquad \downarrow \\ \qquad\qquad\qquad\qquad\qquad\qquad CH_3SCH_2^+ + RCOOH \\ \qquad\qquad\qquad\qquad\qquad\qquad m/e = 61 \end{array} \qquad (5.51)$$

It will be recalled from the preceding section that, in the methoxy-substituted esters, methoxymethyl cation was produced only by the use of methane as the chemical ionisation reactant and it is suggested[32] that under identical conditions the rate constant for the formation of gaseous $CH_3SCH_2^+$ is roughly 500 times greater than that of $CH_3OCH_2^+$. While there is supporting evidence that in the gas phase $CH_3SCH_2^+$ is formed more easily than $CH_3OCH_2^+$, condensed-phase solvolysis results are quite to the contrary and it is predicted that the sulphur esters should suffer hydrolysis at a much lower rate than methoxymethyl esters. Thus a significant difference in behaviour is observed in the gas and condensed phases and further study in both is necessary.

5.5.7 Equilibrium reactions

In almost all of the compounds for which temperature studies have been made (esters of various types) ions with m/e values higher than the molecular weight of the compound have been observed and, for many of these ions, the relative intensities decreased sharply as the ionisation chamber temperature increased. Thus, for example, in benzyl acetate an $m/e = 301$ peak is observed and this must comprise the $(2M+1)^+$ ion, i.e. the protonated dimer of benzyl acetate[47]. At a source temperature of 37 °C the relative intensity of this ion is 0.175 and it becomes negligibly small at 196 °C. $(2M+1)^+$ ions have also been observed with alcohols[25] and for water[54, 55].

The compounds on which the studies under consideration were made were invariably of a high degree of purity, and it is not probable that the ions in question were formed from contaminants. For some ions such as $(2M+1)^+$ ions, one could in principle argue that the observed ions are formed by protonating a neutral dimer molecule present in some equilibrium amount in the vapour of the substance under investigation. However, such dimers are not known to exist in significant amounts in vapours of esters, but beyond that, ions are observed, for example $(M+39)^+$ for which this argument cannot be applied. One must conclude that the ions are being produced by ion–molecule reactions from lighter ions in the spectrum and if this be the case, by use of conventional concepts of ion–molecule reactions, one is hard pressed to understand the observed decreases in intensities of higher ions as the temperature increases. One must postulate that the rate constants for these reactions show a sharp decrease with an increase in temperature; such behaviour has never been observed for ion–molecule reaction rate constants in other contexts; and the idea is really not very tenable.

Consequently, it was postulated[47] that the higher ions are formed by equilibrium reactions or, at least, reversible reactions, and the observed effect of temperature is simply a manifestation of the effect of temperature on the reversible reaction. Thus the ions $m/e = 169$, 189 and 301 observed in the t-butyl spectrum of benzyl acetate at 37 °C are considered to be association complex ions formed by the reactions:

$$C_6H_5CH_2OAcH^+ + H_2O \rightleftarrows C_6H_5CH_2OAc{\cdot}H_3O^+ \quad (5.52)$$

$$C_6H_5CH_2OAc + C_3H_3{}^+ \rightleftarrows C_6H_5CH_2OAc{\cdot}C_3H_3^+ \quad (5.53)$$

$$C_6H_5CH_2OAcH^+ + C_6H_5CH_2OAcH^+ \rightleftarrows (C_6H_5CH_2OAc)_2H^+ \quad (5.54)$$

In reaction (5.52) the $C_3H_3^+$ ion is present in the i-butane plasma at 0.70 torr to the extent of *c.* 3 %, and in reactions (5.53) and (5.54) the protonated benzyl actate ions are produced by reactions of t-$C_4H_9^+$ with benzyl acetate according to reaction (5.32). The H_2O in equation (5.52) is residual water in the mass spectrometer.

For a reaction such as (5.54) the equilibrium constant is

$$K_P = \frac{I_{(BzAc)_2H^+}}{(I_{(BzAcH^+)})(P_{BzAc})} \quad (5.55)$$

where the I values are the ion intensities and BzAc is an abbreviation for benzyl acetate. P_{BzAc} is the partial pressure of benzyl acetate in atmospheres. Analogous expressions can be written for the other equilibria. A criterion of the attainment of equilibrium in a reversible system is the constancy of the equilibrium constant as a function of the concentrations of the components of the reaction and it has been shown that the equilibrium constants for reaction (5.54) are sensibly independent of the pressure of benzyl acetate. Thus on the basis of this criterion equilibrium has been achieved. The equilibrium constants obtained from equation (5.55) vary with temperature according to the van't Hoff relationship and this can be taken as an indication, although certainly not a proof, that a reversible reaction is occurring and possibly that equilibrium is attained.

A kinetic analysis has been made[54] of the reactions involved in reversible reactions occurring under chemical ionisation conditions. Two possible systems are considered. In the first a protonated species is produced by chemical ionisation transfer of a proton from a reactant ion to an additive molecule, which is followed by a reversible association of the protonated additive molecule with a second additive molecule. The reactions involved are

$$RH^+ + A \longrightarrow AH^+ + R \tag{5.56}$$

$$AH^+ + A \rightleftharpoons A_2H^+ \tag{5.57}$$

where RH^+ is the chemical ionisation reactant ion and A is the additive. In the second system an additional reaction is considered, namely,

$$A_2H^+ + A \longrightarrow \text{Products} \tag{5.58}$$

where the products are not specified except that they are not considered to be capable of reverting back to $AH^+ + A$. Reaction (5.58) is used to simulate the possibility that a whole series of higher equilibria may exist between protonated polymeric species of A, as is indeed the case in water, for example. Making certain reasonable assumptions about the rate constants, concentrations and residence times, the conclusion is reached[54] that under appropriate circumstances equilibrium in reaction (5.57) can be achieved in both systems considered, i.e. reactions (5.56) and (5.57) taken alone or taken in conjunction with (5.58). The most important condition for the attainment of equilibrium seems to be that the experimental conditions be adjusted to hold the concentration of A_2H^+ formed in equation (5.57) to a relatively small fraction (perhaps 10% or less) of the concentration of AH^+.

In summary, it seems to be established that reversible reactions occur in the ionisation chamber of a mass spectrometer under chemical ionisation conditions and, in addition, evidence exists which suggests that equilibrium is attained. To the extent that this is in fact the case, from the value of the equilibrium constant and its temperature variation one calculates free energies, enthalpies, and entropies for the reactions. As an illustration of the results obtained we give in Table 5.7 thermodynamic values for ionic

Table 5.7 Experimental thermodynamic values for ionic equilibria in benzyl acetate and t-amyl acetate

*Reaction**	ΔH/ kcal mol^{-1}	$-\Delta G^\circ_{300}$/ kcal mol^{-1}	$-\Delta S$/ cal deg^{-1} mol^{-1}	$K_{p(300)}$†
$HBzAc^+ + BzAc \rightleftharpoons H(BzAc)_2^+$	5.4 ± 0.1	9.7	-14 ± 3	1.12×10^7
$HAmAc^+ + AmAc \rightleftharpoons H(AmAc)_2^+$	9.1 ± 1.5	9.8	-2 ± 5	1.35×10^7
$HBzAc^+ + H_2O \rightleftharpoons HBzAc{\cdot}H_2O^+$	13.7	10	12	3×10^7
$HAmAc^+ + H_2O \rightleftharpoons HAmAc{\cdot}H_2O^+$	13.8	10	12	3×10^7
$C_3H_3^+ + BzAc \rightleftharpoons BzAc \cdot C_3H_3^+$	10	~10	~−7	~9×10^6

*BzAc = benzyl acetate, AmAc = t-amyl acetate.
†Standard state = 1 atm.

equilibria in benzyl and t-amyl acetates. The remarkable aspect of these results is the positive ΔS values obtained for the formation of the protonated dimers of benzyl and t-amyl acetates and the $BzAc{\cdot}C_3H_3^+$ association complex. Reactions involving the association of two entities result in the loss of three translational degrees of freedom, which corresponds to an entropy change of −(30–40) e.u. Thus the positive entropies obtained with benzyl and t-amyl acetates and with substituted benzyl acetates[52] are completely contrary to expectation. The origin and significance of these results are not understood. It is felt that they are not the consequence of some kind of random error and it is stated[52] that if a systematic error is operating it is indeed well hidden. Clearly a very intensive investigation of these equilibrium phenomena is needed and judgment about the significance of the thermodynamic results obtained must be suspended until such investigations are completed.

The ionic equilibria in the water system have been investigated under chemical ionisation conditions by use of methane[55] and propane[54] as reactants. Protonated water forms a series of consecutive equilibria which may be written as

$$(H_2O)_nH^+ + H_2O \rightleftharpoons (H_2O)_{n+1}H^+ \qquad (5.59)$$

where under chemical ionisation conditions n has been observed to range from 1 to 5. Equilibrium constants, enthalpies and entropies for reactions (5.59) have been obtained, and the chemical ionisation values obtained for the equilibria where $n = 3$–5 are in good agreement with those obtained earlier under non-chemical ionisation conditions[56], but the values obtained for $n = 2$ are in poor agreement with the earlier values, and those for $n = 1$ are in very strong disagreement. The cause of the lack of agreement in the last two cases is not known.

References

1. Munson, M. S. B. and Field, F. H. (1966). *J. Amer. Chem. Soc.*, **88**, 2621
2. Munson, M. S. B. and Field, F. H., Process for Chemical Ionisation for Intended Use in Mass Spectrometry and the Like, U.S. Patent No. 3,555,272, Jan. 12, 1971
3. Field, F. H. (1968). *Accounts Chem. Res.*, **1**, 42
4. Field, F. H. (1968). *Advances in Mass Spectrometry*, Vol. 4, 645 (London: Institute of Petroleum)
5. Field, F. H., Munson, M. S. B. and Becker, D. A. (1966). *Advances in Chemistry Series, No. 58*, 167 (Washington, D.C.: American Chemical Society)
6. Field, F. H. and Munson, M. S. B. (1967). *J. Amer. Chem. Soc.*, **89**, 4272
7. Michnowicz, J. (1971). *Thesis* (University of Delaware, Newark, Del.)
8. See Budzikiewicz, H., Djerassi, C. and Williams, D. H. (1967). *Mass Spectrometry of Organic Compounds*, 19 (San Francisco: Holden-Day)
9. Futrell, J. H. and Miller, C. D. (1966). *Rev. Sci. Instr.*, **37**, 1521
10. Futrell, J. H., Abramson, F. P., Bhattacharya, A. K. and Tiernan, T. O. (1970). *J. Chem. Phys.*, **52**, 3655
11. Baldeschwieler, J. D. (1968). *Science*, **159**, 263
12. Bursey, M. M., Elwood, T. A., Hoffman, M. K., Lehman, T. A. and Tesarek, J. M. (1970). *Anal. Chem.*, **42**, 1370
13. Field, F. H. (1961). *J. Amer. Chem. Soc.*, **83**, 1523
14. Futrell, J. H. and Wojcik, L. H. (1971). *Rev. Sci. Instr.*, **42**, 244
15. Michnowicz, J. and Munson, B. (1970). *Org. Mass. Spectrosc.*, **4**, 481
16. Arsenault, G. P., Althaus, J. R. and Divekar, P. V. (1969). *Chem. Commun.*, 1414
17. Dzidic, I., Desiderio, D. M., Wilson, M. S., Crain, P. F. and McCloskey, J. A., private communication
18. Beggs, D., Fales, H. M., Milne, G. W. A. and Vestal, M. L., *Rev. Sci. Instr.*, in the press
19. Field, F. H. (1968). Paper presented at 155th Meeting of American Chemical Society, San Francisco
20. Whitney, T. A., Klemann, L. P. and Field, F. H. (1971). *Anal. Chem.*, **43**, 1048
21. Lindholm, E. (1966). *Advances in Chemistry Series, No. 58*, 4 (Washington, D.C.: American Chemical Society)
22. Franklin, J. L., Dillard, J. G., Rosenstock, H. M., Herron, J. T., Draxl, K. and Field, F. H. (1969). *Ionisation Potentials, Appearance Potentials and Heats of Formation of Gaseous Positive Ions* (Washington, D.C.: NSRDS-NBS26)

23. Fales, H. M., Lloyd, H. A. and Milne, G. W. A. (1970). *J. Amer. Chem. Soc.*, **92,** 1590
24. Haney, M. A. and Franklin, J. L. (1969). *J. Phys. Chem.*, **73,** 2857
25. Field, F. H. (1970). *J. Amer. Chem. Soc.*, **92,** 2672
26. Munson, M. S. B. and Field, F. H. (1966). *J. Amer. Chem. Soc.*, **88,** 4337
27. Munson, M. S. B. and Field, F. H. (1967). *J. Amer. Chem. Soc.*, **89,** 1047
28. Kiryushkin, A. A., Fales, H. M., Axenrod, T., Gilbert, E. J. and Milne, G. W. A. (1971). *Org. Mass Spectrosc.*, **5,** 19
29. Ziffer, H., Fales, H. M., Milne, G. W. A. and Field, F. H. (1970). *J. Amer. Chem. Soc.*, **92,** 1597
30. Milne, G. W. A., Fales, H. M. and Axenrod, T. (1971). *Anal. Chem.*, **43,** 1815
31. Weeks, D. P. and Field, F. H. (1970). *J. Amer. Chem. Soc.*, **92,** 1600
32. Field, F. H. and Weeks, D. P. (1970). *J. Amer. Chem. Soc.*, **92,** 6521
33. Milne, G. W. A., Axenrod, T. and Fales, H. M. (1970). *J. Amer. Chem. Soc.*, **92,** 5170
34. Fales, H. M., Milne, G. W. A. and Nicholson, R. S. (1971). *Anal. Chem.* **43,** 1785
35. Field, F. H. (1968). *J. Amer. Chem. Soc.*, **90,** 5649
36. Field, F. H. (1967). *J. Amer. Chem. Soc.*, **89,** 5328
37. Field, F. H., Hamlet, Peter and Libby, W. F. (1967). *J. Amer. Chem. Soc.*, **89,** 6035
38. Bursey, M. M. and McLafferty, F. W. (1967). *J. Amer. Chem. Soc.*, **89,** 1
39. Fales, H. M., Nagai, Y., Milne, G. W. A., Brewer, H. B., Jr., Bronzert, T. J. and Pisano, J. J. (1970). *Anal. Biochem.*, **43,** 288
40. For a summary of this work (utilising exclusively electron ionisation) see Das, B. C. and Lederer, E. (1970). *Topics in Organic Mass Spectrometry*, 286 (New York: Wiley-Interscience)
41. Biemann, K. (1962). *Mass Spectrometry*, 284 (New York: McGraw Hill)
42. Gray, W. R., Wojcik, L. H. and Futrell, J. H. (1970). *Biochem. and Biophy. Res. Commun.*, **41**, 111
43. Fales, H. M., Milne, G. W. A. and Axenrod, T. (1970). *Anal. Chem.*, **42,** 1432
44. Wilson, M. S., Dzidic, I. and McCloskey, J. A., private communication
45. Fales, H. M., Milne, G. W. A. and Vestal, M. L. (1969). *J. Amer. Chem. Soc.*, **91,** 3682
46. Porter, R. F. and Solomon, J. J. (1971). *J. Amer. Chem. Soc.*, **93,** 56
47. Field, F. H. (1969). *J. Amer. Chem. Soc.*, **91,** 2827
48. McDaniel, E. W. (1964). *Collision Phenomena in Ionised Gases* (New York: John Wiley and Sons, Inc.)
49. Rosenstock, H. M., Wallenstein, M. B., Wahrhaftig, A. L. and Eyring, H. (1952). *Proc. Nat. Acad. Sci.*, **38,** 667
50. Rosenstock, H. M. and Krauss, M. (1963). *Mass Spectrometry of Organic Ions*, 1 (New York: Academic Press)
51. Magee, J. L. (1952). *Proc. Nat. Acad. Sci.*, **38,** 764
52. Field, F. H. (1969). *J. Amer. Chem. Soc.*, **91,** 6334
53. Salomaa, P. (1957). *Acta Chem. Scand.*, **11,** 132; (1960). *ibid.*, **14,** 586
54. Beggs, D. P. and Field, F. H. (1971). *J. Amer. Chem. Soc.*, **93,** 1576
55. Beggs, D. P. and Field, F. H. (1971). *J. Amer. Chem. Soc.*, **93,** 1567
56. Kebarle, P., Searles, S. K., Zolla, A., Scarborough, J. and Arshadi, M. (1967). *J. Amer. Chem. Soc.*, **89,** 6393

6
Ion Cyclotron Resonance Mass Spectrometry

C. J. DREWERY, GILL C. GOODE and K. R. JENNINGS
University of Sheffield

6.1 PRINCIPLES OF OPERATION AND INSTRUMENTATION

6.1.1 Introduction

Since it was first described by Wobschall and his co-workers in 1963[1], ion cyclotron resonance mass spectrometry (ICR) has been developed and applied to problems of physical and chemical interest in many laboratories[2, 3]. This review aims to give examples of the many different types of application which have been reported in the literature up to mid-1971.

6.1.2 Basic principles

The central feature of ICR is the absorption of energy by an ion from a linearly polarised radio-frequency field which is applied perpendicular to a uniform magnetic field. Energy absorption occurs when the angular cyclotron frequency of the ion, ω_c, is equal to that of the r.f. field, and for an ion of mass to charge ratio m/e in a magnetic field of strength B, $\omega_c = eB/m$. For Ar^+ in a field of 8000 G, $\nu_c = \omega_c/2\pi = 307$ kHz. The absorption of energy is independent of the initial velocity of the ion and accelerates the ion so that its cyclotron radius ($r_c = v/\omega_c$) increases; absorption continues until the ion collides with a molecule, a wall or drifts out of the r.f. field. Mass analysis is performed by sweeping ω or, more usually, B, so that ions of different m/e values fulfill the resonance condition and the relative number densities of ions can be obtained from the magnitude of the power absorption at $\omega = \omega_c$. High sensitivity is achieved (1–10 ions cm^{-3}) and permits the use of low electron-trap currents (<0.1 μA) and the observation of collision phenomena at pressures of 10^{-4}–10^{-6} torr, owing to the long path-length of the ions. The application of two or more r.f. fields simultaneously to different regions of the instrument, together with the use of appropriate modulation techniques, greatly extends the capabilities of the technique.

6.1.3 Instrumentation

Two general types of apparatus have been used (i) that developed by Wobschall *et al.*[1,4], and (ii) designs based on that of the commercial instrument produced by Varian Associates[5], or developed from it. Since the latter have found wider application, they will be discussed more fully.

6.1.3.1 Wobschall's instrument

Wobschall described his first apparatus in 1963[1], but in 1965[4] an improved version in which a solenoid replaced an electromagnet was developed and this is illustrated in Figure 6.1. The low-energy ions formed in the source

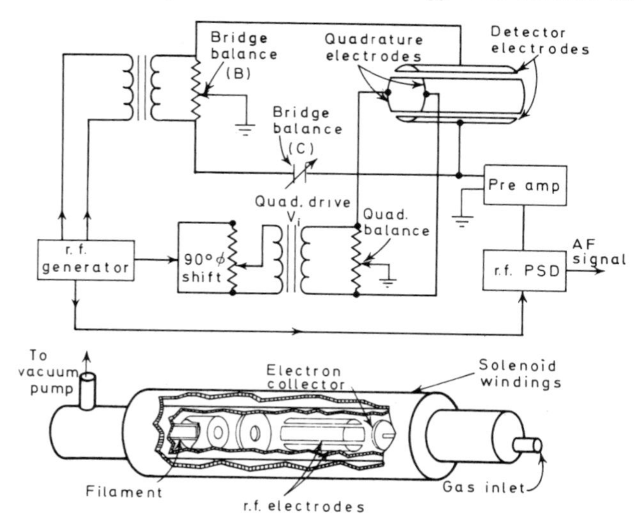

Figure 6.1 Ion cyclotron resonance apparatus. Electrons from a retarding potential difference type electron source travel along the magnetic field direction into the measurement chamber where attachment (or ionisation) occurs. Power absorption at the detector electrodes is sensed by an unbalance of the r.f. bridge and is displayed on a recorder as the derivative of the absorption line
(From Wobschall, D. *et al.*[21], by courtesy of the American Institute of Physics)

drift into the analyser region where they may collide with neutral species. The analyser consisted of a quartz tube on which split cylindrical r.f. electrodes were deposited by evaporating gold; the electrodes formed one arm of a r.f. bridge and any imbalance due to power absorption by resonant ions was detected using a r.f. phase-sensitive detector. Mass analysis was obtained by sweeping the oscillator frequency since diffusion of ions into the analyser was sensitive to the magnetic field strength. The influence of instrumental design on such parameters as line-width, resolution and sensitivity were discussed together with an account of space-charge effects and collision-broadening effects. Theoretical expressions for absorption line-shapes were obtained and compared with experimental curves and from the relationship between line-widths and collision frequencies, collision cross-sections for momentum transfer could be calculated.

6.1.3.2 *The Varian instrument*

In the basic Varian instrument[5], the drift cell is as shown in Figure 6.2 and is placed between the poles of an electromagnet. Ions are produced in the source region and are constrained to move in a circular path in a plane perpendicular to the magnetic field and movement along the magnetic field

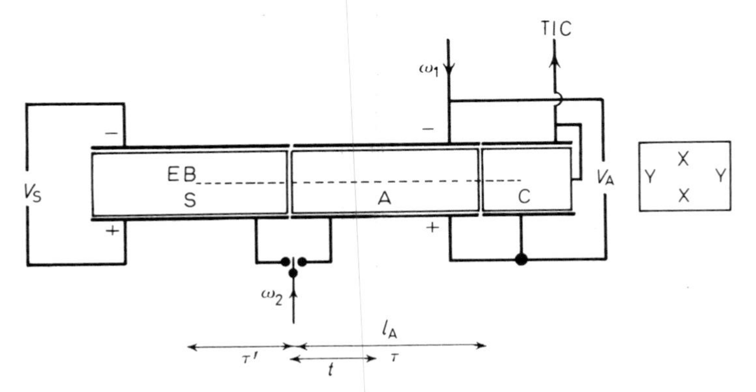

Figure 6.2 Schematic view of ion cyclotron resonance cell showing source S, analyser A and collector C regions. Source and analyser drift voltages V_S and V_A are applied to plates XX. The trapping potentials are applied to plates YY in the source and analyser regions, but not in the collector region. The observing oscillator of angular frequency ω_1 is applied to the analyser and the irradiating oscillator of angular frequency ω_2 is applied to the source or analyser regions. The source and analyser residence times are τ' and τ respectively and t is the time measured from $t = 0$ at the beginning of the analyser region (see Section 6.2.3). The length of the analyser, l_A is typically 6.35 cm

lines of force is limited by applying trapping potentials of a few tenths of a volt to the trapping plates. Small potentials applied to the drift plates, usually split about earth potential, provide drift fields E which are typically 0.2–0.6 V cm^{-1}. Under the influence of the crossed electric and magnetic fields, ions drift along the cell with a drift velocity $v_D = E/B$, independent of the mass of the ion. Ions are detected by power absorption from a marginal oscillator applied to the analyser region and a mass spectrum is obtained by scanning the magnetic field at fixed frequency. The total ion current is measured by an electrometer connected to the collector region trapping plates, on which the trapping potential is removed to allow ions to reach the plates; typical total ion currents are in the range 10^{-11}–10^{-12} A. Under normal operating conditions, ion path lengths of 10–100 m can be achieved. The separation of the source and analyser regions minimises trapping effects and space-charge effects arising from the electron beam.

In several laboratories an additional region, known as the reaction region, has been introduced between the source and analyser regions, principally for use in double-resonance experiments. These cells have the advantage that the time over which ions are observed in the analyser region is a smaller fraction of the total reaction time than in the three-section cell. In order to try to improve on the rather poorly defined electric fields found in standard cells of rectangular cross-section, cells utilising shaped drift plates[6] have been constructed and, for certain types of experiments, they appear to offer some advantages.

In most systems, the signal-to-noise ratio is improved by using a phase-sensitive detector and the number of ions which absorb power in the analyser region is modulated at a frequency referenced to the phase-sensitive detector. Several methods have been employed, but the most common are modulation of the electron beam, electron energy or the magnetic field. An alternative detection system employs a higher r.f. field strength so that resonant ions are removed completely from the cell, and a spectrum is obtained by recording the fall in total ion current (TIC) as ions of different m/e are removed from the cell[7]. For certain applications, this has the advantage that a spectrum may be obtained at a fixed magnetic field strength by sweeping the frequency of the irradiating oscillator, but in general the resolution and sensitivity obtainable are less than that given by power absorption measurements.

6.1.3.3 Double-resonance techniques

Several double-resonance techniques have been described. If $P^+ \rightarrow S^+$ in an ion–molecule reaction, it is often found that the rate constant for the reaction is a function of the translational energy of P^+. If S^+ is observed at a fixed magnetic field strength and a second oscillator, tuned to the resonant frequency P^+, is applied to the source or reaction region of the cell, the yield of S^+ will change[8]. At low irradiating field strengths, positive signals (S^+ increases) are normally observed for charge-transfer reactions, but negative signals (S^+ decreases) are observed for almost all exothermic ion–molecule reactions. Care must be taken to avoid sweeping out primary ions if the signs of double-resonance signals are to be meaningful. By sweeping the frequency of the second oscillator, all precursors of S^+ may be identified from the relationship $\omega_p/\omega_s = m_s/m_p$ at fixed magnetic field strength. The relative importance of each precursor ion in forming S^+ may be determined by sweeping the irradiating oscillator at higher r.f. field strengths at which each precursor ion is swept out of the cell in turn. Standard modulation methods may be used, or the amplitude of the irradiating oscillator may be modulated so as to produce a pulsed double-resonance spectrum. A variation of this technique has been used in the study of translational energy effects[6]. A bunch of ions is produced by a pulsed electron beam; the ions are immediately subjected to an r.f. pulse at their resonant frequency causing them to attain a terminal translational energy determined by the r.f. field strength and the duration of the pulse. Modulation of the magnetic field may be used to give a standard spectrum or modulation of the r.f. field strength may be used to display the effects of translational energy directly.

6.1.3.4 Ion ejection and residence times

The presence of the trapping plates imparts an oscillatory motion to the ions in the direction of the magnetic field[9], the frequency of which is given by $\omega_T = (4eV_T/md^2)^{\frac{1}{2}}$ where V_T is the trapping potential and d is the separation of the trapping plates. Application of an oscillating field of frequency ω_T (typically 10–50 kHz) to the plates increases the amplitude of this motion, thereby ejecting the ions of a given m/e. The resolution obtainable is about five, so that the technique is best suited to ejecting light ions or ions widely separated in mass. Normal spectra may be obtained in the presence and absence of the ejected ion or a pulsed ion ejection spectrum may be obtained directly by modulating the ejection field at the reference frequency of the phase-sensitive detector. Pulsing of the trapping plates can also be used to determine the residence times of ions in different regions of the cell[10]. A bunch of ions is formed in the source region by pulsing the electron beam and is allowed to drift along the cell. If the sign of the trapping voltage in the analyser region is momentarily reversed by applying a short pulse, any ions in the analyser region will be removed, and the TIC will fall. By varying the delay time between the electron beam pulse and the ejection pulse, the residence time in the analyser region can be determined as the difference between the shortest and longest delay times which result in ion ejection. Such measurements show that residence times calculated from the drift velocity, $v_D = E/B$, are only approximately correct owing to the imperfectly defined fields within the cell.

6.1.3.5 The trapped ion cell

A somewhat different type of cell is the trapped ion cell[11] which consists of a single region complete with end plates to form an enclosed box. For the study of positive ions, drift plates, end plates and trapping plates are held at a small positive potential ($\sim +1$V) and a bunch of ions is formed using a pulse of electrons. If required, the translational energy of the ions is increased by immediately applying a r.f. pulse, after which the ions are allowed to react. The marginal oscillator is set so that it is very close to the resonant frequency of the ion of interest and, after the desired reaction time has elapsed, the ions are detected by pulsing either the magnetic field or the trapping voltage so that the change in conditions is sufficient to bring the ions into the resonance condition. The latter method utilises the fact that the effective frequency seen by the ion is given by $\omega_{eff} = \omega_c(1-\omega^2{}_T/2\omega_c)$ and $\omega_T \propto V_T^{\frac{1}{2}}$. The advantages of this cell are that residence times are known accurately, and ions may be retained for several hundred ms, allowing the study of slow reactions and ionic equilibria. After observation, ions are removed from the cell by reversing the polarity of the trapping potential with a quenching pulse; the sequence is then repeated.

6.1.3.6 *Transient ion cyclotron resonance*

An alternative method of detection using the standard three-section cell is the transient or heterodyne technique[12]. A bunch of ions produced by a pulse of electrons drifts into the analyser region where the ions are detected by a rapid-response marginal oscillator tuned to a frequency slightly different from the resonance frequency, ω_0. The resulting signal is an interference pattern of frequency $|\omega_0 - \omega|$, damped by ion–neutral molecule collisions and analysis of this can in principle lead to rate data for both reactive and non-reactive collisions free from uncertainties of residence times and fringing field effects.

6.2 THEORY OF ION CYCLOTRON RESONANCE SPECTRA

6.2.1 Power absorption and interpretation of ion cyclotron resonance spectra

The motion of a chemically unreactive ion of mass m and charge e undergoing cyclotron resonance can be described by equation (6.1)[13]

$$\frac{\mathrm{d}\boldsymbol{v}}{\mathrm{d}t} = \boldsymbol{\varepsilon}(t)\frac{e}{m} + \frac{e\boldsymbol{v}\times\boldsymbol{B}}{m} - \xi\boldsymbol{v} \tag{6.1}$$

where $\boldsymbol{\varepsilon}(t)$ is the time-dependent electric field in V m^{-1}, $\boldsymbol{B}$ is the flux density of the magnetic field in tesla, ξ is the collision frequency for momentum transfer in s^{-1} and $\boldsymbol{v}$ is the average ion velocity in m s^{-1}. The solution of this equation leads to an expression for the power absorbed by an average ion at resonance.

Two sets of limiting condition may be distinguished: (a) the high-pressure limit at which power absorption is collision-limited, and (b) the low-pressure limit at which power absorption is limited by the time which the ion spends in the r.f. field. The high-pressure limit was treated by Beauchamp[14], and Buttrill[15] has considered the low-pressure limit, and this treatment was extended by Buttrill and Marshall[16], and by Marshall[17]. More generalised treatments incorporating both limiting sets of conditions have been given by Dunbar[12] and by Comisarow[18] and the following discussion is largely based on the last treatment.

Equation (6.1) can be written in component form, and the z-component (in the direction of the magnetic field) may be omitted since it does not enter into power absorption considerations:

$$\frac{\mathrm{d}v_x}{\mathrm{d}t} = \frac{e\varepsilon_x(t)}{m} + \frac{ev_yB}{m} - \xi v_x \tag{6.2a}$$

$$\frac{\mathrm{d}v_y}{\mathrm{d}t} = \frac{e\varepsilon_y(t)}{m} + \frac{ev_xB}{m} - \xi v_y \tag{6.2b}$$

The linearly-polarised field $\boldsymbol{\varepsilon}$, of amplitude ε_0, may be split into the sum of two counter-rotating polarised fields such that $\boldsymbol{\varepsilon} = \boldsymbol{\varepsilon}^+ + \boldsymbol{\varepsilon}^- = \varepsilon_0 \cos \omega t_{\boldsymbol{j}}$ where only $\boldsymbol{\varepsilon}^+$ $(= \frac{1}{2}\varepsilon_0 \cos \omega t_{\boldsymbol{j}} + \frac{1}{2}\varepsilon_0 \sin \omega t_{\boldsymbol{i}})$ is effective in influencing the motion of a positive ion. Equations (6.2a) and (6.2b) can therefore be re-written

$$\dot{v}_x = (\tfrac{1}{2}\varepsilon_0 \sin \omega t)\frac{e}{m} + \frac{eB}{m} v_y - \xi v_x \tag{6.3a}$$

$$\dot{v}_y = (\tfrac{1}{2}\varepsilon_0 \cos \omega t)\frac{e}{m} - \frac{eB}{m} v_x - \xi v_y \tag{6.3b}$$

For an ion of resonant frequency $\omega_c = eB/m$, with an initial velocity v_0 and which leads the rotating electric field by a phase angle γ, the complete solution of equations (6.3a) and (6.3b) is

$$v_x = C \cos \omega t + D \sin \omega t + (F \cos \omega_c t + G \sin \omega_c t)e^{-\xi t} \tag{6.4a}$$

$$v_y = D \cos \omega t - C \sin \omega t + (G \cos \omega_c t - F \sin \omega_c t)e^{-\xi t} \tag{6.4b}$$

where

$$C = \varepsilon_0 e[(\xi^2 + \omega_c^2 - \omega^2)(\omega_c + \omega) - 2\xi^2 \omega]/2m[(\xi^2 + \omega_c^2 - \omega^2)^2 + 4\xi^2\omega^2]$$

$$D = \varepsilon_0 e\xi[2\omega(\omega_c + \omega) + \xi^2 + \omega_c^2 - \omega^2]/2m[(\xi^2 + \omega_c^2 - \omega^2)^2 + 4\xi^2\omega^2]$$

$$F = v_0 \sin \gamma - C$$

$$G = v_0 \cos \gamma - D$$

and t is the time during which the ion has been absorbing energy.

The power absorbed by the ion, $A(t)$, from the alternating electric field, $\boldsymbol{\varepsilon}$, is given by the product of the electric force on the ion, $\boldsymbol{F}$, and the velocity of the ion, $\boldsymbol{v}$, so that

$$A(t) = \boldsymbol{F} \cdot \boldsymbol{v} = e\boldsymbol{\varepsilon}^+(v_x + v_y) \tag{6.5}$$

Substituting for $\boldsymbol{\varepsilon}^+$, v_x and v_y gives an expression for the instantaneous power absorption for an ion in an alternating field of amplitude ε_0 as a function of time, and where ω_c is the resonant frequency:

$$A(t) = \frac{\varepsilon_0 e}{2}[-F \sin (\omega_c - \omega)t + G \cos (\omega_c - \omega)t]e^{-\xi t} + \xi \tag{6.6}$$

A simplified expression can be obtained by averaging over the phase angle, γ, since all values are equally probable, and by making the approximations $\xi \ll \omega$ and $\omega_c + \omega \simeq 2\omega$, this is of the form

$$A(t) = \frac{\varepsilon_0^2 e^2}{4m[\xi^2 + (\omega_c - \omega)^2]}[\{(\omega_c - \omega) \sin (\omega_c - \omega)t - \xi \cos (\omega_c - \omega)t\}e^{-\xi t} + \xi] \tag{6.7}$$

This gives the instantaneous *average* power absorption *per ion* of an ensemble of ions.

6.2.2 Non-reactive ions

If the rate of production of ions per second is P_0, then at any point in the analyser there are $P_0\mathrm{d}t$ ions absorbing energy at a rate $A(t)$ and the total power absorption is given by

$$A = \int_{\tau'=0}^{\tau} P_0 A(t)\,\mathrm{d}t \tag{6.8}$$

where τ' is the time at which the ions enter the analyser region (set equal to zero for convenience) and τ is the time taken to drift through the analyser region. Substituting for $A(t)$ from equation (6.7) leads to the general expression for total power absorption

$$A = Z\int_0^{\tau} (\omega_c-\omega)\sin(\omega_c-\omega)t\,\mathrm{e}^{-\xi t}\,\mathrm{d}t - Z\int_0^{\tau} \xi\cos(\omega_c-\omega)t\,\mathrm{e}^{-\xi t}\,\mathrm{d}t + Z\int_0^{\tau}\xi\,\mathrm{d}t \tag{6.9}$$

$$= \frac{Z\mathrm{e}^{-\xi\tau}[-2\xi(\omega_c-\omega)\sin(\omega_c-\omega)\tau + \{(\xi^2-(\omega_c-\omega)^2\}\cos(\omega_c-\omega)\tau]}{[\xi^2+(\omega_c-\omega)^2]}$$

$$-Z[\{\xi^2-(\omega_c-\omega)^2\}/\{\xi^2+(\omega_c-\omega)^2\}] + Z\xi\tau \tag{6.10}$$

where $Z = P_0\varepsilon_0^2e^2/4m\{\xi^2+(\omega_c-\omega)^2\}$. In the low-pressure limit, $\xi\rightarrow 0$ so that the expression reduces to that obtained by Buttrill[15],

$$A = Z\{1-\cos(\omega_c-\omega)\tau\} = P_0\varepsilon_0^2e^2\tau^2/8m \text{ at resonance} \tag{6.10a}$$

Similarly, in the high-pressure limit, $(\xi/t)\rightarrow\infty$ and Beauchamp's expression results[14],

$$A = Z\tau\xi = P_0\varepsilon_0^2e^2\tau/4m\xi \text{ at resonance} \tag{6.10b}$$

Spectra are normally obtained by scanning the magnetic field at fixed frequency; since $\omega_c = eB/m$ and $\tau = l_A B/E$ (where l_A is the length of the analyser region and E is the drift field), it follows that $\tau\propto m$. In an ICR spectrum, therefore, the peak height $A\propto P_0 m$ in the low-pressure limit and $\propto P_0/\xi$ in the high-pressure limit, so that at the two limits, $P_0\propto A/m$ and A/ξ respectively.

6.2.3 Reactive ions

If the primary ions P^+ react with a neutral species M to form a secondary ion S^+, appropriate time-dependent expressions for $P(t)$ and $S(t)$ must be used when evaluating the expression for the total power absorption. A simple treatment suitable for the determination of relative rate constants has been given by Goode *et al.*[7] and is valid under the following conditions (i) low ion densities (ii) total ion current independent of magnetic field strength (iii) limiting low-pressure conditions and (iv) low percentage conversion. The low percentage conversion allows one to simplify the integrated form of the rate equation by taking the first term only of the exponential expansion, so that $P(t') = P_0(1-k[\mathrm{M}]t')$ and $S(t') = P_0k[\mathrm{M}]t'$, at a time t' after formation of primary ions in the electron beam. The power absorption of secondary ions is made up of two contributions, one arising from S^+ ions formed during the time τ' the ions take to traverse the source region and the second arising from ions formed during the time τ which the ions spend traversing the analyser region. Ions formed in the source absorb r.f. power for the full analyser residence time τ, whereas those formed at the time t after entering the analyser absorb r.f. power only for the time $(\tau - t)$. Under the limiting low-pressure conditions, the appropriate expression for power absorption by the secondary ions is

$$A_{S^+} = \frac{e^2\varepsilon^2 k[M]P_0}{4m_S}\int_0^\tau (\tau' + t)(\tau - t)dt \tag{6.11}$$

where m_S is the mass of the secondary ion. On integration, this yields

$$A_{S^+} = \frac{e^2\varepsilon^2 k[M]P_0}{4m_S}(\tau'\tau^2/2 + \tau^3/6) \tag{6.12}$$

Both τ' and τ are functions of m_S when B is scanned, so that

$$\tau' = \beta' m_S \qquad \tau = \beta m_S \tag{6.13}$$

and hence

$$A_{S^+} = e^2\varepsilon^2 k[M]P_0 m_S^2(\beta'\beta^2/2 + \beta^3/6)/4 \tag{6.14}$$

$$= K \times k m_S^2 \tag{6.15}$$

for given operating conditions, where $K = e^2\varepsilon^2[M]P_0(\beta'\beta^2/2 + \beta^3/6)/4$. Hence for competing reactions in which k_1 and k_2 are the rate constants for the formation of S_1^+ and S_2^+ from a single primary ion P^+,

$$k_1/k_2 = A_{S_1^+} m_{S_2}^2 / A_{S_2^+} m_{S_1}^2 \tag{6.16}$$

The treatment presupposes that there are no other sources of S_1^+ and S_2^+ and that these ions do not react further. Any errors arising from working at finite pressures and conversions may be minimised by extrapolating k_1/k_2 to zero pressure and conversion.

6.2.4 Absolute rate constants

If a primary ion P^+ reacts to give a single secondary ion S^+ under the conditions outlined in the previous section, the appropriate expression for power absorption by P^+ is

$$A_{P^+} = \frac{e^2\varepsilon^2 P_0}{4m_P}\int_0^\tau \{1 - k[M](\tau'_P + t)\} t\, dt$$

$$= \frac{e^2\varepsilon^2 P_0}{4m_P}[\tau_P^2/2 - k[M](\tau'_P\tau_P^2/2 + \tau_P^3/3)] \tag{6.17}$$

Substituting $\tau'_P = \beta' m_P$ and $\tau_P = \beta m_P$ gives

$$A_{P^+} = \frac{e^2\varepsilon^2}{8}\cdot\beta^2 P_0 m_P - \frac{e^2\varepsilon^2}{4}k[M]P_0(\beta'\beta^2/2 + \beta^3/3)m_P^2 \tag{6.18}$$

Combining this with equation (6.14) yields

$$A_{P^+} = A_{P^+}(0) - A_{S^+} \cdot m_P^2(3\beta' + 2\beta)/m_S^2(3\beta' + \beta) \tag{6.19}$$

where $A_{P^+}(0)$ is the power absorption due to primary ions at zero conversion and A_{S^+} is the power absorption due to S^+ ions obtained by scanning the magnetic field. Since $P_0 = 8A_{P^+}(0)/e^2\varepsilon^2\beta^2 m_P$, equation (6.14) can be re-written as

$$A_{S^+} = A_{P^+}(0)k[M]\,(\beta' + \beta/3)m_S^2/m_P \tag{6.20}$$

From equations (6.19) and (6.20) and putting $\beta' = \tau'_p/m_p$, $\beta = \tau_p/m_p$, one obtains

$$k = \frac{m_P^2 A_{S^+}}{[M]\,[A_{P^+}m_S^2(\tau'_P + \tau_P/3) + A_{S^+}m_P^2(\tau'_P + 2\tau_P/3)]} \tag{6.21}$$

so that k may be obtained by measuring A_{P^+} and A_{S^+} at fixed drift fields but at varying pressures or at constant pressure as a function of drift fields. In a similar way, if $k_1, k_2 \ldots k_j$ are rate constants for the formation of S_1^+, $S_2^+ \ldots S_j^+$ from a given primary ion P^+, the power absorption due to the primary ion is

$$A_{P^+} = A_{P^+}(0) - \frac{3\tau'_P + 2\tau_P}{3\tau'_P + \tau_P}\sum A_{S_j^+}\frac{m_P^2}{m_S^2} \tag{6.22}$$

and the expression for an individual rate constant is

$$k_j = \frac{A_{S_j^+}m_P^2}{[M]m_{S_j}^2\left[A_{P^+}(\tau'_P + \tau_P/3) + (\tau'_P + 2\tau_P/3)\sum_j A_{S_j^+}\dfrac{m_P^2}{m_{S_j}^2}\right]} \tag{6.23}$$

The individual rate constants may be obtained as described above.

The major uncertainties in obtaining absolute rate data from ion cyclotron resonance measurements are in evaluating residence times and measuring absolute pressures. Non-ideal electric fields within the cell will lead to errors in τ' and τ, but direct measurement[10] of this is possible using a pulsing technique and the difference between experimental and calculated values is usually less than 20%. Absolute pressure measurement is usually accomplished by use of an ionisation gauge which has been calibrated against a capacitance manometer.

A treatment very similar to that given above was developed by Buttrill[15] for the low-pressure limit but the full exponential expression was retained until after integration before expanding it up to the third term. A table of correction factors is given for conversions up to 50% for different values of m_S/m_P. In a further paper, Buttrill and Marshall[16] extended this treatment to cases involving primary, secondary and tertiary ions.

For the limiting high-pressure case, the appropriate expression for the rate constant was derived by Bowers *et al.*[19]:

$$k = 2m_P\xi_S A_{S^+}/[M]\,(m_P\xi_S A_{S^+} + m_S\xi_P A_{P^+})\,(\tau'_P + \tau_P) \tag{6.24}$$

where ξ_P and ξ_S are the collision frequencies for momentum transfer for the primary and secondary ions respectively. The generalised form of this equation is

$$k_j = \frac{2m_P \xi_{S_j} A_{S_j^+}}{[M]m_{S_j}\left[\xi_P A_{P^+} + \sum_j (m_P/m_{S_j})\xi_{S_j} A_{S_j^+}(\tau'_P + \tau_P)\right]} \tag{6.25}$$

and rate constants are again obtained by varying either the pressure or the drift times.

Comisarow[18] obtained a completely general expression for power absorption by primary, secondary and tertiary ions which leads to the above expressions for the limiting high- and low-pressure cases. The general equations require solving by computer, but providing values are available for all the terms in the equations quantitative rate data may be obtained from ICR spectra under all operating conditions.

For the collision-limited high-pressure case, Beauchamp and Buttrill[20] have developed an expression for the intensity of the pulsed double-resonance signal, ΔA_{S^+}, arising from a contribution to S^+ from the irradiated ion. They have interpreted this as arising from a change in the rate constant from k at thermal energies to k^* at higher translational energies and for low conversions,

$$\frac{k^* - k}{k} = \frac{\Delta A_{S^+}}{A_{S^+}}\left[\frac{\tau'_S + \tau_S}{\tau'_S - \tau_S}\right] \tag{6.26}$$

At threshold irradiating energies, therefore, the sign and magnitude of the pulsed double-resonance signal should reflect the behaviour of dk/dE_{P^+}, where dE_{P^+} is the change in translational energy of the primary ion.

6.3 PHYSICAL APPLICATIONS

6.3.1 Resolution, line-widths and ion lifetimes

6.3.1.1 Factors affecting resolving power

The resolving power of a mass spectrometer is defined as $M/\Delta M$, where ΔM is the range of mass about M which is detected simultaneously with M; in a magnetic deflection instrument, this is a function of slit widths and focusing. In an ion cyclotron resonance instrument, however, detection by power absorption makes it convenient to define the resolving power of the instrument as given by $\rho = \omega_c/\Delta\omega_{\frac{1}{2}}$, where ω_c is the cyclotron resonance frequency of the ion and $\Delta\omega_{\frac{1}{2}}$ is the half-width of the peak at half height. In general therefore

$$\rho = \omega_c/\Delta\omega_{\frac{1}{2}} = eB/m\Delta\omega_{\frac{1}{2}} \tag{6.27}$$

In the low-pressure limit, the expression for the total power absorbed may be solved for $\Delta\omega_{\frac{1}{2}}$ and leads to $\Delta\omega_{\frac{1}{2}} = 2.783/\tau_A$, and since $\tau_A = l_A B/E$,

$$\rho = el_A B^2/2.783\, mE \tag{6.28}$$

under these conditions. For low mass ions observed at high magnetic field strengths, a resolving power of several thousand may be obtained. Under normal operating conditions in which spectra of ions up to $m/e = 100$ are obtained by sweeping B, ρ varies linearly with B and is in the range 500–1500.

The above analysis is valid only for the conditions under which no event occurs which interrupts the absorption of r.f. power during the time which the ion spends traversing the analyser region. There are three general cases when this will not be so:

(a) The resolution is predicted to be independent of the r.f. field strength ε; this is valid only for low r.f. field strengths since, at higher values, ions absorb sufficient energy for them to strike the drift plates of the analyser region, thereby reducing τ_A, and hence ρ.

(b) Power absorption may be interrupted by the ion suffering a reactive or non-reactive collision so that even if τ_A is unchanged, the limiting low-pressure power absorption equation is no longer valid.

(c) In the case of negative ions, the ion may decompose by an autodetachment process, $X^- \rightarrow X + e$, and even if the ion is rapidly re-formed by electron capture, the effect is similar to that found for the ion suffering a collision and the limiting low-pressure power absorption equation is again no longer valid.

6.3.1.2 *Non-reactive collision processes*

At low pressures, instrumental parameters such as field gradients and inhomogeneity control the ultimate line-width attainable, but at pressures of $\sim 10^{-5}$ torr upwards, broadening of the lines due to the effects of collisions begins to be apparent under normal operating conditions. When the time between collisions is short compared with τ_A, the half-width of the absorption peak at half height, obtained by sweeping ω at a fixed magnetic field strength, is equal to the collision frequency, ξ, i.e.

$$\xi = \Delta\omega_{\frac{1}{2}} \tag{6.29}$$

This relationship has been used to determine the collision frequencies[1, 21–23] and hence the collision cross-sections for several positive and negative ions in their parent gases as a function of ε/P, where ε is the observing r.f. field strength and P is the pressure. For the pairs $N_2^+ \cdot /N_2$, $Ar^+ \cdot /Ar$ and $H_2^+ \cdot /H_2$, cross-sections of 185, 215 and 275 Å^2 were found for low ε/P, falling to 125, 140 and 68 Å^2 at high ε/P, respectively. This work has been extended to higher ε/P values and results are in good agreement with those obtained from d.c. mobility measurements.

A comprehensive theory of collision broadening resulting from resonant charge transfer has been formulated[14] and has been used to calculate ICR spectra as a function of both pressure and electric field strength. More recently[24], a reduced collision frequency, ξ/n, has been obtained from the slope of a plot of $\Delta\omega_{\frac{1}{2}}$ against pressure for several unreactive hydrocarbon ions, CH_5^+, $C_2H_5^+$, $C_3H_7^+$ and $C_4H_9^+$, in methane. Good agreement was obtained between the experimental values and those calculated from the expression

$$\xi/n = 2.21\ \pi e(\alpha\mu)^{\frac{1}{2}}/m \tag{6.30}$$

where m/e = mass to charge ratio of the ion, α is the polarisability of the molecule and μ is the reduced mass of the colliding pair. The expected dependence on $(\mu^{\frac{1}{2}}/m)$ was also found experimentally. At low translational energies, ξ/n is independent of ion energy and so enables one to calculate diffusion cross-sections and reduced mobilities. The conditions under which collisions lead to ion losses to the drift plates are also discussed.

The transient ICR[12] or heterodyne technique has been used in a study of collision processes occurring in the $N_2^{+\cdot}/N_2$ and $CH_4^{+\cdot}/CH_4$ systems. Six different types of collision process are distinguished, both reactive and non-reactive, and in favourable cases, overall rate constants for all reactive and for all non-reactive processes may be determined. A plot of ξ against P, after correction for the non-thermal energy of the ions, leads to $k = 8 \times 10^{-10}$ cm^3 molecule^{-1} s^{-1} for momentum relaxation in the $N_2^{+\cdot}/N_2$ system and 1.15×10^{-9} cm^3 molecule^{-1} s^{-1} in the $CH_4^{+} \cdot /CH_4$ system. Comparison with values obtained from the charge-induced dipole polarisation theory suggests a substantial non-orbiting charge-transfer rate in the nitrogen system but a negligible rate in the methane system; this is ascribed to the different geometries of the CH_4 and $CH_4^{+\cdot}$ species.

6.3.1.3 Autodetachment by negative ions

The case of autodetachment by negative ions[25] has been shown to be similar in many respects to collisions in its effect on peak shapes. If k is the rate constant for the autodetachment process, the general expression from which the half-width at half-height may be evaluated is

$$\frac{2k^2}{\Delta\omega_{\frac{1}{2}}^3 + \Delta\omega_{\frac{1}{2}}k^2} = \frac{1-(1+k\tau_A)\exp(-k\tau_A)}{\Delta\omega_{\frac{1}{2}} - (k \sin \Delta\omega_{\frac{1}{2}}\tau_A + \Delta\omega_{\frac{1}{2}} \cos \Delta\omega_{\frac{1}{2}}\tau_A)\exp(-k\tau_A)} \quad (6.31)$$

For low values of k, $\Delta\omega_{\frac{1}{2}} = 2.783/\tau_A$ at low pressure, as before, and for high values of k, $\Delta\omega_{\frac{1}{2}} = k$ (cf. equation (6.29)). Hence k may be evaluated from $\Delta\omega_{\frac{1}{2}}$ under appropriate conditions and may be used to calculate line-widths in the general case from equation (6.31). Under these conditions, the resolution obtainable at low pressures is given by equation (6.27) rather than equation (6.28).

Experimental low-pressure plots of resolution against drift fields for a given ion are compared with plots calculated by means of equation (6.31) for various assumed values of k. Good agreement between theory and experiment was obtained for infinite lifetime ions such as SF_5^- and Cl^-, and half-lives of ~ 500 μs and ~ 200 μs were obtained for SF_6^- and $C_4F_8^-$ respectively. These values are much higher than those in the literature and it is suggested that ions formed under ICR conditions are those formed by the re-capture of thermal electrons ejected by short-lived ions, whereas in other instruments stray fields may cause the electron energy to be substantially greater than thermal, thereby leading to the formation of excited ions.

6.3.2 Other physical applications of ion cyclotron resonance

6.3.2.1 Ionisation by metastable molecules

There have been three further applications of ICR mass spectrometry in fields other than the study of ion–molecule reactions. The standard cell has been modified by fitting a multichannel beam source which could produce a well-collimated high-intensity molecular beam directed along the length of the cell[26]. The beam is intersected by an electron beam in a field-free region producing a small proportion of excited molecules in metastable states. An additional inlet system was used to admit a second reactant and the products were analysed in the normal manner. Excited N_2 molecules were produced by using 24 eV electrons and were allowed to react with C_6H_6 to give $C_6H_6^{+\cdot}$. The $a^1\pi_g$ state of N_2 was identified as the species responsible for producing the ions.

6.3.2.2 Threshold excitation spectra

The inelastic scattering of electrons from neutral species has been the subject of two studies[27,28]. Since SF_6 is an efficient scavenger of thermal electrons, it may be used to capture electrons which have transferred essentially all their energy to a neutral species, and this has been used to obtain threshold excitation spectra[27] of several atoms and molecules by monitoring the SF_6^- signal or the total electron and negative ion current whilst sweeping the electron beam energy. In a similar study, the Cl^- ion from CCl_4 was used to monitor the thermal electrons[28]. Very similar spectra were obtained for the N_2 molecule in the two studies and peaks arising from the formation of several triplet states were identified, together with a peak at ~2.1 eV ascribed to the formation of transient negative ions. Spectra given by several olefins were found to contain similar peaks in the 1.5–2.0 eV region, and excitation to the lowest triplet level gave rise to a peak in the 4.4–4.9 eV region.

6.3.2.3 Photodetachment

The fitting of a quartz window on the end of the vacuum envelope allows one to pass a beam of photons along the analyser cell. This has been used[29] to determine the electron affinity of the SH^- radical by observing the threshold wavelength for the occurrence of the photodetachment reaction

$$SH^- + h\nu \rightarrow SH\cdot + e \qquad (6.32)$$

The threshold was at ~584 nm, leading to a vertical detachment energy of 2.28 ± 0.15 eV. More recent experiments[30] using a tunable laser have given values of 1.26 and 0.74 eV for the electron affinities of PH_2 and NH_2, respectively.

6.4 CHEMICAL APPLICATIONS

6.4.1 Ion–molecule studies of simple systems

6.4.1.1 The H_2, D_2 and HD systems

There have been several ICR studies of the H_2, HD and D_2 [6,22,31] systems with a view to comparing experimental and theoretical rate constants, measuring isotope effects and observing translational energy effects. For the reactions

$$H_2^{+\cdot} + H_2 \rightarrow H_3^+ + H^\cdot \qquad (6.33)$$

$$D_2^{+\cdot} + D_2 \rightarrow D_3^+ + D^\cdot \qquad (6.34)$$

Bowers *et al.*[31] reported $k_{6.33} = 2.11 \times 10^{-9}$ cm^3 molecule^{-1} s^{-1} and $k_{6.34} = 1.60 \times 10^{-9}$ cm^3 molecule^{-1} s^{-1}, i.e. $k_{6.33}/k_{6.34} = 1.33$. In the HD system,

$$HD^{+\cdot} + HD \begin{cases} \rightarrow H_2D^+ + D\cdot & (6.35a) \\ \rightarrow HD_2^+ + H\cdot & (6.35b) \end{cases}$$

they found $k_{6.35a} = 0.75 \times 10^{-9}$ cm^3 molecule^{-1} s^{-1} and $k_{6.35b} = 1.05 \times 10^{-9}$ cm^3 molecule^{-1} s^{-1}, so that $k_{6.35a}/k_{6.35b} = 1.40$. These results are in good agreement with literature values[32].

The dependence of the rate constants of the above reactions on the translational energy of the primary ion has been the subject of several investigations[6,31,33]. For reactions (6.33), (6.34) and (6.35a), a positive double-resonance signal is obtained at very low irradiating field strengths, but this becomes negative at higher field strengths. This has been interpreted as a change in sign of $\mathrm{d}k/\mathrm{d}E_{tr}$ due to the change from reaction via a complex at low energies to a stripping mechanism at higher energies[31]. An alternative explanation of the reversal in sign of the double-resonance signal has been given in terms of ion losses due to sweep-out effects[33], as indicated by a fall in the total ion current at higher irradiating field strengths, and the small negative signal reported for reaction (6.35b) was observed only when the total ion current began to fall. Using a double pulsing technique similar to that of Anders[34], Clow and Futrell[6] found that $k_{6.34}$ was almost independent of translational energy of D_2^+ over the range 0–20 eV, but more recent work[35] suggests that it passes through a maximum, as suggested by merging beam studies[36]. The interpretation of results at higher energies is difficult, however, owing to the possible interference from ion losses and the problems associated with calibrating the primary ion energy scale.

Ion–molecule reactions occurring in mixtures of argon and H_2, D_2 or HD[37], and of nitrogen with H_2, D_2 or HD[37,38] have also been examined. In the Ar/H_2 mixture, two reactions may yield ArH^+ at low pressure:

$$Ar^{+\cdot} + H_2 \rightarrow ArH^+ + H\cdot \qquad (6.36)$$

$$H_2^+ + Ar \rightarrow ArH^+ + H\cdot \qquad (6.37)$$

The ejection of H_2^+ in the source region made it possible to obtain an absolute rate constant for reaction (6.36) by varying the drift voltages. This, together with relative rate constants, led to absolute rate constants being determined for some ten reactions in systems containing argon; the experimental values obtained are between 0.19 and 0.67 times those predicted by the Gioumousis and Stevenson theory[39]. In a similar study of systems containing nitrogen[38], on the other hand, the rate constants are all much closer to the values predicted by the theory. Marked differences were found in the behaviour of the double-resonance signals at different irradiating field strengths when observed for several of the reactions in the two systems. For example, the signals for reactions (6.36) and (6.37) are negative and positive, respectively, and this is also the case for the corresponding reactions in the N_2 systems; the magnitudes of the two pairs of signals suggest that there is more non-reactive scattering in the Ar–H_2 system than in the N_2–H_2 system. In a more recent study of the Ar–D_2 system[40], a kinetic analysis of the results suggests that the reaction

$$D_3^{+\dagger} + Ar \rightarrow ArD^+ + D_2 \quad (6.38)$$

occurs only above a threshold, indicating that the proton affinity of $D_2^{\dagger}$ is greater than that of Ar. The $D_3^{+\dagger}$ species is considered to be an excited state which loses energy by successive deuteron transfer reactions, in which the D_2 molecules are formed in vibrationally excited states. Clow and Futrell[6] have studied the translational energy dependence of the two reactions

$$Ar^{+\cdot} + D_2 \longrightarrow ArD^+ + D\cdot \quad (6.39)$$

$$D_2^{+\cdot} + Ar \longrightarrow ArD^+ + D\cdot \quad (6.40)$$

in a modified ICR cell and find $k_{6.39} = 0.9 \pm 0.1 \times 10^{-9}$ cm^3 molecule^{-1} s^{-1} and $k_{6.40} = 1.6 \pm 0.2 \times 10^{-9}$ cm^3 molecule^{-1} s^{-1} respectively, independent of translational energy up to 10 eV in centre-of-mass coordinates. Both values are in good agreement with the Gioumousis and Stevenson theory[39]. The translational energy dependence of HeH^+ formation in H_2/He mixtures[6] was found to be in good agreement with previous observations[41, 42].

6.4.1.2 *Methane and related systems*

The methane system (see Chapter 5) has been studied by several groups by ICR[6, 12, 15, 34, 43], and in several cases it has been suggested that the rate constant for the reaction

$$CH_4^{+\cdot} + CH_4 \longrightarrow CH_5^+ + CH_3\cdot \quad (6.41)$$

is sufficiently well-established for it to be used to calibrate the ICR technique. Clow and Futrell[6] have studied the translational energy dependence of several reactions in the methane system, following earlier work on the system by Anders[34]. For reactions (6.41) and (6.42):

$$CH_3^+ + CH_4 \longrightarrow C_2H_5^+ + H_2 \quad (6.42)$$

they[6] obtain thermal energy rate constants of $k_{6.41} = 1.2 \pm 0.1 \times 10^{-9}$ cm^3 molecule^{-1} s^{-1} and $k_{6.42} = 1.0 \pm 0.1 \times 10^{-9}$ cm^3 molecule^{-1} s^{-1}; each value is a little below that predicted by the theory of Gioumousis and Stevenson[39]. Buttrill[15] has obtained a value of 0.95×10^{-9} cm^3 molecule^{-1} s^{-1} for $k_{6.41}$ by computer analysis of his data, together with values of 0.51×10^{-9} cm^3 molecule^{-1} s^{-1} and 0.41×10^{-9} cm^3 molecule^{-1} s^{-1} for the formation of $CH_3D_2^+$ and $CH_2D_3^+$ in the CH_2D_2 system. Inoue and Wexler[43], on the other hand, find a rather lower value of 0.31×10^{-9} cm^3 molecule^{-1} s^{-1}, analysing their results in terms of the high-pressure formalism for rate constants. These values should be compared with a value of 1.2×10^{-9} cm^3 molecule^{-1} s^{-1} determined in a medium pressure source by Harrison *et al.*[44], and suggest that a 'best' value of $1.1 \pm 0.1 \times 10^{-9}$ cm^3 molecule^{-1} s^{-1} can probably be adopted.

There have been two studies of the $CH_4 - H_2$[43, 45] system by ICR mass spectrometry, but unfortunately there is a considerable disagreement over both the observations and the interpretation of the data. Inoue and Wexler[43] observed the system at relatively high pressures and using 70 eV electrons; their observations were made on the two mixtures CH_4–D_2 and CD_4–H_2. The double-resonance technique was used to infer the occurrence of *c*. 20 reactions in the system, but the complexity of the reaction sequence renders interpretation very difficult. Bowers and Elleman[45] have also studied the CD_4–H_2 system in detail and they observed slightly fewer but somewhat different reactions from those reported by Inoue and Wexler[43]. Both investigations agree that CD_4H^+ is the major product, but whereas Bowers and Elleman find that H_3^+ is almost the sole precursor, Inoue and Wexler find that approximately half comes from $CD_4^{+\cdot}$ and a further 27.5% from CD_5^+. Bowers and Elleman can find no evidence for the formation of CD_4H^+ from CD_5^+ and consider that a deactivated form of H_3^+ gives CD_4H^+ whereas an excited form gives CD_3^+ and CD_2H^+ in small yields. Inoue and Wexler report that the only reaction of H_2^+ with CD_4 is the formation of CD_4H^+, whereas Bowers and Elleman find only the formation of $CD_4^{+\cdot}$ CD_3^+ and CD_2H^+ from these species. It is clear that further work on this mixture is desirable. The C_2H_6 system[46] was studied at electron energies of 75 eV and the double-resonance technique was used to infer the occurrence of many reactions, some of which are endothermic for thermal energy ions. In mixtures with D_2, it was found that $D_2^{+\cdot}$ does not react with C_2H_6 and that the major product arising from D_3^+ ions is $C_2H_6D^+$ which may then fragment to give $C_2H_5^+$ or $C_2H_4D^+$.

Marshall and Buttrill[16] used the CH_3F system to test a very complete analytical scheme for the determination of absolute rate constants. Three reactions were observed and their rate constants determined as follows:

$$CH_3F^{+\cdot} + CH_3F \longrightarrow CH_4F^+ + CH_2F\cdot \quad 1.36 \times 10^{-9}\ \text{cm}^3\ \text{molecule}^{-1}\ \text{s}^{-1} \quad (6.43)$$

$$\longrightarrow C_2H_4F^+ + HF + H\cdot \quad 0.96 \times 10^{-10}\ \text{cm}^3\ \text{molecule}^{-1}\ \text{s}^{-1} \quad (6.44)$$

$$CH_4F^{+\cdot} + CH_3F \longrightarrow C_2H_6F^+ + HF \quad 0.8 \times 10^{-9}\ \text{cm}^3\ \text{molecule}^{-1}\ \text{s}^{-1} \quad (6.45)$$

Two studies have been made of the charge exchange reactions in Xe–CH_4 mixtures[47,48] and the occurrence of two exothermic processes has been confirmed:

$$Xe^{+\cdot}(^2P_{\frac{1}{2}}) + CH_4 \longrightarrow Xe + CH_4^{+\cdot} \qquad \Delta H = -7\ \text{kcal mol}^{-1} \quad (6.46)$$

$$CH_4^{+\cdot} + Xe \longrightarrow CH_4 + Xe^{+\cdot}(^2P_{\frac{3}{2}}) \qquad \Delta H = -23\ \text{kcal mol}^{-1} \quad (6.47)$$

The charge exchange $CH_3^+ \longrightarrow Xe^{+\cdot}$ reported in the earlier study[47], and which is very endothermic, was later shown not to occur[48].

6.4.1.3 The ammonia and ammonia–methane systems

Recently, reactions occurring in the NH_3 system[49] have been studied as a function of the translational energies of the primary ions, and absolute rate constants for the following reactions were obtained at thermal energies:

$$NH_3^{+} + NH_3 \longrightarrow NH_4^{+} + NH_2 \qquad 1.9 \pm 0.2 \times 10^{-9}\ \text{cm}^3\ \text{molecule}^{-1}\ \text{s}^{-1} \tag{6.48}$$

$$\longrightarrow NH_3 + NH_3^{+} \qquad <0.04 \times 10^{-9}\ \text{cm}^3\ \text{molecule}^{-1}\ \text{s}^{-1} \tag{6.49}$$

$$NH_2^{+} + NH_3 \longrightarrow NH_4^{+} + NH \qquad 1.1 \pm 0.2 \times 10^{-9}\ \text{cm}^3\ \text{molecule}^{-1}\ \text{s}^{-1} \tag{6.50}$$

$$\longrightarrow NH_3^{+} + NH_2 \qquad 1.1 \pm 0.2 \times 10^{-9}\ \text{cm}^3\ \text{molecule}^{-1}\ \text{s}^{-1} \tag{6.51}$$

An increase in $k_{6.51}$ was found at higher translational energies (~20–30 eV), $k_{6.50}$ falling so that $(k_{6.50}+k_{6.51})$ rises only slightly. Reaction (6.48) is the major reaction of NH_4^{+} ions at thermal energies, but $k_{6.48}$ falls and $k_{6.49}$ rises as the translational energy increases, the sum of the two being approximately constant.

Because of its interest in connection with the chemistry of the atmosphere of the planet Jupiter, the CH_4–NH_3 [50] system has been studied in detail. In a mixture of CH_4 and ND_3, ND_3H^{+} was shown to arise from $CH_2^{+\bullet}$, CH_3^{+} and $CH_4^{+\bullet}$, the latter two ions also giving $ND_3^{+\bullet}$ by charge transfer. The only reaction involving methane as a neutral species was found to be that in which NH_4^{+} is formed from $NH_3^{+\bullet}$. The only new ion found, which is not observed in the mass spectra of the pure substances, was the $CH_2NH_2^{+}$ ion, shown by the double-resonance technique to arise mainly from the reactions:

$$CH_3^{+} + NH_3 \longrightarrow CH_2NH_2^{+} + H_2 \tag{6.52}$$

$$CH_2^{+\bullet} + NH_3 \longrightarrow CH_2NH_2^{+} + H\cdot \tag{6.53}$$

The yield of this ion rises rapidly with pressure, passes through a maximum and reaches a lower limiting value at higher pressures. The ion cyclotron double-resonance signal was negative below the maximum, changing to positive at higher pressures. Since proton transfer to NH_3 from $CH_2NH_2^{+}$ is endothermic at thermal energies, the fact that it is shown to occur using ion cyclotron double resonance suggests that $CH_2NH_2^{+}$ is formed in an excited state with sufficient energy to overcome the endothermicity. The high-pressure behaviour is ascribed to collisional stabilisation which prevents the ion from reacting further. The isotopic variants of reaction (6.52) were studied using CD_4–NH_3 and CH_4–ND_3 mixtures and the results are discussed in terms of competing decompositions of $(CH_3NH_3)^{+*}$ ions and isotope exchange reactions.

6.4.2 Ion–molecule reactions in unsaturated systems

6.4.2.1 Acetylene and ethylene

These reactions have been the subject of a number of ion cyclotron resonance studies and considerable mechanistic information and rate data have been obtained. In the acetylene system[51, 52], the two major reactions of the molecular ion are (6.54) and (6.55)

$$C_2H_2^{+\bullet} + C_2H_2 \rightarrow C_4H_2^{+\bullet} + H_2 \quad (6.54)$$

$$C_2H_2^{+\bullet} + C_2H_2 \rightarrow C_4H_3^{+} + H^{\bullet} \quad (6.55)$$

and $k_{6.54}/k_{6.55}$ was measured as 0.46–0.47, slightly higher than literature values of 0.36–0.44. Fragment ions react by adding a C_2H unit and eliminating a hydrogen atom from the collision complex and at higher pressures collisional stabilisation of tertiary and quaternary ions takes place with H/D randomisation.

In the ethylene system[19, 53], the molecular ion undergoes two main reactions:

$$C_2H_4^{+\bullet} + C_2H_4 \rightarrow C_3H_5^{+} + CH_3^{\bullet} \quad (6.56)$$

$$C_2H_4^{+\bullet} + C_2H_4 \rightarrow C_4H_7^{+} + H^{\bullet} \quad (6.57)$$

and various estimates of $k_{6.56}/k_{6.57}$ are in the range 9.5 ± 0.5, in good agreement with literature values. Both $C_5H_9^+$ and $C_5H_7^+$ arise from $C_3H_5^+$ at higher pressures, and from the pressure dependence of the relative abundance of the two ions two stabilisation mechanisms for $C_5H_9^{+*}$ have been proposed[19]. Line-width measurements give a lower limit of 1.1×10^{-4} s. for the dissociative life-time. Complete H–D scrambling was observed to occur in the $(C_4H_4D_4)^{+\bullet *}$ complex[19], in agreement with literature results, but at higher primary ion translational energies specific isotope effects were observed. The effect of varying the internal energy of a collision complex was demonstrated by forming the $C_4H_6^{+\bullet *}$ complex from $C_2H_2^{+\bullet} + C_2H_4$ in the ethylene system and from $C_2H_4^{+\bullet} + C_2H_2$ in an ethylene–acetylene mixture[19] when the relative rate constants for

$$(C_4H_6^{+\bullet})^* \rightarrow C_3H_3^{+} + CH_3^{\bullet} \quad (6.58)$$

$$(C_4H_6^{+\bullet})^* \rightarrow C_4H_5^{+} + H^{\bullet} \quad (6.59)$$

were observed to be $k_{6.58}/k_{6.59} = 1.55$ and 4.4, respectively, the internal energy being some 20 kcal mol^{-1} lower in the latter case. In reaction (6.58), H–D scrambling is almost complete, but there is a tendency to lose $H^{\bullet}$ rather than $D^{\bullet}$ atoms in reaction (6.59).

6.4.2.2 Propylene, allene and propyne

In the propylene system, reactions of the molecular ion give rise to the four secondary ions $C_3H_7^+$, $C_4H_7^+$, $C_4H_8^{+\bullet}$ and $C_5H_9^+$ in relative abundances of 0.20:0.10:0.43:0.27, in good agreement with medium-pressure source results at zero field[54]. Higher olefins up to hexene[55] have been studied and several generalisations and relationships between reaction paths and reactant structure were found. For the higher olefins, partial H–D scrambling occurring to different extents for reactions proceeding through common intermediates was observed, but extensive scrambling does not occur.

In the allene and propyne systems[56], there is generally good agreement between results obtained by ion cyclotron resonance and by the medium pressure source method[57]. Differences in reaction rate constants for the ions $C_3H_4^{+\cdot}$ and $C_3H_5^+$ were ascribed to different translational energy distributions of reactant ions in the two techniques. The relative rate constants for the fragmentation of $(C_5H_8^{+\cdot})^*$ complexes were shown to depend upon the reactions leading to its formation and the changes were ascribed to differences in structure or internal energy. In general, reactions could be rationalised in terms of the formation of a four-centre complex, formed by the four unsaturated carbon atoms.

6.4.2.3 *Halogenated olefins*

Several halogenated olefins[53, 58] have been studied by ion cyclotron resonance mass spectrometry. The chloroethylene system[58] was the earliest to be studied and was found to be somewhat different from the ethylene system[19, 53]. The molecular ion gives rise to three secondary ions which corresponds to the loss of $CH_2Cl^\cdot$, HCl and $Cl^\cdot$ from the collision complex and reactions of fragment ions and secondary ions often occur with the elimination of HCl. In the fluoroethylene case[53], the secondary ions $C_3H_3F_2^+$, $C_3H_4F^+$ and $C_3H_5^+$ arise from reactions of the molecular ion with relative rate constants of 0.6:0.25:0.15, respectively. These reactions parallel reaction (6.56) in the ethylene system and the formation of $C_4H_5F^{+\cdot}$ by HF elimination is a very minor process. In reactions of other ions[60], however, HF elimination is common and parallels the HCl eliminations of the chloroethylene system. The kinetic order of reactions occurring in halogenoethylene systems is a subject of controversy: under ion cyclotron resonance conditions, the formation of $C_3H_3F_2^+$ in the fluoroethylene system is second order[53], but other techniques suggest that in both the CH_2CHCl and CH_2CHF systems, the majority of the products are formed in third-order processes[59]. Reactions of the molecular ions of $CHFCF_2$ and C_2F_4 with the parent molecule[60] lead mainly to the elimination of $CF_3\cdot$, but CH_2CF_2 appears to behave quite differently in that the sole reaction of the molecular ion is the formation of the dimer ion, $(CH_2CF_2)_2^+$. In mixtures of C_2D_4 and the fluoroethylenes[61], partial H–D scrambling is observed in reactions in which a methyl or substituted methyl radical is eliminated, but in reactions in which olefinic ions and molecules are formed, a loose four-centre complex best explains the results. For example,

$$CH_2CF_2^{+\cdot} + C_2D_4 \longrightarrow CD_2CF_2^{+\cdot} + CH_2CD_2$$

$$\text{but not } CHDCF_2^{+\cdot} + CHDCD_2 \qquad (6.60)$$

Preliminary results indicate that these generalisations can be extended to the fluoropropenes[62].

6.4.2.4 *Other unsaturated systems*

In the vinyl methyl ether system[63], the molecular ion $[CH_3O{\cdot}CH{=}CH_2]^{+\cdot}$ reacts to give as major products $C_4H_8O_2^{+\cdot}$, $C_5H_8O^{+\cdot}$ and $C_3H_7O_2^+$ by elimination of C_2H_4, CH_3OH and $C_3H_5\cdot$ from the collision complex; the relative rates of formation are 0.70:0.10:0.20 respectively. In mixtures with $CD_3OCH{=}CH_2$, little or no H–D scrambling occurred. A similar lack of H–D scrambling was found in a study of reactions in the C_2H_2–H_2S and C_2H_4–H_2S systems[52] and was attributed to the localisation of the positive charge by the hetero-atom. The major reactions in the C_2H_4–H_2S system were loss of $CH_3^{\cdot}$ or CH_4 from the collision complex $(C_2H_6S^{+\cdot})^*$; in the C_2H_2–H_2S system, $C_2H_2^{+\cdot}$ did not react with H_2S, but H· and $CH_3^{\cdot}$ are lost from the collision complex formed from $H_2S^{+\cdot}+C_2H_2$.

The fractional yield of molecular ions of *cis*-but-2-ene as a function of pressure has been shown to vary much less under ion cyclotron resonance conditions than under medium-pressure source conditions[64]. This is attributed to the different time scales used in the two instruments and to the effective competition between unimolecular fragmentation of excited ions and their collisional stabilisation.

Both the positive and negative ion–molecule reactions in the acetonitrile system have been investigated[65, 66]. The major reaction of the molecular ion is to form the CH_3CNH^+ ion, which deuterium labelling and the double-resonance technique show to arise in a proton transfer reaction. Several fragment ions are found to transfer a proton to the CH_3CN molecule in exothermic processes, but the observation of the reaction:

$$CH_3^+ + CH_3CN \longrightarrow CH_3CNH^+ + CH_2 \qquad (6.61)$$

indicates that the CH_3^+ ions are formed in an excited state by fragmentation of the $CH_3CN^{+\cdot}$ ion since the reaction is endothermic by *c.* 10 kcal mol^{-1} for ground state CH_3^+ ions. For the reaction

$$CH_2CN^+ + CH_3CN \longrightarrow C_3H_4N^+ + HCN \qquad (6.62)$$

the use of ^{15}N and D labelling showed that the N atom of the HCN molecule comes exclusively from the CH_2CN^+ species, whereas the H atom comes entirely from the neutral reactant; this may be rationalised in terms of a cyclic complex. In a similar reaction in which the $C_3H_4N^+$ ion is formed from the CH_2^+ ion, the H atom lost from the collision complex again comes entirely from the neutral reactant.

The major negative ion formed is, not unexpectedly, the CN^- ion which reacts with the parent molecule to form C_3N^- with the elimination of ammonia. In a H_2O–CH_3CN mixture, OH^- ions react to give CH_2CN^- and CNO^-.

In a study of the HCN system[67], the principle reaction of the molecular ion was shown to be the formation of the H_2CN^+ ion with the elimination of a CN radical. Fragment ions react principally by addition of a CN radical and elimination of a hydrogen atom.

6.4.3 Ion–molecule reactions of alcohols and ethers

In one of the earliest studies to make use of ion cyclotron resonance mass spectrometry, the ion–molecule reactions of methanol were extensively investigated[68]. The reaction scheme proposed was in good general agreement with that proposed by Munson[69], using conventional techniques. The dominant secondary ion is the $CH_3OH_2^+$ ion, which arises from reactions of the molecular ion and from the fragment ions CHO^+ and CH_3O^+. The $CH_3OH_2^+$ reacts further to form $(C_2H_9O_2^+)^*$ which is stabilised by collision at high pressures but eliminates H_2O to give $C_2H_7O^+$ at lower pressures. This ion was shown to have the structure $(CH_3)_2OH^+$ rather than $C_2H_5OH_2^+$ by comparing reactions of $C_2H_7O^+$ ions produced from $(CH_3)_2O$ and C_2H_5OH in mixtures with methanol. Both the protonated parent ion of the ether and the $C_2H_7O^+$ formed in the methanol system react to give the $C_3H_{11}O_2^{+*}$ ion which does not fragment, whereas that formed from the methanol–ethanol mixture eliminates a water molecule.

Dehydration reactions[70, 71] have been shown to occur in ion–molecule reactions in several aliphatic alcohol systems. In the t-butanol system[70], the ion at $m/e = 59$, assumed to be the $(CH_3)_2COH^+$ ion, reacts as follows:

$$(CH_3)_2\overset{+}{C}OH + (CH_3)_3COH \longrightarrow (CH_3)_2CO{-}\overset{+}{H}{-}OH_2 + C_4H_8 \quad (6.63)$$

and a similar reaction of the protonated molecular ion yields a butene molecule. A labile proton appears to be necessary for this type of reaction to occur, and similar reactions were observed in the propan-2-ol, butan-2-ol and cyclopentanol systems but not in the ethanol system. Experiments with $(CD_3)_2CHOH$ indicated that HOD is transferred.

The reaction giving rise to the $C_5H_9^+$ ion[71] in the mass spectrum of propan-2-ol was shown, by deuterium labelling to involve the loss of two water molecules:

$$CD_3CHOH^+ + CD_3CH(OH)CD_3 \longrightarrow H_2O + C_5H_2D_9O^+ \longrightarrow D_2O + C_5H_2D_7^+ \quad (6.64)$$

$$CH_3CDOH^+ + CH_3CD(OH)CH_3 \longrightarrow H_2O + C_5H_9D_2O^+ \longrightarrow H_2O + C_5H_7D_2^+ \quad (6.65)$$

Hence both hydroxyl groups in the reactants are involved, as well as two hydrogen atoms from the methyl groups of either the ion or the neutral reactant. A mechanism in which hydrogen transfer precedes nucleophilic displacement is postulated as the main route to the formation of the $C_5H_{11}O^+$ ion, which in turn loses water by passing through a six-membered ring complex, thereby giving the observed isotopic distribution.

A similar dehydration reaction[71] but one involving a structural rearrangement is found in ethers. For example, at high electron energies, the following reactions occur in the 2-propyl ether system, reaction (6.67) predominating:

$$CH_3CH{=}\overset{+}{O}{-}CH(CH_3)_2 \longrightarrow C_3H_6 + CH_3CH\overset{+}{O}H \quad (6.66)$$

$$CH_3CH{=}\overset{+}{O}{-}CH(CH_3)_2 \longrightarrow H_2O + C_5H_9^+ \quad (6.67)$$

6.4.4 Ion–molecule reactions in other systems

6.4.4.1 Nucleophilic and electrophilic displacement reactions

The interpretation of ionic reaction mechanisms in solution is often complicated by the presence of solvation effects and there has been increasing interest in carrying out analogous reactions in the gas phase. For example, in a mixture of HCl and CH_3F, the nucleophilic displacement reaction[72]

$$HCl + CH_3FH^+ \rightarrow CH_3ClH^+ + HF \tag{6.68}$$

is observed to occur with a rate constant of $3.1 \pm 0.3 \times 10^{-10}$ cm^3 molecule^{-1} s^{-1}. When CH_3Cl is mixed with water, a nucleophilic displacement does not take place, but when CH_3Cl is replaced by C_2H_5Cl a reaction is observed:

$$H_2O + CH_3ClH^+ \not\longrightarrow CH_3OH_2^+ + HCl \tag{6.69}$$

$$H_2O + C_2H_5ClH^+ \longrightarrow C_2H_5OH_2^+ + HCl \tag{6.70}$$

This can be rationalised by noting that the proton affinities of CH_3Cl, H_2O and C_2H_5Cl are respectively, 160, 164 and 167 kcal mol^{-1}, so that a proton transfer will occur rather than reaction (6.69). For a displacement reaction to occur, it must be exothermic, and proton transfer to the nucleophile must be endothermic. If nucleophilicity is measured by methyl cation affinity ($M + CH_3^+ \rightarrow MCH_3^+$), then the order in which nucleophilic displacement will occur is:

$$NH_3 > CO > H_2S > CH_2O > HI > H_2O > HBr > HCl > N_2 > HF$$

Gas-phase electrophilic aromatic substitution reactions have also been studied by ion cyclotron resonance mass spectrometry[73]:

$$XE^+ + C_6H_6 \rightarrow C_6H_6E^+ + X \tag{6.71}$$

For example, benzene can be nitrated using NO_2 or the fragment ions $H_2ONO_2^+$ and $CH_2ONO_2^+$ derived from alkyl nitrates. In the case of NO_2 however, nitration does not occur by electrophilic substitution since the precursor of the $C_6H_6NO_2^+$ ion was shown to be the $C_6H_6^{+\bullet}$ ion rather than the NO_2^+ ion. The NO_2^+ ion reacts with benzene to form $C_6H_6O^+ + NO$. In acetonitrile–benzene mixtures, CH_3CNH^+ was found to be the precursor of $C_6H_6CH_2CN^+$, $m/e = 118$, and a Hoesch synthesis was proposed to accommodate the results of deuterium labelling studies, e.g.

$$C_6H_6 + CD_3CND^+ \rightarrow C_8H_5D_3N^+ \ (m/e = 121) + HD \tag{6.72}$$

Reactions in the ethyl nitrate system[74] have recently been investigated and this study was noteworthy in that the CHO^+ and $C_2H_5^+$ ions could be resolved at a magnetic field of 16.5 kG by use of an observing frequency of 691 kHz. The protonated molecular ion was shown to be formed from $C_2H_5^+$, CHO^+ and $C_2H_3^+$ and to undergo collision-induced decomposition when irradiated, giving the $H_2NO_3^+$ ion.

6.4.4.2 Diborane

A study of the diborane system was the first dealing solely with reactions of an inorganic species[75, 76]. The negative ions BH_4^-, $B_2H_7^-$, $B_3H_8^-$, $B_4H_9^-$, $B_5H_{10}^-$ and $B_6H_9^-$ were all observed, but some ions, e.g. the $B_4H_9^-$ and $B_6H_9^-$ ions, were formed by pyrolysis at the filament. The only strong ion cyclotron double-resonance signal was obtained for the reaction

$$BH_4^- + B_2H_6 \rightarrow B_2H_7^- + BH_3 \qquad (6.73)$$

Isotopic studies indicate that full equilibration of the boron atoms occurs in the collision complex. When water vapour is present in the system, no ions containing both boron and oxygen were found, but OH^- is shown to give rise to excited BH_4^- ions which react further to give $B_2H_5^-$ and $B_2H_7^-$ ions. Several new ions were observed in the H_2S–B_2H_6 system; the major ion was BS_2^-.

A large number of positive ions are found in the B_2H_6 system and reactions could be grouped into one of three types:

(i) $$B_xH_y^+ + B_2H_6 \rightarrow B_{x+2}H_{y+2}^+ + 2H_2 \qquad (x = 1\text{–}4) \qquad (6.74)$$

(ii) $$B_2H_5^+ + B_2H_6 \rightarrow B_4H_n^+ + \text{hydrogen} \qquad (n = 4\text{–}6) \qquad (6.75)$$

(iii) $$B_2H_x^+ + B_2H_6 \rightarrow B_3H_y^+ + BH_3 + \text{hydrogen} \quad \begin{matrix}(x = 2\text{–}6)\\(y = 4\text{–}6)\end{matrix} \qquad (6.76)$$

6.4.4.3 Phosphine

More recently, reactions occurring in the PH_3 and PH_3–PD_3 systems have been investigated as a function of electron energy and pressure[77, 78]. The major reaction of the molecular ion was the formation of PH_4^+ which is formed by several other routes at higher pressures. Several $P_2H_x^+$ ions were observed, ($x = 0$–5) and isotopic labelling indicated that proton transfer is favoured over hydrogen abstraction. The rate constant of $1.05 \pm 0.2 \times 10^{-9}$ cm^3 molecule^{-1} s^{-1} which was obtained for the reaction

$$PH_3^{+\cdot} + PH_3 \rightarrow PH_4^+ + PH_2^{\cdot} \qquad (6.77)$$

is in excellent agreement with literature values obtained by other techniques. Absolute rate constants were also obtained for the formation of cross-condensation products such as POH^+, PNH_2^+ and PCH_2^+ in PH_3 mixtures with H_2O, NH_3 and CH_4. Unlike the ammonia system, charge-transfer and proton-transfer reactions do not predominate, the PH_3 behaves more like CH_4 in its ion–molecule chemistry.

6.4.4.4 Reactions of O^- ions

Both NO_2 and N_2O were used as sources of O^- ions in a study of their reactions with acetylene[79]. The major products are C_2H^- and C_2OH^-; by comparing the rate constants of these reactions with those for NO^- production with N_2O and NO_2^- production with NO_2, absolute rate constants could be determined. If $k = 1.2 \times 10^{-9}$ cm^3 molecule^{-1} s^{-1} is taken for the reaction[80], $O^- + NO_2 \rightarrow NO_2^- + O$, the rate constants for the reactions producing C_2H^-, C_2OH^- and NO^- are respectively, 4×10^{-10}, 2×10^{-11} and 3×10^{-11} cm^3 molecule^{-1} s^{-1}.

6.4.5 The determination of ion structures

The determination of ion structures by conventional mass spectrometry makes use of techniques such as deuterium labelling, intensity ratios of metastable ions, and fragmentation patterns. Such techniques give information only for ions which have sufficient energy to fragment and may therefore be unreliable if this energy is also sufficient to cause ground-state ions to isomerise prior to fragmentation. An alternative technique, applicable to ground-state ions, is to study their ion–molecule reactions on the assumption that ions of different structure will undergo different reactions.

6.4.5.1 The $C_3H_6O^+$ ion

This technique was used in a tandem mass spectrometer to distinguish between the $C_3H_6^{+\bullet}$ ions given by propylene and cyclopropane[81], but ICR mass spectrometry was first used in a study of the structures of the $C_3H_6O^+$ isomers (1–3)[82]:

$$\underset{(1)}{(H_3C)_2C{=}O^{+\bullet}} \qquad \underset{(2)}{H_2C{=}C(CH_3){-}\overset{\bullet +}{O}{-}H} \qquad \underset{(3)}{H_2C{\overset{\bullet}{\cdots}}C(CH_2){-}\overset{+}{O}H_2}$$

The molecular ion of acetone is presumed to be of structure (1) and the $C_3H_6O^+$ ion produced from the molecular ion of hexan-2-one by a McLafferty rearrangement is presumed to be of structure (2). Several reactions were found in various mixtures which were characteristic of the keto and enol forms of the $C_3H_6O^+$ ion; for example, they may be distinguished by their reactions with nonan-5-one:

$$(CH_3)_2C{=}O^{+\bullet} + CH_3(CH_2)_3CO(CH_2)_3CH_3 \rightarrow (CH_3)_2C{=}\overset{+}{O}{-}C(OH)(CH_3)(CH_2^{\bullet}) + 2C_3H_6 \qquad (6.78)$$

$$CH_2{=}C(CH_3){-}\overset{+\bullet}{O}H + CH_3(CH_2)_3CO(CH_2)_3CH_3 \rightarrow C_4H_9{-}\underset{\underset{\bullet + OH}{\|}}{C}{-}C_4H_9 \; (+\; C_3H_5O) \qquad (6.79)$$

The $C_3H_6O^+$ ion formed by fragmentation of the molecular ion of 1-methylcyclobutanol was found to undergo reactions identical to those of the ion given by hexan-2-one, suggesting that the two sources give rise to ions of common structure, i.e. the enol form. Structure (3) has been suggested as the structure of the $C_3H_6O^+$ ion formed by a double McLafferty rearrangement in the mass spectrum of nonan-5-one, but the ion produced in this way was found to exhibit all the reactions of the enol form (2) and none of those of the keto form, (1). This suggested that the ion given by nonan-5-one was probably the enol form, although it is possible that the keto form is produced initially but isomerises rapidly before reaction. Further evidence concerning the structure of the $C_3H_6O^+$ ion formed in a double McLafferty rearrangement was obtained from a study of reactions of ions produced from partially deuteriated nonan-4-one[83]. The two possible fragmentation pathways are illustrated below, where R^1 and R^2 are protons in the 1- and 7-positions:

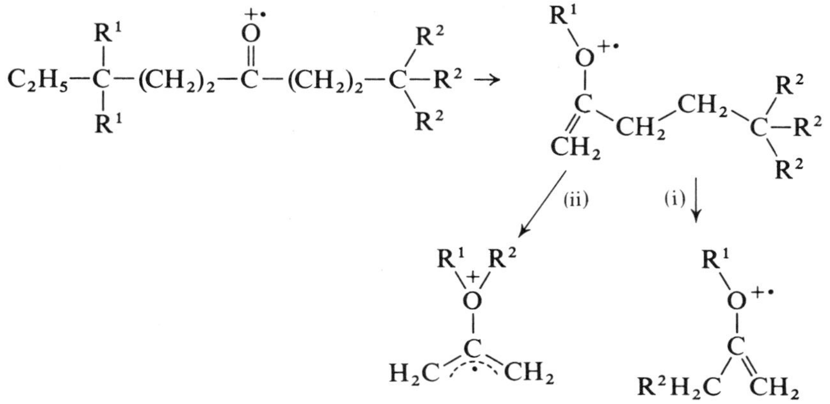

$$C_2H_5-C(R^1)_2-(CH_2)_2-\overset{+\bullet}{C}(=O)-(CH_2)_2-C(R^2)_3 \rightarrow R^1-\overset{+\bullet}{O}-C(=CH_2)-CH_2-CH_2-C(R^2)_3$$

(ii) $\rightarrow$ $R^1R^2\overset{+}{O}-C(\cdots CH_2)(\cdots CH_2)$, i.e. $H_2C\cdots\overset{\bullet}{C}(-\overset{+}{O}R^1R^2)\cdots CH_2$

(i) $\rightarrow$ $R^1-\overset{+\bullet}{O}-C(=CH_2)-CH_2R^2$, i.e. $R^2H_2C-C(=CH_2)-\overset{+\bullet}{O}R^1$

Metastable ion characteristics appear to support route (i) but the ion cyclotron resonance results suggest that fragmentation occurs via route (ii). In the nonan-4-one-1,1,1-d_3 system R^1 = H, R^2 = D, the protonation reaction (6.79) can be formulated either as

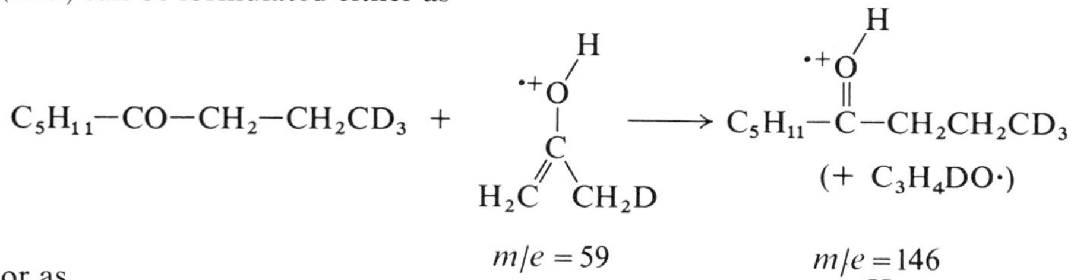

$$C_5H_{11}-CO-CH_2-CH_2CD_3 + H_2C{=}C(CH_2D)-\overset{\bullet+}{O}H \longrightarrow C_5H_{11}-C(=\overset{\bullet+}{O}H)-CH_2CH_2CD_3 \quad (+\ C_3H_4DO\cdot)$$

$m/e = 59$ $\qquad$ $m/e = 146$

or as

$$C_5H_{11}-CO-CH_2-CH_2CD_3 + H_2C\cdots\overset{\bullet}{C}(-\overset{+}{O}HD)\cdots CH_2 \longrightarrow C_5H_{11}-C(=\overset{\bullet+}{O}H)-CH_2CH_2CD_3 \quad (+\ C_3H_4DO\cdot)$$

$m/e = 59$ $\qquad$ $m/e = 146$

or

$$C_5H_{11}-C(=\overset{\bullet+}{O}D)-CH_2CH_2CD_3 \quad (+\ C_3H_5O\cdot)$$

$m/e = 147$

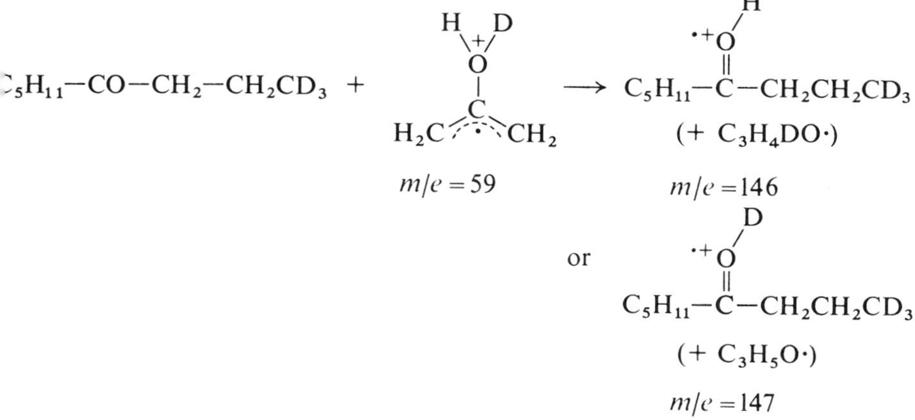

Use of the double-resonance technique showed that $m/e = 59$ is a precursor of the $m/e = 146$ ion but not of the $m/e = 147$ ion, indicating that fragmentation pathway (iii) occurs. In order to eliminate the possibility of the production of an anomalous result by a large isotope effect, reactions of the nonan-4-one-7,7-d_2 system were investigated. Pathway (iii) would lead to $m/e = 59 \rightarrow m/e = 146$, whereas pathway (iv) would lead to $m/e = 59 \rightarrow m/e = 145$ and 146; ion cyclotron double resonance showed that $m/e = 59$ is a precursor only of the $m/e = 146$ ion, again supporting pathway (iii). This observation also rules out the route

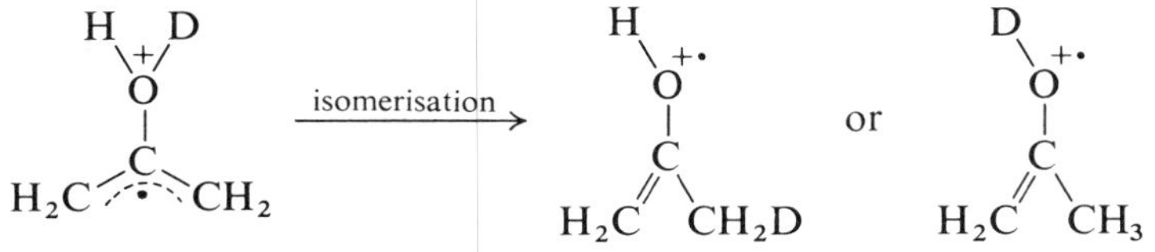

$$\mathrm{H_2C\overset{\cdot}{\cdots}C(\cdots CH_2)\text{–}\overset{+}{O}HD} \xrightarrow{\text{isomerisation}} \mathrm{H_2C{=}C(CH_2D)\text{–}\overset{+\cdot}{O}H} \text{ or } \mathrm{H_2C{=}C(CH_3)\text{–}\overset{+\cdot}{O}D}$$

(v) (vi)

since (v) and (vi) would lead to $m/e = 145$ and 146 respectively, both having $m/e = 59$ as a precursor. These observations therefore indicate that the structure of a $C_3H_6O^+$ ion formed in a double McLafferty rearrangement is best represented by the enol structure, (2).

6.4.5.2 The $C_2H_5O^+$ ion

The three most stable forms of the $C_2H_5O^+$ ion are structures (4)–(6)[84]

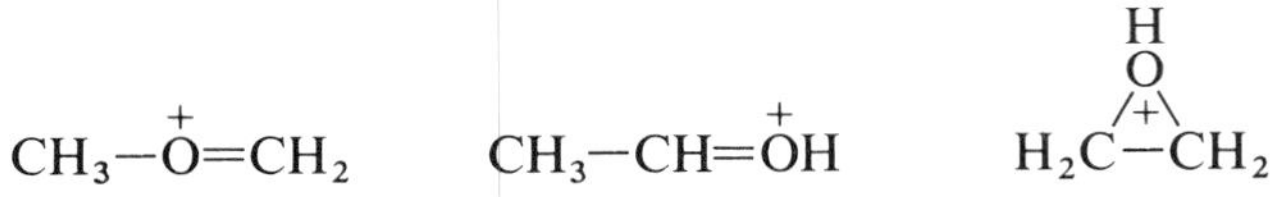

$$\mathrm{CH_3\text{–}\overset{+}{O}{=}CH_2} \qquad \mathrm{CH_3\text{–}CH{=}\overset{+}{O}H} \qquad \mathrm{H_2C\text{–}CH_2\ (\text{bridged by }\overset{+}{O}H)}$$

(4) (5) (6)

and a similar type of ion cyclotron resonance study has been made of this ion. Ions of structure (4) were generated from methyl ethyl ether and their major reaction with the parent molecule was to form the $CH_3\overset{+}{O}{=}CHCH_3$ ion in a hydride abstraction reaction. Ions of structure (5) were generated from propan-2-ol and undergo four main reactions with the parent molecule; the principal reaction is proton transfer to give $(CH_3)_2CH\overset{+}{O}H_2$ and no hydride abstraction reaction is observed. Ions of structure (6) were formed by ion–molecule reactions in ethylene oxide and reacted further to give only the proton-bound dimer ion. When ethylene oxide-d_4 was mixed with propan-2-ol, or ethylene oxide with $(CD_3)_2CHOH$, the same four reactions with very similar product distributions were found as for ions of structure (5), and it is suggested that structure (6) rapidly isomerises to structure (5) before reaction. Twelve different sources of $C_2H_5O^+$ were examined and in all cases, the structure assigned from ion–molecule reactions agreed with that derived from fragmentation reactions.

6.4.5.3 *The $C_3H_6^+$ ion*

Similar studies[85] have confirmed the conclusions obtained from thermochemical and high pressure source studies that the $C_3H_6^{+\cdot}$ ions formed from propene and cyclopropane differ structurally[81]. In a mixture of cyclopropane and ammonia, ions of $m/e = 30$ and 31 were found and were ascribed to the occurrence of the reactions

$$\begin{aligned} C_3H_6^{+\cdot} + NH_3 &\rightarrow CH_4N^+ + C_2H_5^{\cdot} \\ &\rightarrow CH_5N^{+\cdot} + C_2H_4 \end{aligned} \tag{6.80}$$

No such reactions were observed in a propene–ammonia mixture, nor between $C_3H_6^{+\cdot}$ ions formed from pentane or cyclopentane when mixed with ammonia. On the other hand, as expected from thermochemical data, $C_3H_6^{+\cdot}$ ions formed from tetrahydrofuran react in the same way as those produced from cyclopropane. When the ions were formed from cyclopropane, low yields of $m/e = 30$ and 31 were rationalised by assuming that ions of both structures were formed, since metastable ion transitions indicated that both $C_4H_8O^{+\cdot}$ and $C_5H_{10}^{+\cdot}$ are precursors of $C_3H_6^{+\cdot}$ ions.

6.4.5.4 *The $C_8H_8^{+\cdot}$ ion*

Recently, ion cyclotron resonance mass spectrometry of the styrene and cyclo-octatetraene[86] systems have shown that the molecular ions of the two isomers react quite differently. Styrene was found to form a dimer ion and, by use of deuterium labelling and comparison with mass spectra of $C_{16}H_{16}$ isomers, the structure of the dimer was shown to be consistent with that of 1-phenyltetralin.

This technique appears to be very promising as a method of investigating ion structures although at present there are few guiding principles to indicate general types of reaction which are most suitable for use in structural investigations.

6.4.6 Thermochemical applications

6.4.6.1 *Proton, hydrogen and electron affinities*

The translational energies of ions produced in an ion cyclotron resonance instrument are essentially thermal so that only thermoneutral or exothermic reactions occur. Use can be made of this in the study of competitive reactions of the type

$$M_1H^+ + M_2 \rightleftharpoons M_1 + M_2H^+ \tag{6.81}$$

in order to put limits on such quantities as proton affinities $P(M)$, hydrogen affinities $H(M^+)$ and electron affinities $E(M)$. A given compound is mixed with other species having proton affinities above and below that of the compound, and the occurrence or absence of a reaction enables one to put limits on the values of $P(M)$ and $H(M^+)$. The success of the method depends on the assumption that ground state ions are produced and on the availability of systems leading to a relatively low uncertainty in the bracketed values.

Reactions observed in mixtures of hydrocarbons with H_2O and H_2S were used to obtain data for these compounds[20]. For example, in the case of H_2S:

$$H_2S^{+\bullet} + CD_4 \rightarrow H_2DS^+ + CD_3^\bullet \quad (6.82)$$

$$H_3S^+ + C_3H_6 \rightarrow H_2S + C_3H_7^+ \quad (6.83)$$

These respectively put limits of $\leqslant 184$ kcal mol^{-1} and $\geqslant 181$ kcal mol^{-1} for $\Delta H_f(H_3S^+)$, leading to values of $P(H_2S) = 178 \pm 2$ kcal mol^{-1} and $H(H_2S^{+\bullet}) = 106 \pm 2$ kcal mol^{-1}. A similar bracketing technique led to $P(H_2O) = 164 \pm 4$ kcal mol^{-1} and $H(H_2O^{+\bullet}) = 141 \pm 4$ kcal mol^{-1}.

In a further application of the technique, reactions in mixtures containing ethyl nitrate[74] led to $P(C_2H_5ONO_2) = 180 \pm 3$ kcal mol^{-1} and $H(C_2H_5ONO_2^{+\bullet}) = 125 \pm 3$ kcal mol^{-1}. From a study of the relative acidity of phosphine[78, 87] in mixtures with other simple hydrides, it was possible to show that $P(H_2O) \leqslant P(PH_3) \leqslant P(NH_3)$, i.e. $164 \leqslant P(PH_3) \leqslant 207$ kcal mol^{-1}, but much closer limits could be set from reactions observed in mixtures of PH_3 with acetaldehyde and acetone:

$$CH_3CHOH^+ + PH_3 \rightarrow CH_3CHO + PH_4^+ \quad (6.84)$$

$$PH_4^+ + (CH_3)_2CO \rightarrow (CH_3)_2COH^+ + PH_3 \quad (6.85)$$

These put lower and upper limits of 182 and 189 kcal mol^{-1} so that $P(PH_3) = 185 \pm 4$ kcal mol^{-1}. Similar studies of reactions occurring in mixtures of NF_3 with CH_4 and HCl have led to $P(NF_3) = 151 \pm 10$ kcal mol^{-1}, and $H(NF_3^{+\bullet}) = 138 \pm 10$ kcal mol^{-1} [88]. From an analysis of the reactions observed in a mixture of argon and deuterium[89], it was possible to show that $P(D_2) > P(Ar)$, and the double-resonance technique was used to obtain a threshold energy for the reaction

$$(D_3^+)^\dagger + Ar \rightarrow ArD^+ + D_2 \quad (6.86)$$

suggesting $P(D_2)^\dagger \simeq P(Ar) + 7 \pm 5$ kcal mol^{-1}. These observations set limits of $79 \pm 4 \leqslant P(H_2)^\dagger \leqslant 85.2$ kcal mol^{-1} when combined with literature data.

The above method of determining proton affinities is rarely capable of giving values to better than ± 2–5 kcal mol^{-1}, but a recently developed equilibrium technique[90] provides relative proton affinities which are accurate to better than ± 0.2 kcal mol^{-1}. The equilibrium constant for reaction (6.81) is given by $K = [M_2H^+][M_1]/[M_1H^+][M_2]$ and by measuring the ratios of the ion concentrations and neutral concentrations at equilibrium, K and hence ΔG° for reaction (6.81) can be calculated. For fairly large molecules, ΔS° for a proton-transfer reaction will be very small so that ΔG° is approximately equal to ΔH°. Long reaction times and pressures of $\sim 1 \times 10^{-4}$ torr ensure the establishment of equilibrium and the ratio of neutral molecules is adjusted to give a reasonable ion current ratio. Equilibrium data for mixtures of pairs of the compounds piperidene, trimethylamine, pyrrolidine and azetidine led to proton affinities of 3.1, 2.1 and 1.8 kcal mol^{-1}, respectively for the first three compounds, relative to the proton affinity of azetidine. Possible interference from further reactions of the protonated species, to form proton-bound dimers, was ruled out by working at much lower pressures in a trapped ion cell.

6.4.6.2 Stabilities of bicyclic ions

Proton-transfer reactions have also been used in the study of the gas-phase stabilities of bicyclic olefin and ketone ions[91]. If the reactions

$$M + AH^+ \longrightarrow MH^+ + A \qquad (6.87)$$

$$B + MH^+ \longrightarrow BH^+ + M \qquad (6.88)$$

are observed to occur, then $\Delta H_f(AH^+) - \Delta H_f(A) \geqslant \Delta H_f(BH^+) - \Delta H_f(B)$, and if these values are known they can be used to bracket $\Delta H_f(MH^+) - \Delta H_f(M)$. This method was used to show that the bicyclo(2,2,1)heptyl cation compared to its saturated neutral molecule is 6 kcal more stable than the bicyclo-(2.2.2)octyl cation, compared to its saturated neutral molecule. Rather larger differences were observed in olefinic systems but the differences in the keto systems were much lower and this was attributed to localisation of the charge on the oxygen atom. The larger differences found in hydrocarbon systems were attributed to partial σ-delocalisation of the positive charge.

6.4.6.3 Gas-phase acidities and basicities

In a series of papers, Blair and Brauman[92–98] have used a very similar bracketing technique to obtain relative gas-phase acidities from observations on negative ion–molecule reactions. The general method is exemplified for alcohols[93, 95]:

$$ROH + R'O^- \rightarrow RO^- + R'OH \qquad (6.89)$$

The double-resonance technique was used to demonstrate that reactions of this type proceeded in one direction only in most cases, thereby allowing one to place the alcohols in order of decreasing acidity: $(CH_3)_3CCH_2OH > (CH_3)_3COH > (CH_3)_2CHOH > CH_3CH_2OH > CH_3OH > H_2O$, and $(CH_3)_3COH \simeq n\text{-}C_5H_{11}OH \simeq n\text{-}C_4H_9OH > n\text{-}C_3H_7OH > CH_3CH_2OH$. Phenol was shown to be a much stronger acid than any of the alcohols, but $C_2H_5OH > C_6H_5CH_3 > CH_3OH$. The gas-phase order for the alcohols is the reverse of that found in solution, possibly due to steric hindrance in the larger alkoxides. The stabilising influence of methyl groups is attributed to the polarisability of CH_3— or —CH_2— groups, compared with hydrogen, and this results in an increase in electron affinity of alkoxide radicals as the size of the alkyl group increases. If the strength of the bond broken in each case is assumed to be constant within a given series, the relative acidities will reflect the relative electron affinities of the radicals.

Similar studies have been made of the gas-phase acidities of amines[94, 96] by use of reactions of the type:

$$R_1R_2NH + R_3R_4N^- \rightarrow R_1R_2N^- + R_3R_4NH \qquad (6.90)$$

The order of acidities was found to be diethylamine > neopentylamine ⩾ t-butylamine ⩾ dimethylamine ⩾ isopropylamine > n-propylamine > ethylamine > methylamine > ammonia, and diethylamine > water > t-butylamine. As in the case of alcohols, larger alkyl groups lead to higher acidity and the same trend is oberved for secondary amines. The results are interpreted in terms of stabilisation of the charge on the nitrogen by the alkyl group acting as a polarisable entity. If one assumes a constant D(N—H) bond strength in a given series, the above order also reflects the decreasing electron affinity of the amine radicals. On the further assumption that D(N—H) for primary amines is *c.* 8 kcal mol^{-1} greater than that for secondary amines, the similar acidities found for t-butylamine and dimethylamine suggests that the electron affinity of the $(CH_3)_3CNH\cdot$ radical is *c.* 8 kcal mol^{-1} greater than that of the $(CH_3)_2N\cdot$ radical.

In a parallel study of the basicities of amines[97], reaction (6.81) was used to establish an order of basicity for primary, secondary and tertiary amines. The general trends observed were that for primary amines, an increase in the size of the alkyl group increases the basicity, and it was also observed that an increase in alkyl substitution gives increased basicity for a given type of alkyl group, e.g. $(C_2H_5)_3N > (C_2H_5)_2NH > C_2H_5NH_2$. The observed trends are discussed in terms of differences in the dissociation energies of the N—H bonds and the polarisability effect of the alkyl groups.

Gas-phase acidity studies have also been made on PH_3[78, 87] and on a number of carbon acids[92]. In the PH_3 study[78, 87], it was found that $H_2S > PH_3 > n\text{-}C_3H_7OH$ in acidity, and an approximate electron affinity of 35 ± 11 kcal mol^{-1} was found for PH_2. In the carbon acid study[92], the order of acidity is acetylacetone > acetyl cyanide > hydrogen cyanide.

6.5 CONCLUDING REMARKS

The above sections indicate the many different types of problem to which ion cyclotron resonance mass spectrometry has already been applied. Several studies have been in the nature of prototype experiments and the full potential of the technique in these areas has yet to be realised. The main features of ion cyclotron resonance that are likely to prove useful are as follows:

(i) It is possible to study both reactive and non-reactive collisions of ions at near-thermal translational energies.

(ii) The open construction of the instrument minimises mass discrimination effects and allows the use of several different ionising techniques.

(iii) Used with care, ion cyclotron double resonance offers a simple method of establishing reaction sequences and this can be extended using multiple pulsing and modulation techniques.

(iv) The high intrinsic sensitivity makes possible the study of processes involving negative ions if scattered thermal electrons are ejected.

There are, however, several areas in which improvements in the technique are desirable. The analysis of power absorption and of data to obtain rate constants currently outstrips experimental technique. In the standard cell, poorly defined electric fields lead to an uncertainty in the reaction time of at least $\pm 10\%$, although this uncertainty is much reduced in the trapped ion cell. A similar uncertainty arises in the determination of absolute pressures, so that errors in *absolute* rate constants are likely to be ± 10–20%; *relative* rate constants can usually be obtained to within $\pm 5\%$.

Although the ion cyclotron double resonance technique is very valuable, it has to be used with care even in qualitative studies. At high ion densities, spurious signals can be obtained which appear to suggest that most ions are coupled to most others; these arise from artifacts such as de-tuning effects, ion losses, etc. and are usually fairly small, but several early publications contain references to signals which can probably be ascribed to these effects. The use of this method as an ejection technique by sweeping out the ions is less open to misinterpretation, but used in this way it can give no information on translational energy effects in ion–molecule reactions. If a mixture is introduced into the instrument the ability to eject an ion is an advantage which is absent in the normal high-pressure source technique. The remaining ions may react with either of the neutral species, however, and it is necessary to use a tandem instrument to avoid this.

Attempts to use ion cyclotron double resonance for the study of quantitative translational energy effects look promising, but ion losses limit the usefulness of the technique at higher energies. In addition, it has proved difficult to devise a definitive method of calibrating the energy scale and hence of estimating experimentally the energy spread of the irradiated reactant ions. It is hoped that improvements in cell design may improve this situation.

The ion cyclotron resonance technique is essentially a low-pressure technique and in many applications the best operating conditions are at pressures below 2×10^{-5} torr. It is therefore ideally suited to the study of bimolecular processes, but it is much less well-suited for the quantitative study of termolecular processes and the observation of collisionally stabilised products is rare under normal operating conditions.

The electron energy spread appears to be comparable with that found in conventional instruments; for example the energy resolution in the electron scattering experiments was 0.2–0.5 eV. Although ionisation potentials, appearance potentials and threshold excitation spectra can be determined fairly readily at this energy resolution, there is little room in the standard instrument to incorporate a more sophisticated electron gun to improve on this. Similarly, mass resolution and range is limited by the lack of availability of magnets capable of producing a homogeneous field in excess of 15 kG over a sufficiently wide pole gap and area.

While some of the drawbacks outlined above are inherent in the technique, it must be borne in mind that the first ion cyclotron resonance paper appeared only eight years ago (1963) and that the bulk of the work in this field has been published during the last three years. One can therefore predict with confidence that many of the difficulties outlined above will be overcome within the next few years.

References

1. Wobschall, D., Graham, J. R. and Malone, D. P. (1963). *Phys. Rev.*, **131,** 1565
2. Baldeschwieler, J. D. (1968). *Science,* **159,** 263
3. Gray, G. A. (1971). *Advan. Chem. Phys.*, **19,** 141

4. Wobschall, D. (1965). *Rev. Sci. Instrum,* **36,** 466
5. Varian Associates, Palo Alto, California
6. Clow, R. P. and Futrell, J. H. (1970). *Int. J. Mass Spectrom Ion Phys.,* **4,** 165
7. Goode, G. C., O'Malley, R. M., Ferrer-Correia, A. J., Massey, R. I., Futrell, J. H., Jennings, K. R. and Llewellyn, P. (1970). *Int. J. Mass Spectrom Ion Phys.,* **5,** 393
8. Anders, L. R., Beauchamp, J. L., Dunbar, R. C. and Baldeschwieler, J. D. (1966). *J. Chem. Phys.,* **45,** 1062
9. Beauchamp, J. L. and Armstrong, J. T. (1969). *Rev. Sci. Instrum.,* **40,** 123
10. Beauchamp, J. L. (1971). Private communication
11. McIver, R. T. (1970). *Rev. Sci. Instrum.,* **41,** 555
12. Dunbar, R. C. (1971). *J. Chem. Phys.,* **54,** 711
13. Beauchamp, J. L. (1967). *Ph.D. Thesis,* Harvard University, Cambridge, Mass.
14. Beauchamp, J. L. (1967). *J. Chem. Phys.,* **46,** 1231
15. Buttrill, S. E. (1969). *J. Chem. Phys.,* **50,** 4125
16. Marshall, A. G. and Buttrill, S. E. (1970). *J. Chem. Phys.,* **52,** 2752
17. Marshall, A. G. (1971). *J. Chem. Phys.,* **55,** 1343
18. Comisarow, M. (1971). *J. Chem. Phys.,* **55,** 205
19. Bowers, M. T., Elleman, D. D. and Beauchamp, J. L. (1968). *J. Phys. Chem.,* **72,** 3599
20. Beauchamp, J. L. and Buttrill, S. E. (1968). *J. Chem. Phys.,* **48,** 1783
21. Wobschall, D., Graham, J. R. and Malone, D. (1965). *J. Chem. Phys.,* **42,** 3955
22. Wobschall, D., Fluegge, R. A. and Graham, J. R. (1967). *J. Chem. Phys.,* **47,** 4091
23. Wobschall, D., Fluegge, R. A. and Graham, J. R. (1967). *J. Appl. Phys.,* **38,** 3761
24. Ridge, D. P. and Beauchamp, J. L. To be published
25. Henis, J. M. S. and Mabie, C. A. (1970). *J. Chem. Phys.,* **53,** 2999
26. Huntress, W. T. and Beauchamp, J. L. (1969). *Int. J. Mass Spectrom. Ion Phys.,* **3,** 149
27. O'Malley, R. M. and Jennings, K. R. (1969). *Int. J. Mass Spectrom. Ion Phys.,* **2,** App. 1
28. Ridge, D. P. and Beauchamp, J. L. (1969). *J. Chem. Phys.,* **51,** 470
29. Brauman, J. I. and Smyth, K. C. (1969). *J. Amer. Chem. Soc.,* **91,** 7778
30. Smyth, K. C., McIver, R. T. and Brauman, J. I. (1971). *J. Chem. Phys.,* **54,** 2758
31. Bowers, M. T., Elleman, D. D. and King, J. (1969). *J. Chem. Phys.,* **50,** 4787
32. Reuben, B. G. and Friedman, L. (1962). *J. Chem. Phys.,* **37,** 1636
33. Goode, G. C., Ferrer-Correia, A. J. and Jennings, K. R. (1970). *Int. J. Mass Spectrom. Ion Phys.,* **5,** 229
34. Anders, L. R. (1969). *J. Phys. Chem.,* **73,** 469
35. Futrell, J. H. (1971). *A.S.T.M. Meeting on Mass Spectrometry and Allied Topics,* Atlanta, Georgia
36. Neynaber, R. H. and Trujillo, S. M. (1968). *Phys. Rev.,* **167,** 63
37. Bowers, M. T. and Elleman, D. D. (1969). *J. Chem. Phys.,* **51,** 4606
38. Bowers, M. T., Elleman, D. D. and King, J. (1969). *J. Chem. Phys.,* **50,** 1840
39. Gioumousis, G. and Stevenson, D. P. (1958). *J. Chem. Phys.,* **29,** 294
40. Bowers, M. T. and Elleman, D. D. (1970). *J. Amer. Chem. Soc.,* **92,** 7258
41. Chupka, W. A. and Russell, M. E. (1968). *J. Chem. Phys.,* **49,** 5426
42. von Koch, H. and Friedman, L. (1963). *J. Chem. Phys.,* **38,** 1115
43. Wexler, S. and Inoue, M. (1969). *J. Amer. Chem. Soc.,* **91,** 5730
44. Gupta, S. K., Jones, E. G. Harrison, A. G. and Myher, J. J. (1967). *Can. J. Chem.,* **45,** 3107
45. Bowers, M. T. and Elleman, D. D. (1970). *J. Amer. Chem. Soc.,* **92,** 1847
46. Wexler, S. and Pobo, L. G. (1971). *J. Amer. Chem. Soc.,* **93,** 1327
47. King, J. and Elleman, D. D. (1968). *J. Chem. Phys.,* **48,** 4803
48. Clow, R. P. and Futrell, J. H. (1969). *J. Chem. Phys.,* **50,** 5041
49. Huntress, W. T., Mosesman, M. M., and Elleman, D. D. (1971). *J. Chem. Phys.,* **54,** 843
50. Huntress, W. T. and Elleman, D. D. (1970). *J. Amer. Chem. Soc.,* **92,** 3565
51. O'Malley, R. M. and Jennings, K. R. (1969). *Int. J. Mass Spectrom. Ion Phys.,* **2,** 257
52. Buttrill, S. E. (1970). *J. Amer. Chem. Soc.,* **92,** 3560
53. O'Malley, R. M. and Jennings, K. R. (1969). *Int. J. Mass Spectrom. Ion Phys.,* **2,** 441
54. Herod, A. A., Harrison, A. G., O'Malley, R. M., Ferrer-Correia, A. J. and Jennings, K. R. (1970). *J. Phys. Chem.,* **74,** 2720
55. Henis, J. M. S. (1970). *J. Chem. Phys.,* **52,** 282
56. Bowers, M. T., Elleman, D. D., O'Malley, R. M. and Jennings, K. R. (1970). *J. Phys.* Chem., **74,** 2583

57. Myher, J. J. and Harrison, A. G. (1968). *J. Phys. Chem.*, **72,** 1905
58. Beauchamp, J. L., Anders, L. R. and Baldeschwieler, J. D. (1967). *J. Amer. Chem. Soc.*, **89,** 4569
59. McAskill, N. A. and Harrison, A. G. (1970). *Int. J. Mass Spectrom. Ion Phys.*, **5,** 193
60. O'Malley, R. M. and Jennings, K. R. Unpublished work
61. Ferrer-Correia, A. J. and Jennings, K. R. Unpublished work
62. Drewery, C. J., Goode, G. C. and Jennings, K. R. Unpublished work
63. Drewery, C. J. Unpublished work
64. Clow, R. P., Futrell, J. H. and Wisniewski, L. (1970). *J. Phys. Chem.*, **74,** 2234
65. Gray, G. A. (1968). *J. Amer. Chem. Soc.*, **90,** 2177
66. Gray, G. A. (1968). *J. Amer. Chem. Soc.*, **90,** 6002
67. Huntress, W. T., Baldeschwieler, J. D. and Ponnamperuma, C. (1969). *Nature (London)*, **223,** 468
68. Henis, J. M. S. (1968). *J. Amer. Chem. Soc.*, **90,** 844
69. Munson, M. B. S. (1965). *J. Amer. Chem. Soc.*, **87,** 5313
70. Beauchamp, J. L. (1969). *J. Amer. Chem. Soc.*, **91,** 5925
71. Lehman, T. A., Elwood, T. A., Bursey, J. T., Bursey, M. M. and Beauchamp, J. L. (1971). *J. Amer. Chem. Soc.*, **93,** 2108
72. Holtz, D., Beauchamp, J. L. and Woodgate, S. (1970). *J. Amer. Chem. Soc.*, **92,** 7484
73. Benezra, S. A., Hoffman, M. K. and Bursey, M. M. (1970). *J. Amer. Chem. Soc.*, **92,** 7501
74. Kriemler, P. and Buttrill, S. E. (1970). *J. Amer. Chem. Soc.*, **92,** 1123
75. Dunbar, R. C. (1968). *J. Amer. Chem. Soc.*, **90,** 5676
76. Dunbar, R. C. (1971). *J. Amer. Chem. Soc.*, **93,** 4167
77. Eyler, J. R. (1970). *Inorg. Chem.*, **9,** 981
78. Beauchamp, J. L., Holtz, D. and Eyler, J. R. (1970). *J. Amer. Chem. Soc.*, **92,** 7045
79. Goode, G. C. Unpublished work
80. Ferguson, E. E., Fehsenfeld, F. C. and Schmeltekopf, A. L. (1969). *Advan. Chem. Ser.*, **80,** 83
81. Sieck, L. W. and Futrell, J. H. (1966). *J. Chem. Phys.*, **45,** 560
82. Diekman, J., MacLeod, J. K., Djerassi, C. and Baldeschwieler, J. D. (1969). *J. Amer. Chem. Soc.*, **91,** 2069
83. Eadon, G., Diekman, J. and Djerassi, C. (1969). *J. Amer. Chem. Soc.*, **91,** 3986
84. Beauchamp, J. L. and Dunbar, R. C. (1970). *J. Amer. Chem. Soc.*, **92,** 1477
85. Gross, M. L., and McLafferty, F. W. (1971). *J. Amer. Chem. Soc.*, **93,** 1267
86. Wilkins, C. L. and Gross, M. L. (1971). *J. Amer. Chem. Soc.*, **93,** 895
87. Holtz, D. and Beauchamp, J. L. (1969). *J. Amer. Chem. Soc.*, **91,** 5913
88. Beauchamp, J. L., Holtz, D., Henderson, W. G. and Taft, R. W. (1971). *Inorg. Chem.*, **10,** 201
89. Bowers, M. T. and Elleman, D. D. (1970). *J. Amer. Chem. Soc.*, **92,** 7258
90. Bowers, M. T., Aue, D. H., Webb, H. M. and McIver, R. T. (1971). *J. Amer. Chem. Soc.*, **93,** 4314
91. Kaplan, F., Cross, P. and Prinstein, R. (1970). *J. Amer. Chem. Soc.*, **92,** 1445
92. Brauman, J. I. and Blair, L. K. (1968). *J. Amer. Chem. Soc.*, **90,** 5636
93. Brauman, J. I. and Blair, L. K. (1968). *J. Amer. Chem. Soc.*, **90,** 6561
94. Brauman, J. I. and Blair, L. K. (1969). *J. Amer. Chem. Soc.*, **91,** 2126
95. Brauman, J. I. and Blair, L. K. (1970). *J. Amer. Chem. Soc.*, **92,** 5986
96. Brauman, J. I. and Blair, L. K. (1971). *J. Amer. Chem. Soc.*, **93,** 3911
97. Brauman, J. I., Riveros, J. M. and Blair, L. K. (1971). *J. Amer. Chem. Soc.*, **93,** 3914
98. Brauman, J. I. and Blair, L. K. (1971). *J. Amer. Chem. Soc.*, **93,** 4315

7
Time-of-flight Mass Spectrometry

R. S. LEHRLE
University of Birmingham
and
J. E. PARKER
Heriot-Watt University, Edinburgh

7.1 TIME-OF-FLIGHT MASS SPECTROMETRY (TOFMS): PRINCIPLES AND TECHNIQUES

7.1.1 General reviews and bibliographies

The reviews by Joy[1] and Price[2] provide a clear introduction to the subject of time-of-flight mass spectrometry (TOFMS). A quite comprehensive list of early papers is contained in two publications of the Bendix company[3,4]. More recent work is described in the proceedings of three European TOFMS symposia[5,6] published in 1969, 1970 and 1971. A very comprehensive list of all published papers on TOFMS which are readily available was compiled in 1970 by Joy[7], who grouped the papers according to application. This list includes references to a number of reviews published in various journals. The Mass Spectrometry Bulletin[8], published monthly, provides a continuing source of information on MS, including TOFMS studies. The Scientific Documentation Centre[9] also provides a mass spectrometry information service.

7.1.2 Basic design and theory

The essential principle of TOFMS is very simple: a bunch of ions, usually created by an electron pulse passing through the source, is accelerated by a potential gradient so that all ions acquire the same (kinetic) energy. If the bunch contains ions of different mass, it will therefore contain ions of different velocities, and while it passes down a field-free drift tube it will separate into several smaller bunches each containing ions of the same mass. These smaller bunches strike the detector target (e.g. an electron multiplier) at times which are mass-dependent. If the detector signal is displayed as a single sweep on an oscilloscope, peaks are observed at positions on the time-base according to their mass (more precisely their m/e ratio), and the amplitude of the peaks is proportional to the abundance of the species. Thus an 'instantaneous' record of a mass spectrum is obtained. If the oscilloscope time-base is synchronised with the frequency of the electron impact pulses (e.g. 10 kHz), the oscilloscope displays a 'continuous' record of changes in the mass spectrum. With oscilloscope display, the lower limit of amplitude sensitivity for a specified ionic species is when the abundance of this species corresponds (on average) to one ion per cycle; smaller abundances merely provide a 'single ion peak' less frequently. This problem may be overcome (at the expense of the speed of obtaining information) by scanning the 'continuous' spectrum with a gate which effectively integrates the whole of the signal corresponding to any mass number while it passes through that mass number. Control of the width of this gate and the rate of its scan therefore provides variable control of the sensitivity, resolution and speed of obtaining a single mass spectrum, which (for intermediate scanning rates) can be displayed on a fast u.v.-galvanometer recorder or on a conventional pen recorder (for slow scanning rates). This is the 'analogue output' mode of operation. At slow scanning rates, the sensitivity and resolution are comparable with those obtained by conventional single-focusing magnetic deflection instruments.

Thus a TOFMS is rather more versatile than a conventional magnetic deflection MS in that the operator can choose anywhere between the extremes of recording mass spectra (at low sensitivity) at the rate of one every 100 μs, or (at high sensitivity) at the rate of one in several minutes. There is, however, another important advantage of TOFMS machines: conventional instruments depend for their resolution on precisely defined slit geometry in the source, whereas the resolution of TOF instruments is primarily dependent on the application of appropriate electrical pulses to comparatively large grids in the source unit. Thus the ionisation region in a TOFMS source can be larger and more open than in a magnetic deflection instrument, and this facilitates the introduction of various beams or probes. For these reasons TOF instruments have been much developed for special research applications (see Section 7.3) as well as used for conventional mass analysis (Section 7.2).

Although the TOF principle was first discussed 25 years ago[10], the first detailed analysis of essential design requirements was published in 1955 by Wiley and McLaren[11], and this was the basis of the first commercial instrument produced by the Bendix Corporation. Moorman and Bonham[12], in discussing factors influencing TOFMS resolution, conclude that (allowing for a reasonable signal spread in the detector) a resolution of *c.* 2000 should be achievable, although this has not yet been attained. The paper describes design improvements resulting from a study of the factors causing loss of resolution; with these improvements a resolution ($M/\Delta M$ at 10% peak height) of over 600 was achieved. Damoth[13] has noted Moorman's realisation that the drift-tube length factor cancels out in the basic TOF equation, and that consequently the energy focusing of the TOFMS does not depend primarily on the length of the drift tube. Thus a Bendix instrument (MA-1) has been designed in which there is a flight path of only 20 mm, and the resolution still exceeds 200. Damoth points out that there is a definite relationship between the accelerating and drift voltages; the 1000 V ion energy of the MA-1 limits its resolution to *c.* 250, whereas to achieve a resolution of 1000, 4000 V would be required. The MA-1 instrument incorporates design developments[13] such as solid-state circuitry (giving greater stability), and several features which are especially useful in TOFMS–g.l.c. applications (Section 7.3.9), e.g. automatic repetitive analogue scanning of a selected mass range, a total output integrator, which every eighth cycle integrates the entire mass spectrum (or, if desired, those peaks above a specified mass number, so that He or other carrier gas can be eliminated from the integration), and an optional logarithmic amplitude output. (Dyson[14] has recently described a simple linear-to-logarithmic conversion circuit, so that those instruments with a linear (intensity) response of the analogue will display a boosting of the smaller signals at the expense of the larger. This is useful for the preliminary scan over the complete mass range of a new sample.) The necessity for standardising the definitions of resolution and sensitivity for mass spectrometers working on different principles (TOF, magnetic deflection, quadrupole) has been pointed out by Damoth[15], who also discusses the optimisation of TOFMS resolution and sensitivity. He concludes that the principal factors affecting TOF resolution are time-lag focusing, the flatness of the source grids (e.g. stainless steel grids are better than the original molybdenum ones), the rise-time of the draw-out pulse, and the stability of the accelerating voltage. The principal factors affecting TOF sensitivity are considered to be the time duration available for measurement ('the statistics problem'), the variation of sensitivity with mass number, the dynamic range required, and the residual cycle. In practice the sensitivity can be increased by (i) increasing the duty cycle, (ii) improving the ion transmission percentage and (iii) decreasing the noise and drift in the electrometer and the electronic amplifier outputs. The performance characteristics of a TOFMS (Bendix 3012) have been reported by Bonham *et al.*[16]. This instrument has *c.* 2 × the resolution of previous models and 10 × the sensitivity. It also offers a number of alternative sources, i.e. an ion–molecule reaction source, a nude source which may be cantilevered into a vacuum system for residual gas analysis work, a surface ionisation source, an atmospheric sampling system, and a source featuring direct line-of-sight sampling into the ionisation region for flame and plasma analysis. Wiley and McLaren's original analysis[11] of the focusing action of a TOFMS has been extended by Franklin *et al.*[17–20] to permit the measurement of the initial translational energies of ions from an analysis of their peak shapes. (Ions with excess translational energy exhibit peak broadening.) This technique is valuable in precise determinations of appearance potentials, where an accurate assessment of kinetic energy is required.

Sazone[21] has theoretically analysed the ability of TOF instruments to compensate for the initial ion velocities. It is demonstrated that the Wiley–McLaren two-stage ion acceleration is the limit of energy resolution for

TOF mass spectrometers which employ time-independent fields. An alternative explanation is advanced[21] for the increased resolution obtained by operating the TOFMS in the Studier mode, see Section 7.1.3.1 (and References 41 and 42).

Electron multipliers often used in TOFMS instruments are of the resistive strip magnetic type[22], since these are less susceptible to contamination. A technique to improve signal-to-noise ratio of a multiplier has been described by Young[23], and the dependence of the secondary electron yield (gain per stage) on the electron energy for coated-glass dynode resistance strips has been examined by Streeter *et al.*[24]. These workers found that for electron energies over the range 20–70 eV, the measured mean yield increases almost linearly from about 1.2 to 1.65. The mass-discrimination of electron multipliers has been discussed by Hunt *et al.*[25] and Thomas[26]. A discussion of the merits of various ion detectors is included in the MS instrumentation review by Honig[27].

Various TOFMS instruments have been designed to analyse negative as well as positive ions. Thus the earlier Bendix instruments 12-107 and 14-107 can deal with negative ions, but since the analogue output on these instruments is designed for positive ions only, oscilloscope display must be used for negative ion measurements. The Bendix 3015 is a more versatile instrument[13], since analogue recordings of both positive or negative ions is possible, and incidentally the resolution exceeds 1000. (The model 3012 is the 'positive ions only' version of 3015.) The negative ion operation is achieved by floating the electron multiplier and the analogue electrometers; the ion source is operated at earth potential. The 3015 has been used for negative ion work by Thynne and co-workers[29] (see also Section 7.3.1).

Whilst Bendix instruments are the ones principally used in North America and Western Europe, Elliot Automation[28] manufacture a 37 cm flight tube TOFMS. This instrument cannot be easily modified for specialist applications (e.g. Section 7.3), and the small unit mass resolution (up to m/e 85) and absence of analogue output appear to limit its application to leak detection and residual gas analysis. Several instruments have been designed and constructed in Eastern Europe[30–33]. More specialised designs are discussed in later sections of this review.

7.1.3 Basic techniques and instrumentation

Recent developments in basic techniques and instrumentation may be conveniently classified under the following headings: (1) appearance potential measurements and source modifications, (2) multiplier modification, (3) display and recording techniques (including those suitable for single-scan and time-resolved mass spectrometry), (4) data processing, and (5) unconventional TOF instruments.

7.1.3.1 Appearance potential measurements and source modifications

Kiser and Gallegos[34] have compared ionisation potential results obtained by conventional methods (i.e. linear extrapolation method, logarithmic plot method, extrapolated voltage difference method and energy compensation method) (see Chapter 5). More recently there has been interest in the retarding potential difference (RPD) technique[35]. Delwiche[36] has pointed out that the TOFMS is well suited to the RPD technique because (i) ionisation occurs in a field-free region, (ii) the ion source is near earth potential so that insulation problems are simplified, (iii) the source magnets are outside the vacuum envelope (this simplifies the collimation of the electron beam); and (iv) constant focusing of a particular mass is facilitated by the absence of an analysing magnet. The appropriate modifications to the electronics and the electron and ion optics are described by the author, and the result of RPD measurements on SF_6 are compared with those obtained by photoelectron spectroscopy. The Bendix TOFMS (model 3015) is equipped with an RPD source consisting of five electrodes. However, it is probable that the RPD method does not produce as sharp an energy cut-off as was once assumed[37]; it is more probable that the difference in ion currents measured by RPD is not due to an almost monoenergetic electron slice, but to the difference between two slightly-displaced energy distributions of unequal intensity. This idea is supported by the work of Winters *et al.*[38], who have devised another method (energy-distribution-difference method, EDD) of removing the effect of a large energy spread. Here the width of the distribution is reduced by subtracting from it a fixed fraction of the distribution at a slightly higher acceleration energy. A further method ('deconvolution') has been described and applied[39,40]. This method essentially removes the effect of the electron energy distribution by a computational procedure, but it can be used only if the shape of the distribution is known.

A continuous ion source for a TOFMS has been described by Studier[41,42], this provides a 300 × increase in sensitivity over the pulsed mode of operation, without decrease in resolution or signal-to-noise ratio.

Miller *et al.*[43] have modified their TOFMS so as to improve the base-line stability at source pressures of *c.* 1 torr. The noisy base-line is due to ions drifting into the flight tube during periods when the ion-withdrawal pulse is off. This problem is overcome by replacing the simple ion-withdrawal grid by a gating arrangement which consists of two grids on either side of a slit, which are electrically connected to the top half of the slit while the bottom half of the slit has a standing − 125 V potential (the maximum of the drawout pulse in their instrument). This arrangement deflects all ions from entering the flight tube until the ion drawout pulse is applied to the top half of the gate. Although this modification was intended for use at high pressures, the reviewers feel that it might be usefully incorporated for work at conventional pressures, thereby eliminating one source of noise in the final signal (see Section 7.1.3.2).

7.1.3.2 Multiplier modification

The basic properties of electron multiplier ion detection and pulse counting methods in mass spectrometry have been discussed by Dietz[44], and a technique for improving the signal-to-noise ratio of the electron multiplier has been described by Young[23].

A modification of magnetic electron multipliers involves the application of a synchronised gate pulse to the acceptance end of the multiplier so that superfluous portions of the mass spectrum are blanked out[45]. This technique has been developed by Studier and co-workers[46, 47], who point out that dirtying of the dynode strip is thereby retarded, multiplier gain is increased and non-linearity of the multiplier is minimised. (In normal multiplier operation, a large increase in the abundance of mass 28 could reduce the intensity of a peak at 200 by a factor of three[46].) The modification is especially valuable in linked g.l.c.-TOFMS equipment, where very large ratios of carrier gas: sample may be present (see Section 7.3.9).

7.1.3.3 Display and recording techniques

Fast display and recording techniques have been comprehensively reviewed by Meyer[48], who stresses that although the various techniques may have been devised originally for specific applications (e.g. time-resolved mass spectrometry in flash photolysis work), they must all be considered to be more generally applicable. Meyer's classification may be summarised as follows:

(a) *Oscilloscope z-axis gating method*—This displays ion intensity (y axis) versus time (x axis) for any chosen mass in the spectrum. Use of dual-beam oscilloscopes and multiple gate-marker circuits allows the recording of several masses simultaneously. The disadvantages are that a transient event may cause a slight displacement of the peak, and that precise positioning of the gate for intermediate species such as atoms or radicals may be very difficult.

(b) *Synchronised multiplier anode gating*—The display here is the same as in (a), but it is achieved by conventional gating of the electron multiplier. The advantages of (b) are that the total ion current for a given mass is measured, rather than the peak height only, and that by the use of oscilloscope-input filter capacitors, the statistical fluctuations of the ion beam can be filtered out to produce a continuous line recording.

(c) *Three-dimensional intensity-modulated raster*—The x- and y-axes correspond to mass and time respectively, and the z-axis (i.e. the intensity of the electron beam) corresponds to ion intensity. All the mass peaks are thus intensity modulated, and a qualitative record of the time-variation of the entire mass spectrum is obtained. This method is useful for the preliminary detection of peaks sharing interesting kinetic behaviour.

(d) *Two-dimensional linear intensity-modulated display*—In this method the mass spectrum at $t = 0$ is displayed vertically as a series of intensity-modulated spots (i.e. base-line eliminated). These spots are moved in the x direction at a speed commensurate with the duration of the experiment. Changes in the abundance of any species are displayed as vertical displacements of the spots. The advantage of this method over (c) is that quantitative abundances can be estimated. The method is limited by peak overlap; this restricts the portion of a spectrum which can be displayed on a single oscilloscope.

(e) *Multiple trace raster display*—A standard oscilloscope display (i.e. ion abundance *v.* mass number) is successively displaced along the vertical dimension by means of raster operation. Clearly the number of spectra is limited by overlap of the larger peaks; Moulton[49, 50] reduced this problem by using five identical circuits and oscilloscopes in series to obtain 35 consecutive mass spectra for each shock-tube experiment. The Bendix Corporation have also tackled the problem by horizontally offsetting the mass spectrum each time it is displayed; the number of traces can then be increased to a maximum of 64. A limitation of this method is that the maximum peak height resolution is only about 1 part in 20, due to the finite size of the electron beam spot.

(f) *The drum camera method*—This technique, first applied to TOFMS by Kistiakowsky and Kydd[51], involves the time-resolution of successive traces by the rapid vertical motion of a strip of film past the oscilloscope. Photographic sensitivity is the main problem; improved legibility of the film records is achieved by using a 24 kV accelerating potential for the electron beam of the oscilloscope. The technique has been widely used in flash photolysis work (see Section 7.3.8) where it is necessary to synchronise the flash lamp with the pulsed spectrometer and suitably control the shuttering action of the camera. The advantage of this method is that up to 120 consecutive cycles of the TOFMS output can be recorded with only one oscilloscope and camera.

Synchronisation problems in time-resolved mass spectrometry have been discussed by Kende[52] and Greene[53]. Kende devised a circuit to start a pre-selected number of TOF scans at a pre-selected time during a reaction. Greene's procedure is to make the mass spectrometer a slave to the process and cycle the mass spectrometer only at the time of interest. The only additional equipment required is a delay generator. He used the technique to measure the velocity distributions of species in molecular beams, but points out that it can be applied to any repetitive process in which time-resolved spectra are required.

The ultimate single-cycle sensitivity in time-resolved mass spectrometry has been discussed by Meyer and Freese[54], who obtained a value of 1 part in 100 000 for pulsed operation at 20 kHz, corresponding to an absolute abundance of reactive species of 0.5×10^{-7} mol l^{-1}. Bimolecular rate constants approaching 10^{11} l mol^{-1} s^{-1} may be studied with this sensitivity and time resolution. The results were achieved in flash photolysis experiments and the sensitivity was enhanced by suitable geometrical coupling of the flash lamp and reaction vessel to permit efficient transfer of the u.v. radiation to the absorbing gases.

As mentioned in Section 7.1.2, a u.v. galvanometer recorder may be used to display the TOFMS analogue output at intermediate scanning rates. D'Oyly-Watkins and Gaythorpe[55] have recently described an amplitude-limiting device for such recorders. This is desirable because ions in high abundance can produce currents greater than the maximum safe limit of the galvanometer recorder. Their design can be switched easily to give limits of $\frac{1}{3}$, $\frac{1}{2}$ and full scale.

7.1.3.4 Data processing

Meyer[48], in a review of data recording and processing systems, has listed the following characteristics of an ideal single system: broad mass range coverage, mass identification, large dynamic amplitude range, high amplitude resolution, high frequency response, high signal-to-noise ratio, ability to record a large quantity of data, capability of dealing rapidly with successive experiments, direct conversion of amplitudes and masses to digital information, and computer data processing. He then refers in detail to three aspects which may be summarised as follows:

(a) *Magnetic tape recording* – Here all masses and amplitudes can be recorded rapidly, and recovered at a later time. The restricted frequency response of available tape recorders limits the scope of this technique.

(b) *Analogue-to-digital conversion* – A semi-automatic device ('Telereader magnifier') requires manual alignment of a pair of x- and y-axis cross-wires with both the base and peak of a mass signal; the peak height and mass are then recorded on a punched card. An improved automatic film scanning and digiting device ('Image digitiser') has a resolution of one part in 500 (vertical) and one part in 660 (horizontal). The CHLOE system provides a method of digitising data recorded in the form of a waveform of ion intensity versus time, but is restricted to data recorded on 35 mm film. The analogue to digital converter (ADC) plus computer memory system has been used by several workers for slow analysis applications; for very fast time-resolved mass spectra the principal limitations are the low frequency response of ADC units and the cost of the system.

(c) *Computer data processing* – For time-resolved mass spectrometry, it is desirable that the computer program should refine the raw data (e.g. eliminate time-dependent systematic variations and impurity signals and compensate for the changing magnitude of dissociative ionisation contributions to mass signals as the parent molecule concentrations change due to the chemical reaction). Such a program (CDC 3600 Fortran) has been described[56]. It can accommodate 50 masses, 250 equally-spaced time points, and amplitudes varying by a factor of 10 000. More recently Fallgatter and Hanrahan[57] have written a program which can accommodate 250 peaks plus 150 background peaks (the background signal being subtracted) up to mass 300. As mentioned in Section 7.1.2, instruments are now being designed with mass scales which are stable for long periods of time; this should allow the use of programs using an optimum-delay jump-scan mode of operation, thereby both increasing the sensitivity and reducing the amount of data to be handled.

7.1.3.5 Unconventional TOF instruments

Carrico *et al.*[58, 59] have developed a novel TOFMS in which a series of grids along the flight tube are pulsed in such a manner that ions of different mass (which will have arrived at grids at different distances from the source) are turned back and arrive at an electron multiplier at different times. Further development of this prototype instrument will help to increase TOFMS resolution by reducing the spatial spread and translational energy distribution of the ions.

French and Locke[60] have developed a TOF analyser for determining the velocity distributions of molecular beams of velocity up to 10 000 m s^{-1}. Crosby and Zorn[61] have built a residual gas analyser and leak detector of similar design to that of French and Locke; both instruments use the TOF of neutral metastable atoms and molecules. The metastable atoms are formed[61] by a 10 μs electron pulse and move with gas kinetic velocity to a windowless photomultiplier cathode which feeds an electron multiplier. Any ions which are formed are removed by a transverse electric field. The ion and electron pulses are counted by a delayed-coincidence circuit, and by varying the delay the mass spectrum can be scanned. Clampitt[62] has designed a similar TOF instrument which is capable, however, of resolving atoms with varying amounts of excitation, e.g. the 29 eV electron bombardment of H_2 produces photons, two groups of H^* and H_2^*. Clampitt[62] has been able to obtain threshold energies for the production of these metastable states.

7.2 APPLICATIONS TO CONVENTIONAL MASS SPECTROMETRIC ANALYSIS

7.2.1 Inorganic analyses

Many inorganic compounds and minerals are involatile, and the study of these requires the special method of Section 7.3. Here we consider a few representative analyses which may be classified as inorganic.

Both continuous and pulsed source TOFMS have been used to observe the mass spectra of the gaseous oxides and oxyacids of xenon and iodine[63]. XeO_4, XeO_3, HIO_4 and I_2O_5 are all sufficiently volatile for mass spectra to be observed by appropriate techniques. TOFMS has also been used[64] to follow the purification of the elusive xenon dioxide difluoride, XeO_2F_2; both positive and negative ion mass spectra were recorded. The mass spectrum of xenon trioxide difluoride has been reported more recently[65].

Some astatine compounds (HAt, CH_3At, AtI, AtBr and AtCl) have been observed and identified using a TOFMS with its source in continuous operation[66]. The stabilities of alane, gallane and indane have been compared by TOFMS study of these gaseous hydrides formed by atom reactions in a fast flow system[67]. The bond energetics and implications of these results are discussed.

TOFMS has also contributed to studies in silicon–fluorine chemistry. These include work on silicon difluoride[68, 69] and perfluorosilanes[68], silicon–boron fluorides (Si_2BF_7, Si_3BF_7, Si_3BF_9, Si_4BF_{11})[70] and the reaction of silicon difluoride with perfluoroaromatics[71], aromatics[71], iodine[71a] and trifluoromethyl iodide[71b]. The chemistry of BF and some of its derivatives[72] and the synthesis of thioborine[73] have also been examined.

A number of reactions have been investigated by TOFMS, for example the reactions of oxygen atoms with hydrogen and ammonia[74], the role of the nitroxyl molecule in the reaction of hydrogen atoms with nitric oxide[75] and the thermal decomposition of ammonium perchlorate[76]. More specialised investigations include a TOFMS study of gases associated with explosions of burning zirconium droplets[77] (here a fast recording oscilloscope display was triggered by the appearance of the mass peak of an inert gas released from a quartz capillary crushed simultaneously with the sample), an investigation of the vapour effusing from a cell of yttrium chloride[78], an examination of the volatile products of the surface reaction between chlorine and heated polycrystalline yttrium[79], and a study of the high-temperature reaction between nickel and chlorine[80].

7.2.2 Organic and organometallic analyses

An obvious use of mass spectrometry in organic chemistry is to identify the products of reactions. TOFMS has been applied in this way to study the products from reactions of some fluorinated and non-fluorinated aromatic compounds with silicon difluoride[71], the exhaust products from an air-aspirating diesel fuel burner[81], the products and intermediates formed by discharge in ammonia–carbon tetrachloride mixtures[82], the products from the liquid-phase oxidation of paraffins[83], the products of the radiolysis of pure liquid cyclopentane[84] and the products from the ionic addition of HCl, HBr and HI to allene and methyl acetylene[85]. It is of interest in the last case that substituted cyclobutane products were observed in addition to conventional Markovnikov adducts.

In some cases TOFMS has been used to study reactions or fragmentation processes in more detail. Thus there has been discussion of the fragmentation behaviour of unsaturated methyl esters[86], of alkyl and aromatic isocyanates[87] and of substituted (methyl, benzoyl, sylyl or syloyl) propionic and butyric acids[88]. In the latter work the proposed fragmentation mechanisms were supported by work with selectively deuteriated compounds. TOFMS has also been used to clarify the rearrangement–dissociation processes of neopentyl esters[89]; such processes are evident in most alkyl esters of oxygen acids. The rearrangement of some piperidine *N*-oxides to hexahydro-1,2-oxazepines was studied by MS analysis of the products from the reaction in solvent at 170 °C [90]. The acid-catalysed (45 % acetic acid) cleavage of some cyclopropyl ketones related to lumisantonin has also been examined[91].

An interesting technique involving a combination of paper chromatography and MS has been developed by Studier *et al.*[92]. The possible presence of purine and pyrimidine bases in meteorites and terrestrial deposits would be of biological significance. G.L.C. of these compounds would be impractical due to their low volatility. A mixture of eight bases was analysed by paper chromatography in two dimensions, and the bases were extracted with 0.1 M HCl and evaporated on to a thin platinum ribbon. This filament was placed in a TOFMS source near to the electron beam, and the filament was heated to distil off the base. It was found that the HCl salts of the bases had the same spectrum as the free bases, which have large molecular peaks. It should be possible to use a similar technique with thin-layer chromatography.

Studies of organometallic compounds include an investigation of the fragmentation behaviour of benzene cyclopentadienylmanganese(I)[93], and of dimethylzinc, trimethylaluminium, and trimethylantimony[94]. The doubly charged transition metal-carbonyl ions have also been examined[95].

Some natural products contain volatile components which have been identified by TOFMS. Thus the volatile components in lipids[96], fish protein concentrate[97], oranges[98] and grapes[99] have been characterised. The fate of volatile hop constituents in beer has also been studied[100].

7.2.3 Medical analyses

Medical applications of TOFMS have been mainly concerned with the effects of breathing abnormal atmospheres. Thus ten years ago a study was reported on the changes in lung volume and alveolar gas concentration during breath holding after breathing oxygen at reduced pressures[101]. There have also been observations of the effects of the exposure of rats to a space cabin atmosphere (98 % O_2 at 258 mm Hg) for two weeks[102]. There have been several papers concerned with the identification of thermal decomposition products from fluorocarbon polymers and the assessment of their toxicity towards laboratory animals[103]. In this connection, the products from polytetrafluoroethylene have been most widely studied; for example, when this polymer is heated below 500 °C the principal toxic component is a particulate material which may have other toxicants absorbed on it (e.g. perfluoroisobutylene)[104]. However, when the polymer is pyrolysed in air above 500 °C, COF_2 is the predominant product in the range 500–650 °C, whilst CF_4 and CO_2 are the major products above 650 °C [105]. In this work, experiments with rats showed COF_2 to be the most toxic component. In further experiments with dogs, rabbits, guinea pigs and mice[106], pathology revealed changes in the lungs and livers of exposed animals; clinical changes such as inhibition of metabolism have also been described[107]. A further topic of interest is the pyrolysis of tobacco. In some work[108] sponsored jointly by the Swedish Tobacco Monopoly and the Swedish Cancer Society, attention is drawn to the problem of secondary pyrolysis (i.e. pyrolysis products initially condensing in the unburnt tobacco are pyrolysed again when the combustion zone approaches). For model studies, work on fractionated tobacco tar is in progress.

7.3 SPECIAL APPLICATIONS

7.3.1 Basic ionisation processes and energetics

The following topics are included under this heading: (1) ionisation potentials, appearance potentials, and energetics, (2) studies of multiple ionisation, (3) ion collision processes and ion decomposition processes in the TOFMS (including dissociation in the drift tube), and (4) studies of negative ions.

7.3.1.1 Ionisation potentials, appearance potentials and energetics

Appearance potential and ionisation potential measurements have been mentioned in Section 7.1.3.1. The appearance potentials of some cobalt carbonyl hydrides[91] and of the dimetallic and complex carbonyls[110] have been measured; in the latter case there appears to be a stepwise loss of CO groups, and mechanistic processes were proposed on the basis of energetic arguments. In some work on arsenic compounds[111, 112] an estimation of the As—F bond energy and of the heats of formation of AsF_3 and AsF_5 were made from appearance potential measurements in electron impact studies of AsF_3 and AsF_5 [111] and estimates of the $H_2As—SiH_3$ bond dissociation energy and heat of formation of $AsSiH_5$ were made from appearance potential measurements on the latter compound[112]. Appearance potentials of ions from a series of trimethylsilanes have been measured and molecular and ionic heats of formation and bond dissociation energies obtained[113]. The appearance potentials and standard heats of formation of a number of methoxy- and halogeno-substituted methyl ions have been determined[114] and the stabilisation energies of some substituted methyl cations have been derived from a systematic investigation of their appearance potentials[115]. TOFMS has also been applied to estimate dissociation energies of the gaseous monoxides of the rare earths[116] and heats of formation and bond dissociation energies have been determined from appearance potentials for parent and fragment ions of ruthenium and osmium tetroxides[117]. Franklin *et al.*[17–20] have evaluated heats of reaction and proton affinities by methods which involve estimation of the excess translational energy of the ions from their TOFMS peak shapes. Dyson[118, 119] has modified a Bendix 14-101 in order to obtain neutral fragment mass spectra of simple polyatomic molecules. A short pulse of electrons produces molecular ions with neutral and ionic fragments. The ions are removed from the source region by a clearance pulse and a further electron pulse is applied. The second beam of electrons produces a normal mass spectrum plus a contribution from the neutral fragments. Dyson[119] has obtained the ionisation cross-sections (*c.* 0.5×10^{-16} cm^2 $atom^{-1}$) for the formation of H^+ from hydrogen atoms produced by the 100 eV radiolysis of ammonia, water and methane. If there is any release of kinetic energy upon fragmentation, however, these cross-sections will be lower limits due to the diffusion of the neutral species away from the source region. Many other examples of energetic studies are listed by Joy[6].

7.3.1.2 Multiple ionisation

Shadoff[120] has discussed the general problem of the analysis of ionisation efficiency curves with respect to multiple ionisation processes. In a study of multiple ionisation in the rare gases (Ne, Ar, Kr, Xe) [121] electron impact energies up to 200 eV were employed; ions with charges up to 9 were observed and corresponding ionisation potentials reported. The relative yields of multiply charged krypton ions have also been investigated as a function of the maximum energy of the x-rays used[122]. A TOFMS study of the decomposition of multiply charged ions into singly-charged fragments was carried out using electron impact energies of 1 keV [123]. In this work the ions and electrons were accelerated in opposite directions to multiplier detectors and the masses of the ions were found by measuring the time interval between the electron and ion pulses by delayed coincidence. The formation of positive ion pairs was demonstrated.

7.3.1.3 Ion collision, decomposition and dissociation processes

Some years ago, Ferguson *et al.*[124] described the observation of the products of ionic collision processes and ion decomposition in a TOFMS. The technique used was to apply a retarding field near the end of the flight path; neutral species and fragment ions formed during flight were thereby separated from their parent ions. Metastable ion decomposition, collision-induced decomposition and charge-exchange processes were studied in this way. More recent studies have been reported by Hunt *et al.*[125, 126] and others[127–132]. Hunt discusses the observation and identification of ion dissociation processes occuring in the drift tube[125] and the recognition of spurious fragment peaks when using the TOF potential-barrier technique[126]. In the latter case the method used was to make flight-time shift measurements at more than one value of the distance from the ion source to the potential barrier. Fromont and Johnsen's[130] method involved the admission of gas to the flight tube, which could be differentially pumped relative to the source. TOFMS has also been used to investigate the bimolecular formation of Ar_2^+ [133] and He_2^+ [134]. In both cases excited states of the monatomic molecule are involved; in the helium case there is the possibility that three families of excited states take part in the process and radiative lifetimes and rate constants were derived for each family. The rate constants for Ar_2^+ formation, and the cross-section for He_2^+ formation, were compared with theoretical predictions. Other examples of ion–molecule interactions in the source are given in Section 7.3.2. The mode of decomposition of doubly charged transition metal-carbonyl ions has also been investigated[135]. In this work the inlet to the TOFMS was modified in order to increase the sample pressure in the ion source.

7.3.1.4 Studies of negative ions

A detailed discussion of the experimental aspects of negative ion studies has been published by Thynne *et al.*[40, 136]. These authors point out that negative ions are usually formed much less abundantly than positive ions and often with excess kinetic energy, which reduces their collection efficiency. However, negative ion mass spectra are often simpler than those of positive ions, since large negative molecular ions do not undergo extensive decomposition, whereas positive ion mass spectra are complicated by rearrangement processes. Studies on negative ions can provide data on electron affinities of atoms, free radicals and molecules, and are essential to an understanding of flames and combustion processes.

Several problems have been encountered in TOFMS studies of negative ions[136]. Thus many of the compounds which most readily form negative ions lead to rapid corrosion of the heated filament used for electron impact studies. Also, when working with a TOFMS (Bendix 3015), the automatic trap current control maintains the trap current constant only if the electron energies exceed 7 eV; at lower energies the emission must be controlled manually or correction must be made for the variation in trap current. The energy spread of the electron beam creates the same problems as in positive ion spectroscopy (see Section 7.1.3.1). Harland *et al.*[40] have studied negative ion formation at low electron energies and removed the effect of energy spread in the electrons by a deconvolution procedure (see Section 7.1.3). This was tested by convoluting known distributions and then deconvoluting. The electron energy distribution was obtained by measuring the SF_6^- peak as a function of electron energy. A resonant electron-capture ionisation curve is then deconvulated (using the experimental electron energy distribution) by means of 15 smoothing and iteration cycles. Using this technique Thynne and co-workers have obtained a wealth of thermodynamic data mostly for compounds containing the more electronegative elements.

The lifetime of a negative ion is the time between its formation by a resonance capture process and the re-emission of the captured electron with the return of the ion to the neutral state. If the lifetime is large enough (*c.* 10 μs) the charged species may be observed mass spectrometrically[136]. Thus Edelson *et al.*[137] have separated negative ions and neutral species in a TOFMS by applying a retarding potential to the ion beam just before it reaches the ion detector. The ions are thereby slowed down and appear at higher mass number, whilst the neutrals are unaffected. Compton *et al.*[138] have achieved this objective in a modified TOFMS (Bendix 14-206) by applying a flat-topped retarding potential to the 'ion time-of-flight adjust' cylinder 170 cm from the ion source.

There is evidence that slow secondary electrons (produced by electron impact processes) may interact with molecules to form negative ions[136]. This depends upon the fact that these product ions are second order in total gas pressure, whereas the ionisation efficiency curves for the product ions cannot be related to the curves for any primary product ions in the system.

Winters and Kiser[139] studied negative ions produced by electron bombardment of metal carbonyls. Using a TOFMS (Bendix 12-100) they measured the relative abundances of the negative ions produced as a result of ion-pair processes at 70 eV and the energy of the dissociative capture maxima for each ion. They proposed a reaction mechanism and discussed it in terms of the quasi-equilibrium theory.

Because of the different modes of formation of negative ions, the ions observed and their abundances in negative ion mass spectra are markedly dependent upon electron energy. Thynne *et al.*[136] discuss the negative ion mass spectra of the following compounds: C_2F_6, CF_3H, CCl_4, $CHCl_3$, CH_2Cl_2, CH_3Cl, CH_3I, $CHCl_2CHCl_2$, CCl_3CH_2Cl, CCl_2CCl_2, CF_3COCF_3 and SF_6. They conclude that negative ion formation is more extensive in Cl compounds than in F or I compounds and that fragmentation tends to favour the formation of ions containing an odd number of halogen atoms but not to produce mixed ions such as CHX^-. In general, large symmetrical multiple-bonded molecules appear to undergo less fragmentation than molecules without these characteristics.

Negative ion studies by TOFMS have been reported for several compounds, e.g. allyl alcohol and acrolein[140], and hexafluoroethane, 1,1,1-trifluoroethane and fluoroform[141]. The ion–molecule reactions of negative ions of sulphur have also been investigated[142].

7.3.2 Gas-phase ion–molecule interactions

In Section 7.3.1.1 it was mentioned that ion–molecule collisional processes may occur in the source region of a TOFMS. Further examples of TOFMS studies of such collisional processes may be quoted; these include those leading to the formation of Ar_2^+ [143,144], N_3^+ [145]; those involving ethanol[146], methanol[146], cyclohexane[147], nitrogen[148], α-particle-irradiated methane and water vapour[149], various ions in argon[150] and some negative ion–molecule reactions[151,152]. The general problems of observing the products of ion collision processes in a TOFMS[153] and of flight-time analysis of scattered particles[154] have been discussed.

Two basic experimental approaches have been used in TOFMS ion–molecule studies (but see also Section 7.3.1). The first of these is similar to the technique originally used with magnetic deflection instruments, i.e. to increase the source pressure in order to encourage ion–molecule collisions during the residence time of ions in the source. Most of the work mentioned above was performed using this 'high pressure method'. The ion–molecule reaction source introduced by the Bendix Corporation and described by Tiernan *et al.*[155] operates essentially on this principle.

Hughes and Tiernan[156] have studied the high pressure (up to 1 torr), ion–molecule reactions of cyclobutane at thermal energies (up to 0.3 eV). $C_4H_8^+$ and $C_4H_7^+$ are apparently unreactive with cyclobutane molecules, although the molecular ion reacts with various added molecules by what appears to be a rearrangement to the linear butene ion. Potzinger and Lampe[156a] have studied the ion–molecule reactions of Me_2SiH_2, Me_3SiH and Me_4Si. Beggs and Lampe[157–160] have examined the high pressure (10^{-3} torr) ion–molecule reactions of silane mixed with methane, ethylene, acetylene and benzene. The cross-reactions between silane ions and hydrocarbons and between hydrocarbon ions and silane were identified and the thermal energy rate constants were measured. Beggs and Lampe[159] have evidence, from deuterium labelling, that in the SiH_4^+/C_2H_2 reaction a symmetrical cyclic complex is formed, whereas in the $C_2H_2^+/SiH_4$ reaction a linear complex is formed. In the silane–ethylene system[158] there is evidence for a long-lived complex in which complete randomisation of the hydrogens has occurred. This situation may be compared with the CH_4^+/SiH_4 system[160] where the hydrogens retain their identity; this result is consistent with the report of Herman *et al.*[161] that the CH_4^+/CH_4 reactions proceeds via a direct mechanism. Beggs and Lampe's[160] results for the $C_6H_6^+/SiH_4$ system indicate non-randomisation of the hydrogens, which they explain as the formation of a silicon-containing side group on the ring and not the formation of an analogue of tropylium ion. Fluegge[149] has measured the lifetime of the complex $(C_2H_7^+)^*$ from the collision of CH_3^+ and CH_4, as 0.18 μs. Methane at 0.2 torr was subjected to α-particle radiolysis and the product ion and the corresponding α-particle were delay-coincidence counted. These 'high pressure methods' permit kinetic and mechanistic information to be obtained from an examination of the variation in mass spectrum with pressure and time. Thus in a study of ion–molecule reactions in acetonitrile and propionitrile[162], is was shown that the principal product is the $[M+1]$ ion in each case; that with CH_3CN, $[M+CH]^+$ is formed in moderate abundance, and that with C_2H_5CN the ions in moderate abundance are $C_2H_5^+$, $C_2H_6^+$ and CH_4CN^+. The rate constants were calculated and fully deuteriated compounds were used in order to establish the composition of the product ions. Deuterium itself has been used in the study of ion–molecule reactions in a mixture of CH_4 and D_2 [163]. The use of a very sensitive TOFMS to study ion–molecule reactions at high pressures has been discussed by Conway[164]. Young, Edelson and Falconer[165] have used a TOFMS to investigate the clustering reactions of H_3O^+ to form $H_3O^+ (H_2O)_n$ in a *c.* 1% mixture of H_2O in Ar. They have been able to compute the first three rate constants as $k_1 = 6.9 \times 10^{-28}$, $k_2 = 1.2 \times 10^{-27}$ and $k_3 = 4.2 \times 10^{-28}$ cm^6 s^{-1} at 337 K. The principle limitations of the high-pressure method are that it is not easy to specify which of the components of a mixture will be the reactant ion and which will be the reactant molecule and that there is only limited control of the relative translational energy of the collisions. The latter is quite important in that mechanistic information about the type and energetics of the reaction process can often be obtained from the dependence of the cross-section on translational energy (see later).

The other basic experimental approach is the beam method. The minimum requirement here is to produce a focused beam of specified reactant ions by means of an auxiliary source and to allow these to interact with the reactant gas in the source of the TOFMS. Greater control of translational energy is then available, since the reactant ions can be electrostatically accelerated to any chosen velocity before entering the interaction zone[166–169]. Greater versatility is achieved if two mass spectrometers are set up in tandem; the first selects as the reactant ion any molecular or fragment ion, these are then accelerated into the reactant gas in the source of the second mass spectrometer which analyses the products[170]. The modification of a TOFMS to achieve selection of reactant ions for mass and energy by a repeated time-of-flight principle has been described by Knewstubb and Field[171]; it is projected that ion pulses produced by this technique will be used for the study of ion–molecule reactions.

At the low translational energies (thermal to several tens of eV) available with the high-pressure method, the most common reaction mechanism involves atom or group transfer. Reactions of this type have also been studied by the beam method in a TOFMS, for example the variation of cross-section with translational energy for the reactions $Ar^+ + D_2 \rightarrow ArD^+ + D$ and $N_2^+ + D_2 \rightarrow N_2D^+ + D$ are discussed in references 168 and 172, and the reactions of O^-, OH^- and NO^- with a large number of organic and inorganic molecules have been reported[170]. In the latter study, the general conclusions were that whilst NO^- is usually unreactive except with alkyl halides, O^- usually abstracts H to give OH^-, and OH^- reacts with alcohols by abstracting an H^+ from the hydroxyl group of the alcohol. The results of a high pressure source TOFMS study of proton-transfer reactions in ethanol have been compared with those obtained using a tandem mass spectrometer technique[173]. The kinematics of some D-atom transfer reactions in the energy range 1–100 eV have also been studied[174]. In atom-transfer work by the beam method the translationally energetic reactant ion 'picks up' the transferred atom (stripping process or orbiting complex mechanism) and continues on its way as the product ion. Both reactant and product ions may be deflected into the TOF analyser, and the ratio of the intensity of product to intensity of reactant is found as the ratio of the two corresponding peaks.

The situation in charge transfer reactions (i.e. $A^+ + B \rightarrow A + B^+$) is rather different in that the maximum cross-section may be observed at high relative translational energies (several orders of magnitude greater than atom transfer reactions), provided that the process is not a resonant one. Such electron jumps can occur at impact parameters considerably in excess of the sum of the van der Waals radii; the result of this is that the product molecule (A) proceeds with the translational energy of the reactant ion (A^+), whereas the product ion (B^+) will be essentially stationary (i.e. thermal velocity), as was the reactant molecule (B). Such processes are conventionally studied by the beam method, in which the reactant beam (A^+), previously accelerated to high velocity, passes through the TOF source (containing molecules B) in a direction perpendicular to the flight axis. The reactant intensity is then measured by a receiver Faraday cup in line with the beam A^+, and the low velocity product ions B^+ measured by the TOF detector. This approach has been used to measure relative cross-sections for several charge-transfer reactions and the energetics have been interpreted in terms of the 'near adiabatic' theory[168, 175]. An extension of the technique allows absolute cross-sections to be measured (by the so-called condenser method); full details of this TOFMS modification, and discussion of the energetic implications of the results for the system N_2^+/N_2, N_2^+/Ar, Ar^+/N_2, and Ar^+/Ar, have been published[176]. The development of a TOFMS working at 10^{-8} torr, and its use for the study of a charge transfer reaction, has also been described[177].

Charge-transfer reactions may be much more complicated if the reactant molecule is polyatomic. Dissociative charge-transfer reactions such as $A^+ + XYZ \rightarrow A + XY^+ + Z$ or $A^+ + XYZ \rightarrow A + X^+ + YZ$ may then occur, in many cases more readily than the direct charge-transfer process. For example in a study of dissociative charge-transfer reactions of propane by a TOFMS technique, 13 fragment ions were observed in large or moderate abundance, and others could be detected, in addition to the propane ion[178]. In this work the cross-sections of the individual fragmentation processes were deduced from condenser-method measurements of the overall cross-section for all processes combined with TOFMS analysis of all the products. Tentative mechanisms were proposed on the basis of the translational energy dependence of these cross-sections. This approach ('Translation energy ion spectroscopy', TES) was later coupled with another technique ('Time-resolved ion spectroscopy', TRS) in a more detailed examination of the dissociative charge transfer of propane by Xe^+, Kr^+ and Co^+ [179]. In this study the results indicate that the mechanism of each of the dissociative charge-transfer systems involves the participation of several excited precursor ions, some of which exist in several excited states of different energy. This implies a rather more complex assembly of reactions than has formerly been assumed.

7.3.3 Surface phenomena

The following topics are covered in this section, (i) gas–surface reactions, (ii) sputtering of solids, and (iii) combined field-ion microscope TOFMS.

Knudsen cell and heated inlet work is discussed in Section 7.3.4; flash pyrolysis and laser–TOFMS are covered in Section 7.3.5.

7.3.3.1 Gas–surface reactions

McKinley[79, 180–182] has studied the surface reactions of heated nickel with F_2[182], Cl_2[181] and Br_2[180] and heated yttrium with Cl_2[79]. Gas pressures in the range 10^{-7} to 10^{-4} torr were used and the volatile products (the temperature range was 300–1600 K depending on the metal–gas system) were sampled by TOFMS. For the nickel systems the species NiX and NiX_2 are formed at rates first order in gas pressure, the formation of the NiX being more important in the higher temperature range. The yttrium–chlorine system yields the gaseous products Cl, YCl_2 and Cl_2, whilst YCl and YCl_3 were not detected. McKinley[79] has interpreted the results as consistent with a strongly chemisorbed layer of Cl atoms stable over the entire temperature range, upon which more chlorine adsorbs as Cl atoms. These either evaporate or react with YCl to form volatile YCl_2. An activation energy of 55 kcal mol^{-1} for YCl_2 evaporation is apparently associated with dissociative adsorption of Cl_2 onto the chemisorbed Cl layer.

Several groups of workers[183–185] have studied the positive ion emission from rapidly heated tungsten filaments by photographing the TOF mass spectrum at 300 frames per second. The emission was found to be dependent upon the heating rate and temperature and it occurred in ion bursts[184]. At sufficient O atom concentration desorption occurs in two stages: low-temperature desorption as oxides and high-temperature desorption as atomic O[183]. Ageev and Ionov[185] found that the oxygen was adsorbed in two different states with desorption activation energies of 7.8 ± 0.6 and 4.1 ± 0.1 eV. In agreement with Ptushinskii and Chuikov[183] the oxygen was desorbed in the atomic state[185] at high temperatures (2400 K), whereas, for greater than 1% coverage, hydrogen and nitrogen are desorbed in the molecular form.

7.3.3.2 Sputtering of solids

The sputtering (i.e. ion bombardment) of solid surfaces in a TOFMS has been reviewed by Thomas[186]. An experimental arrangement[166] consisting of an external Nier source producing an ion beam (continuous or pulsed) with a chosen translational energy (2 keV or less) was used to bombard a metal, polymer or inorganic salt target. The target was placed in the source region of a Bendix 14-101 instrument enabling the spectrum of the sputtered positive ions to be observed. The total electron and positive ion yields were also measured by a total charge collection arrangement. A retarding potential electrode system allowed the kinetic energy distribution of the positive sputtered ions to be measured. The experimental problems have been reviewed by Thomas[186] and Dillon[187]. For example, the continuous ion bombardment of polymer sheets or powder leads to charging of the polymer which causes a pulsating spectrum; this has been overcome by using one or more of the following techniques (i) pulsed ion beam, (ii) neutral beam[188], (iii) mixing of graphite with the polymer powder, and (iv) use of thin sample films (50–500 Å) deposited from a polymer solution. The possibilities of using sputtering of metals to determine surface concentrations of impurities has been assessed through studies of samples of known bulk composition, and a series of intensity coefficients has been determined for some ferrous alloys[186, 187]. The use of the ion-bombardment techniques for investigating surface reactions, such as the synthesis of ammonia on pure iron, has been recently developed[189]. It is possible in this way to obtain information on the relative concentrations of chemisorbed species that occur on a catalyst surface. From this study two feasible kinetic schemes were analysed by use of a steady-state approximation.

These sputtering experiments and the gas–solid studies of the previous section are obviously relevant when considering heterogeneous catalysis and adsorption mechanisms.

Woolley, Warner and Poole[190] have designed and built a TOFMS capable of examining cascade damage in nuclear material. A mass-analysed ion beam is chopped electrically and accelerated to a selected energy of up to 100 keV. The reactant ion beam strikes either a thin solid target or a gaseous target, and the scattered reactant ions, together with any knocked-on target ions, pass into a TOFMS through a collimating slit. The flight tube is 175 cm long and is connected by flexible bellows so that the TOF instrument may be rotated from 0–90 degrees relative to the reactant ion beam. This very versatile instrument should greatly increase our understanding of sputtering processes as well as that of gas-phase ion–molecule reactions.

7.3.3.3 Combined field-ion microscope TOFMS

Muller *et al.*[191–195] Brenner and McKinney[196] and Turner and Southon[197] have combined a pulsed field-ion microscope (FIM) with a TOFMS. By tilting the metal tip both the atomic structure of the solid (microstructure in metals and surface phenomena) and the chemical identity of an individual atom as seen in the fully resolved image, can be obtained. The field-emission ions pass through a collimating slit into the MS which consists of a 2 m flight tube[195] with an electron multiplier detector. The collimating hole is in the middle of an image intensifier, which by means of a prism allows field-ion micrographs to be recorded concurrently with mass analysis. If the potential applied to the specimen is increased above the level required for image formation, a well-defined critical value can be reached at which the surface atoms of the specimen itself are removed as positive ions (i.e. 'field evaporation', a 'thermal' process). Brenner and McKinney[196] have shown for ferrous (and other) materials that temperatures below 80 K and pressures less than 10^{-8} torr are required to reduce field-induced complications to a negligible level. Multiply charged ions are very common, e.g. triply and quadruply charged ions are the predominant species for tungsten, molybdenum and iridium samples. If the noble gases or H_2 are used as imaging gases, they are found to be adsorbed on the tip at 78 K. These gases are desorbed mostly as singly or doubly charged ions (with iron, hydrogen gives H_2^+ and H_3^+) and also as noble gas–metal complex ions. The presence of the imaging gas affects the probability of field ionisation above a surface due to the possibility of charge transfer occurring between the image gas and the metal atom. Therefore, it is not just the geometrical factors of the crystal which determine the local field strength. Also when H_2 is added to the He imaging gas, atomic hydrogen can be adsorbed between the metal atoms forming hydride-like bonds; this removes electronic charge from the metal permitting the ionisation of helium at lower fields.

The atom-probe FIM is extremely sensitive for examining surfaces and surface reactions and the above work indicates that care must be employed in interpreting field-ion micrographs which were taken with an imaging gas present.

7.3.4 Knudsen cell and high-temperature inlet work on substances of low volatility

This section is divided into (a) Knudsen cell and (b) heated-inlet studies.

7.3.4.1 Knudsen cell work

A Knudsen cell design compatible with a TOFMS has been described by Rauh *et al.*[198] and a review of Knudsen cell–TOFMS has been written by Bowles[199]. Inside a Knudsen cell, an equilibrium is set up between the vapour and the condensed phase of the sample. The vapour effuses from a small orifice in the top of the cell and passes into the MS source where it is ionised. The cell is usually heated by radiation from filaments up to *c.* 1170 K; higher temperatures up to 2600 K are achieved with a combination of radiation and electron bombardment heating. Bowles[199] has described the pumping, degassing and pressure control arrangements necessary for use of a Knudsen cell. They are used to determine[199] vapour pressures, enthalpies of vaporisation and dissociation energies of compounds of low volatility, e.g. YCl_3 [200], BeF [201], LaS [202,203], SiCN [204], several titanium tellurides[205], GeNi [206], polymeric Ge [207], Cr_2 [208], copper(II) fluoride[209] and graphite[210,211]. For situations in which a compound evaporates and then decomposes in the gas phase, the residence time of the gaseous species must be sufficiently long to fully establish the decomposition equilibrium. To determine the equilibrium constant K_p for a reaction of the type A(solid, liquid) $\rightleftharpoons$ B(gas) + C(gas), the equilibrium vapour pressure of B and C must be measured. Although P_C may equal P_B in the Knudsen cell, their partial pressures in the TOFMS source may not be equal because of their different volatilities at room temperature. This effect was studied in a Model 14 TOFMS[199]. Early Knudsen cell studies usually assumed that the condensed phase – vapour equilibrium was achieved. The work of Cater *et al.*[212] on US/UO_2 mixtures has shown that equilibrium conditions may not be sampled by the MS if a direct line of sight exists between the liquid surface and the ionisation region. Under typical conditions the mean free path of gaseous molecules is larger than the dimensions of the cell, thus a gaseous equilibrium is not set up. Nearly all the molecules ionised by the MS come directly from the surface which may vary in phase or composition i.e. an average situation is being sampled. Cater *et al.*[212] suggest a better way of examining the equilibrium would be with a redesigned cell, e.g. a spherical cell in which there is no direct line of sight between the surface and the source. Kant[213,214] has shown that for several binary alloys of silver or gold, ideal (Raoultian) behaviour is followed, i.e. the ratio of the ion currents is independent of composition of the liquid at a given temperature.

In studies of the thermodynamic properties of solutions existing in a Knudsen cell (e.g. Ref. 215) one of the main difficulties was the change in instrumental sensitivity between successive runs. This problem has been overcome by Belton and Fruehan's integrated ion-current ratio method[216, 217]. This method assumes there is a direct proportionality between the ion-current ratio and the partial pressure ratio. It is then a simple matter to obtain activity coefficients and partial molar heats of solution by measuring the ion-current ratio as a function of liquid composition. This approach may be applied to binary and ternary mixtures and to complex vapour species. The validity of this approach has been demonstrated by Alcock *et al.*[218] and Hager *et al.*[219] who both used TOFMS (Bendix type 12-101A) spectrometers but different Knudsen cell designs, and whose results for the Cu–Sn and Ag–In alloys overlap within experimental error. The ion-current ratio method has been used to determine relative ionisation cross-sections of transition metals in binary alloys effusing from a Knudsen cell[212–220]. The use of Knudsen cells is now a well-established method for obtaining thermodynamic data for complex high-temperature equilibria.

7.3.4.2 Heated-inlet studies

The heated-inlet system is complementary to that of the Knudsen cell i.e. equilibrium is not established between the condensed and gas phase, but the gas immediately enters the source region of the spectrometer. Zitomer[221] has developed 'thermogravimetric–MS analysis', which can be applied to study a variety of problems involving thermal degradation, structure elucidation and the determination of volatiles. Thus polymethylene sulphide in an atmosphere of helium was heated[221] at 15 °C min^{-1} and the volatiles were passed through an interface which was similar to a GC–MS separator, and then into the TOFMS. The first degradation product is thioformaldehyde; on continued heating CS_2 is also produced. Langer and Gohlke[222] have suggested that the results of thermogravimetric–MS should enable a three-dimensional plot of mass *v.* intensity *v.* temperature to be constructed.

Gohlke[223, 224] has improved the sample introduction device for a TOFMS so that mass spectra may be obtained of compounds with sufficient thermal stability to give vapour pressures of *c.* 10^{-7} torr. Thermal degradation studies have generally been concerned with polymeric material. Such work on the pyrolysis of fluoropolymers has already been mentioned in Section 7.2.3. Other polymer systems examined by thermal degradation–TOFMS are cellulose ethers[225] and epoxy resin systems[226]. The effect of the interface between Al and Fe powder with polycondensate polymers[227] has also been examined.

Several interesting non-polymer pyrolysis–TOFMS studies have been reported. Pai Verneker and Maycock[228, 229] have pyrolysed ammonium perchlorate in an isothermal constant-volume chamber which is connected to a TOFMS by a variable leak. At 230 °C and 10^{-3} torr the major products are O_2, Cl_2 and N_2O in the ratio 3:2:2 and with an activation energy for formation of O_2 and Cl_2 of 28 kcal mol^{-1}. Using the same technique (100–160 °C), Maycock and Pai Verneker[230] have decomposed nitronium perchlorate. An activation energy of 15 ± 1 kcal mol^{-1} is required to produce the three major decomposition products i.e. O_2, NO and Cl_2. Pottie[231] has pyrolysed C_2F_4 in a graphite reactor connected by a pin hole to the TOFMS source. He has measured $I(CF_2) = 11.86 \pm 0.1$ eV, $AP(CF_2^+)_{C_2F_4} = 15.26 \pm 0.05$ eV and $\Delta H_f^0(CF_2) = -36.8$ kcal mol^{-1} (assuming no excess energy) thus demonstrating the thermodynamic stability of CF_2. Using a TOFMS, methylamine has been detected[232] from the gas-phase pyrolysis of azomethane.

7.3.5 Flash pyrolysis and laser–TOFMS work

It is perhaps convenient here to explain why flash pyrolysis is a special case of flash photolysis. If, during flash photolysis, the light energy is converted to heat by (for example) chemical reaction or collisional relaxation, a sudden jump in temperature will occur. This is flash pyrolysis. If, however, a large excess (e.g. one hundredfold) of noble gas diluent is added, then the temperature rise is less than 10 °C, and the photochemistry of the system may be studied as in conventional flash photolysis. Other techniques, such as resistive heating, may be used to obtain a sudden temperature increase of the sample. This section is therefore subdivided as follows:
(1) resistive heating method, (2) flash-tube method, and (3) laser–TOFMS.

7.3.5.1 Resistive heating method

Goldstein[233, 234] has pyrolysed bacteria and organic polymers on thin carbon rod in the source of a TOFMS. Samples were heated at linear rates between 200 and 2000 °C s^{-1} in the temperature range from room temperature to 1000 °C or rapidly (several seconds) to constant temperature in the range of 300–1000 °C and so maintained. Goldstein[233] has noted that the macrostructure of the sample is as important as the rate of heating in producing variations in the pyrolysis products. Korobeinichev *et al.*[235] have pyrolysed (220–280 °C) pyroxyline by the resistive heating of a titanium plate. The temperature was attained after *c.* 10^{-3} s and the following primary decomposition products were detected: NO_2, CH_2O, CO, CO_2 and H_2O. They have derived an Arrhenius expression for the maximum rate, the mechanism of which appears to be autocatalytic. Korobeinichev[235] has stressed that the thickness of the sample (3 μm) may affect the derived kinetic data, presumably through diffusion effects. Meyer[236] has measured the vapour pressure of metals, their oxides and nitrides (e.g. ZrO and ZrO_2 from oxidised zirconium) by resistively heating them in the source of a TOFMS. Koch *et al.*[237] have developed a slightly different approach. This involved the vacuum pyrolysis of solid samples (e.g. CO_2 from heated $CaCO_3$, ethane from coal pyrolysis and the combustion of coal grains in air) which were supported on wire gauze cloth. The gauze was heated electrically at a linear rate of several thousand degrees per second. A plate valve connects the pyrolysis chamber to the TOF ion source. The gas formed during the heating must be pumped out as quickly as possible so that the decisive factor for the readings of the spectrometer is the formation of the gas and not the pressure surge in the reaction chamber.

7.3.5.2 Flash-tube method

Lincoln[238–241] and Friedman[242] have developed a flash pyrolysis apparatus in which a high intensity xenon flash tube is used to heat the sample. Black substances are heated most efficiently; for transparent compounds such as cellulose, a small amount of graphite must be added[238]. Friedman[242] has flash pyrolysed phenol–formaldehyde resins using the xenon-lamp technique. The major products of the decomposition were H_2, CO, H_2O and CH_4. Because of the higher intensities available, lasers are now displacing flash tubes for studies of this kind.

7.3.5.3 Laser–TOFMS

The use of a laser in flash pyrolysis has been reviewed by Joy[243,244] and Bowles[245]. Conventional heating techniques are limited to *c.* 3000 K; thermodynamic data up to 4000 K (required for space research, for example) may be obtained with a pulsed laser. A laser heating source has the advantage that only a small portion of surface is vaporised; the rest acts as a container for the heated portion, thus eliminating any reaction between the container and the specimen. A disadvantage of the method is that the vapour may be in a state of thermal non-equilibrium[246] and may contain highly excited species such as ions. Vastola and co-workers[247,248] use the light from a small ruby laser focused on the target just below the electron beam, giving a rapid heating of a target area as small as 100 μm^2. The area to be volatilised can be selected and viewed through a laser optical system, thus specific regions of heterogeneous samples can be analysed. By using an oscilloscope with adjustable delay time, spectra may be recorded at various times during the pyrolysis. As well as neutrals, ions are produced in the laser pyrolysis of polymeric compounds[248] and thus an ion spectrum may be obtained with the electron beam switched off. Vastola[249] has noted that for thermally unstable salts such as organic sulphates, sulphonates and phosphates, simple mass spectra are produced, i.e. no fragments resulting from C—C bond breakage were found, and peaks representative of the original structure were found in all cases. Bernal, Levine and Ready[246,250] have measured the kinetic energies and mass spectra of adsorbed species on tungsten using a Q-switched ruby laser. The observed species were H^+, C^+, H_2O^+, Na^+, K^+, Co^+, CO_2^+ and W^+. Some of the alkali metal ions have kinetic energies >170 eV which is far too high for a thermal equilibrium process (compare with Knudsen cells, Section 7.3.4). Knox[251] has used the laser microprobe to study the effect of crystal orientation on the vapour species produced by metal oxides[251]. In his studies[252,253] of various selenium compounds he has evidence for two different mechanisms of laser–solid interaction. It is suggested that the ionic products can give information on the short-range interactions of amorphous materials, whereas the excited neutral products are characteristic of high-temperature phase changes. Zavitsanos[211, 254] has vaporised graphite using a ruby laser at low angle in order to reduce the path length of the laser beam through the vapour phase. The species C_1–C_{10} were ejected normal to the surface irrespective of the direction of the beam. Joy and co-workers[1, 243, 244, 255, 256] have flash-pyrolysed coals, aromatic compounds and polymers in order to try and understand the complex processes of coal pyrolysis. Also some substances containing no carbon were laser heated to elucidate the effects of contamination by adsorbed gases, and attempts were made to eliminate the latter by use of a sample holder which could be pre-heated to 150 °C. A Bendix Model 3015 was modified to admit a beam from a ruby laser fired vertically downwards through a quartz window in the flange above the source. The laser light was focused on to the sample, the incident energy being adjustable by the degree of focusing of the beam. The sample was in the form of a thin layer of powder on a quartz rod *c.* 10 mm below the electron beam (thus reducing any secondary reactions). By rotating the rod it was possible to obtain 10 separate pyrolyses. The spectra were displayed on an oscilloscope and recorded with a drum camera. The ion intensity increased during *c.* 1 ms and then decreased. When attempting to detect radical products the initial pyrolysis spectrum must be used in order to avoid the complications of secondary reactions. The use of short delay times also avoids the discrimination against condensable gases[257]. Joy concludes[255] that the flash pyrolysis mass spectrum is characteristic of the solid, e.g. aromatic compounds give molecular and fragment ions, whilst polymers have peaks attributable to the number of monomer units in the fragment. For coal, the products are of low molecular weight (<100) and are dependent upon the laser-beam intensity. At low intensities, coal gives H_2O, C_2H_4 and CH_4 with no radicals or ions; at medium intensities the products (H_2, C_2H_2, $(C_2H)_2$, Na^+ and K^+) are consistent with the sample having reached a higher temperature; at high intensities there is a considerable quantity of radicals and unstable products, e.g. H, C, CH_3, C_2H_2 and polyacetylenes.

Bykovskii, Silnov and co-workers[258, 259] have used laser–TOFMS to investigate impurities in metal films. They have been able to detect 10^{-5}% impurity and to investigate the non-uniformity of the impurity distribution.

Laser–TOFMS is not only a very useful and elegant analytical tool, but it should be possible to study diffusion processes (e.g. in semi-conductor materials), zone refining and etching.

7.3.6 Flame and discharge studies

The use of TOFMS to study flames[1] provides evidence both of the species participating and of the flame reaction mechanism. Agnew and Agnew[260], in trying to understand the self-ignition process of stable low-temperature reactions, used a quartz microprobe to withdraw samples from various positions in a stabilised cool flame of diethyl ether and air. Twenty-seven products were traced quantitatively through the reaction by means of g.c.–TOFMS. Milne and Greene[261] have developed a method of sampling

1-atm flames for the constituent radicals by means of TOFMS. Sheathed burners are used to produce a flat flame, the gases of which are passed through an orifice and a skimmer (both differentially pumped) so as to form a supersonic molecular beam. The molecular beam is mechanically chopped and passed through the ion source of a TOFMS. As well as mass separation, polymer formation[262] is observed in the molecular beams from 1-atm flames. Quantitative sampling[261] of the radicals H, O, OH and Cl (obtained when HCl is added to a H_2–O_2–N_2 flame) is achieved throughout the reaction zones. King and Scheurich[263] have modified the general method of Milne and Greene[261] in order that flame ions may be observed by TOFMS. The modifications consisted of a series of ion lenses for collimating the incoming beam and a grid system for decelerating and bunching the ions as they enter the MS. Individual mass peaks can be resolved to mass 200, although the signal is noisy due to the inability to stop and bunch all the ions. When the alkali metals Na, K and Cs are added to fuel-lean H_2–O_2–N_2 flames the reaction proceeds through a rate-determining step such as $Na + O_2 + M \rightarrow NaO_2 + M$, the rate constants for which have been determined[264]. It is believed[264], however, that the above are not the final products as CsO_2^+ was not observed in the Knudsen cell study of CsO_2 (in O_2 at 1 torr) at 600–1300 K. Much of the work of Milne and Greene on 1-atm flames and molecular beams has been summarised (see Reference 265) including the major phenomena which accompany direct molecular beam sampling, i.e. (i) orifice–system interaction, (ii) mass separation, (iii) nucleation and (iv) influence of the internal energy state on the fragmentation pattern.

TOFMS has been used to analyse the products of gas discharges. Dymshits *et al.*[266] have studied the gaseous discharge products of CF_4. Brinkman and Gordon[267] have examined some intermediate species, such as S—O—Si, Si—O—B, B—O—B, Si—O—Ge, Ge—O—Ge and Ge—O—B, present in discharges. This work was undertaken to examine the synthetic possibilities of gas discharges and some relative stability data have been derived. Ashby *et al.*[268] have linked a TOFMS to a quasi-steady-state accelerator in order to estimate the impurity present in the accelerator. Using D_2, H_2 or He in the accelerator, ions of mass 12–16 were present at *c.* 10% of the intensity of the main ion.

The design of a coaxial d.c. discharge reactor, used in conjunction with a Bendix TOFMS, has been described by Dyson[269]. This was used to study the electrical discharge decomposition of ammonia gas; both NH_2 and other ammoniacal complexes were detected and measured as a function of discharge pressure.

7.3.7 Shock-tube work

The use of TOFMS with shock tubes has been discussed by Robinson[270]. The working pressure of a mass spectrometer is *c.* 10^{-9} of that in a shock-heated gas and there are two general methods of coupling the two pieces of apparatus. First by means of either a pin-hole leak (or small nozzle) in the thin gold foil end wall of the shock tube, or alternatively expanding the shock-heated gas through a nozzle system to a molecular beam which enters the MS. The pin-hole method was developed by Bradley and Kistiakowsky[271–273]. As the shock wave is reflected from the gold end plate it is sampled through the pin-hole and the TOF mass spectrum is displayed on an oscilloscope. A drum camera records spectra every 50 μs. Bradley[273] has used this apparatus to look at the thermal decomposition of nitromethane in argon at temperatures in the range 1145–1460 K. The reaction was first order throughout the temperature range; however, the activation energy became progressively lower as the temperature was raised. This was explained in terms of a reaction of overall stoichiometry $2CH_3NO_2 \rightarrow CH_4 + CO + H_2O + 2NO$, but involving a double activation step. The basic apparatus of Kistiakowsky was greatly improved by Dove and Moulton[50]. In this modified apparatus ions present in the shock wave (chemi-ions) could be analysed, and the spectra could be recorded at intervals of 20 μs by means of three oscilloscopes so that 21 consecutive spectra were obtained. This method gave better light intensity and greater resolution than the drum-camera method (see Section 7.1.3). The pumping speed of the ion-source region was increased, a large ballast volume was added to the source region to reduce the pressure increase in the source, the drift tube was differentially pumped to keep its pressure low (10^{-8} torr), and the filament and high voltage was automatically cut off when the pressure reached a predetermined value. Dove and Moulton[50] examined the oxidation of acetylene and methane in a mixture of 95% Kr carrier gas. In the oxidation of acetylene[50, 272], H_2O and CO were produced simultaneously suggesting that both were formed directly in the main chain-branching cycle. A reaction mechanism was proposed in which the important step leading to regeneration of H atoms is $C_2H + O_2 \rightarrow 2CO + H$. In the oxidation of CH_4 at 1850–2050 K it was observed[50] that significant amounts of ethylene and acetylene are formed early in the main chain-branching steps. It was found that the primary ion in the chemi-ionisation (see Chapter 5) was $m/e = 39$ which is suggested to be cyclo-$C_3H_3^+$. Dove and Moulton[50] discuss some of the experimental difficulties of shock tube–TOFMS, such as ion–molecule reactions and inadvertent analysis of the cold thermal boundary layer in the shock tube. It has been suggested that chemi-ionisation occurs in flames (for example, see Reference 274) by the process $CH^* + O \rightarrow CHO^+ + e$, all other ionic species arising by charge transfer with CHO^+. Kistiakowsky[51, 275–278] has found, however, that in the shock-tube oxidation (and presumably also in flames) of methane, ethylene and acetylene, the first ion observed is cyclo-$C_3H_3^+$, which should be a relatively stable aromatic type species, and the ion CHO^+ was not observed. From measurements[277] of the chemiluminescence of CH ($A^2\Delta$) formed by the reaction $C_2H + O \rightarrow CH(A^2\Delta) + CO$ and of the total ionisation (Langmuir probe method), it was found that the chemiluminescence and ionisation had the same time constants. Thus the reaction $CH(A^2\Delta) + C_2H_2 \rightarrow C_3H_3^+ + e$ is not responsible for the ionisation. Kistiakowsky has suggested three alternative mechanisms:

$$CH_2 + C_2H_2 + O \rightarrow C_3H_3^+ + OH^-$$

$$CH_2 + C_2H \rightarrow C_3H_3^+ + e$$

$$CH(X^2\Pi) + C_2H_2 \rightarrow C_3H_3^+ + e$$

At the moment the correct mechanism is in doubt. The species $C_3H_3^+$ had been observed[276] under non-oxidising conditions. CH_3I, shock heated at 2500 K and 0.2 atm, readily produces the radicals CHI_2, CHI and CH. When acetylene is added, $C_3H_3^+$ is observed and is believed to be due to reaction between CHI and CH with C_2H_2 to give iodocyclopropene and cyclopropene radicals respectively. Due to the high electron affinities of I atoms the following ion-pair formation processes occur,

$$C_3H_3I \rightarrow C_3H_3^+ + I^-; C_3H_3 + I \rightarrow C_3H_3^+ + I^-$$

Free-radical mechanisms for the shock-tube oxidations of formaldehyde[279], ethylene[278], acetylene[50, 277] and methane[50] have been proposed. Kistiakowsky and co-workers have also suggested mechanisms and measured Arrhenius parameters for the pyrolysis of formaldehyde[279], ethylene[280] and acetylene[281].

Modica[282] used the previously studied[271] decomposition of N_2O to compare the pin-hole leak with a slender Pyrex-nozzle sampling technique. Below 3000 K the results agree within experimental error, but at higher temperatures the effect of the cold end wall of the shock tube on the hot gas introduces a systematic error in the measured rate constant. The error is slightly less for the nozzle (see also Reference 277). Modica and La Graff[283] have shock heated C_2F_4 at temperatures of 1200–1800 K and shown that C_2F_2 exists in equilibrium with CF_2. Using C_2F_4 as a source of CF_2, Modica[283, 284] has studied the reactions of the latter with O_2 and also with NO. The heat of formation of CF_2 has been found[285] to be -40.2 ± 4.0 kcal mol^{-1} from the pyrolysis of CHF_3 in Ar at 1600–2200 K. The main decomposition route[285] is

$$CHF_3 \xrightarrow{Ar} CF_2 + HF$$

which has an activation energy of 58.4 ± 2.2 kcal mol^{-1}.

Diesen[286, 287] has studied the thermal decomposition of F_2 and Cl_2 by shock tube–TOFMS, and obtains apparent activation energies of 27 and 41 kcal mol^{-1} respectively. The observed temperature dependence is less than that predicted by simple collision theory and it is felt[287] that at the higher temperatures the Boltzmann distribution is perturbed and the observed rate constants are 'falling off' due to depopulation of higher levels.

The pyrolysis of hydrazine between 1200 and 2500 K has been examined by shock tube–TOFMS[288]. The primary step involves breakage of the N—N bond to give NH_2 radicals which at the higher temperatures react further. The weak N—N bond (20 kcal mol^{-1}) in N_2F_4 makes it a good source of NF_2 radicals; Diesen[289] has studied the decomposition of NF_2 at 2200–3000 K by this means. The experimental data support the following mechanism $NF_2 \rightarrow NF + F$; $2NF \rightarrow N_2F_2^* \rightarrow N_2 + 2F$, where $N_2F_2^*$ has an excitation energy of *c.* 100 kcal mol^{-1}. Below 1900 K NF_3 is observed and is believed to arise from endothermic disproportionation of $2NF_2$ (however, see References 290 and 291). The Arrhenius parameters for the reaction $NF_2 + F_2 \rightarrow NF_3 + F$ have also been measured[292].

Price and co-workers[293] have used shock tube–TOFMS to follow the loss of HCl from 1,2-dichloroethylene to give chloroacetylene. The loss of Cl atoms does not appear to be an important step.

Kistiakowsky *et al.*[294] have studied the isotopic exchange reactions in mixtures of O_2 and CO at 1700–2600 K. The observed rates of O atom attack on O_2 and CO were equal over this temperature range; the concentrations of O atoms, however, is 30 times higher than calculated from literature values for the reaction $CO + O_2 \rightarrow CO_2 + O$. It is suggested that some chain branching process is occurring, possibly involving trace impurities such as H_2.

The expanded-flow technique has been used by Skinner[295] and others[296, 297]. In this method a very thin diaphragm ruptures upon arrival of the shock wave, which passes as a supersonic flow through expansion and skimmer cones. There is then a transition from continuum flow to free molecular flow, and the sample passes through a collimating cone into the TOFMS. One of the disadvantages of this method is that the heavier components become enriched relative to the lighter ones along the axis of the beam[296, 297].

7.3.8 Photochemistry and flash photolysis

A general review of radiation and photochemistry has been given by Ausloos[298], and Meyer[299, 300] has reviewed the combination of flash photolysis with TOFMS. Kistiakowsky and Kydd[51] in 1957 studied the flash photolysis of ketene and NO_2 in the presence of inert gas, by use of a pin-hole leak from the reaction vessel into the source of the TOFMS. Using the drum-camera technique spectra were obtained at 50 μs intervals. Unfortunately Kistiakowsky's[51] apparatus suffered from the disadvantages of low resolution (1/40), low sensitivity (1/700) and a low-intensity flash (leading to a low concentration of intermediates). Meyer[299, 301] has used a higher-intensity flash lamp (10^{18} quanta cm^{-2}) and increased the TOFMS sensitivity to 1 in 100 000. In the NO_2 sensitised flash photolysis of hydrogen–oxygen mixtures[302], evidence was obtained that the primary photochemical step is $NO_2 + h\nu \rightarrow NO + O$. Meyer[303, 304] has flash-photolysed CH_3I in the presence of several added gases. All the systems studied showed that I_2 formation was occurring by two processes, one fast and one slow. The fast reaction $I(^2P_{\frac{1}{2}}) + CH_3I \rightarrow I_2 + CH_3 + 3$ kcal occurred within 76 μs, and was associated with about 50% of the I_2 produced. The slow reaction was accounted for in terms of a termolecular atom recombination. Some interesting investigations have been carried out by Berry *et al.*[305–307] on aromatic systems. The *o*-, *m*- and *p*-carboxylate of diazobenzene was flash photolysed and the time-resolved MS were observed. In all cases a species of $m/e = 76$ was detected, which when formed from the *ortho* isomer dimerised to diphenylene at a rate greater than 7×10^8 l mol^{-1} s^{-1}. This is the transient species benzyne, which differs in structure from the $m/e = 76$ peak obtained from the *meta*- and *para*-derivative, because, for example, the $m/e = 76$ peak derived from *p*-carboxylate persisted for as much as 2 min after the flash photolysis. Appleby *et al.*[293, 308] have noted that attempts to study the decomposition of $Pb(CH_3)_4$ by optical techniques are handicapped by the deposition of metallic Pb on the windows. The advantage of time-resolved MS is that such fogging does not affect the quantitative accuracy of the measurements. When the temperature rise is low (e.g. 50 K using excess of Ar), the only new species detected were methyl radicals and ethane. On the other hand when the temperature rise was 1000 K (i.e. flash pyrolysis occurring concurrently with flash photolysis) ethylene and acetylene were produced. It is suggested that the ethylene arises by a 'disproportionation' reaction of two $Pb(CH_3)_3CH_2$ radicals.

Knewstubb and Reid[309, 310] have modified a TOFMS by replacing the field-free drift tube with an assembly by defining plates fed from a potential divider, with the objective of detecting metastable ion transitions (see Section 7.3.1). A windowless discharge lamp of repetition frequency 8–20 kHz is used. The shape of the metastable ion peak was analysed to give the decomposition function. The conclusions with respect to the decompositions $(CH_3O)_3PO^+ \rightarrow (CH_3O)_2HPO^+ + HCHO$ and $C_4H_6^+ \rightarrow C_3H_3^+ + CH_3$ were compared with the quasi-equilibrium theory of MS.

Reed *et al.*[311] have recently developed a coincidence TOFMS. A He resonance (584 Å) photoionisation source is used, and the product positive ion and electron are separately accelerated and counted by a delayed-coincidence technique. By varying the delay time between the electron pulse and ion pulse, a TOF mass spectrum may be produced. The resolution of this instrument is extremely low (*c.* 10); however, with improved resolution it is hoped to study some simple ion production phenomena.

Kamaratos and Lampe[312] have carried out some elegant work on the nitric oxide inhibited mercury-photosensitised reactions of silane and methyl silane. The photocell was connected by pin-hole to the source of the TOFMS, and continuously irradiated with 2537 Å light. Evidence was found for the existence of H_3SiON which reacts further to give $H_3SiONNO$, the latter decomposing mainly to give $H_3SiO + N_2O$. A full reaction scheme is proposed for both the inhibited and uninhibited system.

7.3.9 Gas chromatography—TOFMS coupling

The direct coupling of a mass spectrometer to gas chromatography apparatus facilitates the characterisation of the GLC peaks. At an early stage[313] the rapid response of TOFMS instruments was recognised as a valuable asset for such work and by 1964 there had been many reports of the analysis and characterisation of complex mixtures using GLC–TOFMS systems (see for example References 313–318).

The large excess of carrier gas is a problem present in all such work and one attempt to overcome this difficulty involves the trapping of a given peak while stopping the carrier-gas flow; this component is then introduced into the mass spectrometer whilst the remaining components are held stationary in the column. This 'stopped-flow' technique[314, 319] not only reduces the effects of carrier gas, but also reduces the demand for very fast scanning when this would require a compromise on sensitivity. An alternative approach to the carrier gas problem is to modify the TOFMS to incorporate a 'blanking generator' to eliminate this superfluous part of the mass spectrum[45]. Saturation of the multiplier by carrier gas, and other problems such as those involving solvents, may be avoided in this way, as mentioned in Section 7.1.3.2. The advantages of various collection methods compared with the various direct coupling methods have been discussed by McFadden[320]. The use of the TOFMS as the actual detector for the GLC apparatus has been described in detail by D'Oyly-Watkins *et al.*[321]. In this work the g.l.c. effluent is split, and no attempt is made to separate the carrier gas because such 'separators' (see below) depend for their operation on mass discrimination and therefore reduce quantitative accuracy. (The performance of this TOFMS detector[321] was tested by injecting known mixtures into the g.l.c. apparatus, and it is claimed that the linear range of the TOFMS detector approaches 10^4 for virtually all compounds and is superior to that available with other selective detectors.) In most current g.l.c.–MS work however, the carrier-gas problem is solved by the use of separators[322–326]; the sample enrichment obtained in this way permits peak characterisation without carrier-gas problems in the mass spectrometer and accuracy of quantitative component analysis is sacrificed or measured where necessary by an independent method (e.g. by the independent g.l.c. detector). Luchte and Damoth[327] have recently described a silver membrane separator which was specifically developed for use with a TOFMS. The device was connected so that it eliminated the need for a splitter valve at the exit to the g.l.c. column.

Papers involving the application of g.l.c.–TOFMS to flame analysis[328] radiation chemistry[329] and photochemistry[330, 331] have been published and many other examples are listed by Joy[7].

7.3.10 Other applications

Certain specialised applications of TOFMS are of importance and have not fitted into the above sections. One such sphere of interest is molecular beams, where for example a TOFMS has been used as a modulated molecular beam detector[332]. In this work the molecular beam was chopped and phase-sensitive detection was employed in the mass spectrometer output circuitry. This principle improves the signal-to-noise ratio over ordinary d.c. detection and discriminates against interfering ions which have the same nominal mass but are formed from residual gases in the ion source.

A development of the beam technique used in ion–molecule reactions (see Section 7.3.2) permits the creation of neutral beams with translational energies of many orders of magnitude greater than thermal energies. This depends upon the fact that in a symmetrical charge transfer reaction ($X^+ + X \rightarrow X + X^+$), the electron transfer occurs over a long range, so that if the reactant ion X^+ is previously accelerated electrostatically to high energy, the product molecule X proceeds with essentially the same energy. Thus if an energetic beam of X^+ is passed through a chamber containing X, a high-energy neutral beam, together with unreacted X^+, emerges. The latter ions may be removed by an electrostatic field, leaving the high-energy neutral beam to pass into the source region of the TOFMS where it may interact with gaseous or solid samples. Some preliminary TOFMS studies of this kind have been published[168, 333].

Other TOFMS applications relate to studies of the synthesis, reactivity and energetics of compounds that are stable only at low temperatures, such as O_2F_2, O_3F_2, H_2O_4, N_2H_2 and BH_3 (see references 334–336). Such work requires the use of cryogenically refrigerated mass spectrometer inlet apparatus[335]. A detailed TOFMS study of the products obtained from fast cryogenic quenching of reactions involving atomic hydrogen or atomic oxygen has been performed[337].

7.4 APPENDIX

The proceedings of the Third International TOFMS Symposium held at Salford University in July 1971 are to be published in *Dynamic Mass Spectrometry*, Volume III, by Heyden and Son Ltd. (London), in 1972. The following papers were presented at the Symposium:

1. Flash photolysis in the vacuum ultraviolet; Johnston, G. R. and Price D.
2. Flash-photolytic decomposition of nitrosyl chloride studied by TOFMS; Fornstedt, L. and Lindquist, S. E.
3. The kinetics of methyl radical reactions at high temperatures by MS sampling behind reflected shock waves; Clark, T. C.
4. Some studies of chemical reactions initiated by shock waves; Heald, P., Lippiatt, J. H., Madeley, J. D. and Myers, P.
5. Gas evolution from molten glass; Rigby, L. J.
6. TOFMS of molecular gas laser discharges; Austin, J. M. and Smith, A. L. S.
7. Mass assignment of TOF mass spectra; D'Oyly-Watkins, C.
8. Thermal volatilisation analysis; Murdoch, I. A. and Rigby, L. J.
9. Application of a TOFMS to the study of interactions between ions and molecules; Lavigne, J. M.
10. A coincidence TOFMS; Jardine, I. and Reed, R. I.
11. Negative-ion formation by some perfluorocarbons; Harland, P. and Thynne, J. C. J.
12. The formation of secondary ion peaks in the TOFMS; Di Valentin, M. A.
13. Uses of inhomogeneous oscillatory electric fields in ion physics; Carrico, J. P.
14. Laser volatilisation of organic contaminants on metal surfaces; Kumar, M. and Rigby, L. J.
15. Non-linear behaviour and recovery time from saturation of the electron multiplier; Di Valentin, M. A.
16. Direct coupled GC–MS with automated readout; Damoth, D. C.

References

1. Joy, W. K., (1967). *British Coal Utilization Research Assoc. Monthly Bulletin,* **31,** 581
2. Price, D. (1968). *Chem. Brit.,* **4(b),** 255
3. *Abstracts and Bibliographies on the Bendix TOFMS,* published by the Cincinnati division of the Bendix Corporation, 3625 Hauk Rd., Cincinnati, Ohio 45211, U.S.A.
4. *Abstracts and Bibliographies on the Bendix TOFMS,* published by the Bendix Scientific Instrument and Equipment Division, 1775 Mt. Read. Blvd., Rochester, N.Y. 14603, U.S.A.
5. *Time-of-flight Mass Spectrometry.* (1969). *(Proceedings of the first European TOFMS Symposium,* Univ. of Salford, 1967), Ed. by Price, D. and Williams, J. E. (Oxford: Pergamon Press)

6. *Dynamic Mass Spectrometry* (1970). *(Proceedings of the second European TOFMS Symposium,* Univ. of Salford, 1969), Ed. by Price, D., and Williams, J. E., (London: Heyden and Son Ltd.)
7. Joy, W. K., (1970). Reference 6, p.225
8. *Mass Spectrometry Bulletin,* published by the Mass Spectrometry Data Centre, A.W.R.E., Aldermaston, Berkshire, England.
9. Scientific Documentation Centre, Ltd., Halbeath, Dunfermline, Scotland
10. Stephens, W. E., (1946). *Bull. Amer. Phys. Soc.,* **21,** 22
11. Wiley, W. C., and Mclaren, I. H., (1955). *Rev. Sci. Inst.,* **26,** 1150
12. Moorman, C. J., and Bonham, R. W., (1967). Factors influencing the resolution of a TOFMS. (*15th Conference on MS,* Denver, Colo., May 1967. *ASTM E-14*)
13. Damoth, D. C. (1969). Reference 5, p.1
14. Dyson, K. O. (1970). Reference 6, p.211
15. Damoth, D. C. Reference 6, p.199
16. Bonham, R. W., Damoth, D. C. and Moorman, C. J. (1966). *Performance Characteristics of a new TOFMS.* (The Bendix Corporation, Scientific Instrument Division, Cincinnati, Ohio, 1966)
17. Haney, M. A. and Franklin, J. L. (1968), *J. Chem. Phys.,* **48,** 4093
18. Haney, M. A. and Franklin, J. L. (1969), *J. Phys. Chem.,* **73,** 4328
19. Franklin, J. L. and Haney, M. A. (1969), *J. Phys. Chem.,* **73,** 2857
20. Bafus, D. A., De Corpo, J. J. and Franklin, J. L. (1971). *Advan. Mass Spect.,* **5,** in the press
21. Sazone, G. (1970), *Rev. Sci. Instr.,* **41,** 741
22. Goodrich, G. W. and Wiley, W. C., (1961), *Rev. Sci. Instr.,* **32,** 846
23. Young, J. R., (1966), *Rev. Sci. Instr.,* **37,** 1414
24. Streeter, J. K., Hunt, W. W. and McGee, K. E., *Secondary electron yield v. Primary Energy for Commercial coated-glass Resistance Strips.* (A.F.C.R.L. Technical Report)
25. Hunt, W. W., McGee, K. E., Streeter, J. K. and Maughan, S. E. (1968). *Rev. Sci. Instr.,* **39,** 1793
26. Thomas, D. W. (1970). Reference 6, p.207
27. Honig, R. E. (1971), *Advan. Mass Spect.,* **5,** in the press
28. Wilson, D. B. (1969). *Vacuum,* **19,** 323
29. Thynne, J. C. J., MacNeil, K. A. G. and Caldwell, K. J. (1969). Reference 5, p.147
30. Anufriev, G. S. and Mamyrin, B. A. (1964). *Prilsurg i. Tekh. Eksper.,* **1964,** 150
31. Matus, L., (1965). *Magyar Tud. Akad. Kosp. Fiz. Kut. Int. Kozlemen,* **13,** 251
32. Rafal'son, A. E., and Shereshevskii, (1968). *Nucl. Sci. Abstr.,* **22,** 4384 N. 43160
33. Pavlenko, V. A., Ozerov, L. N. and Rafal'son, A. E. (1968). *Sov. Phys. Tech. Phys.,* **13,** 431
34. Kiser, R. W. and Gallegos, E. J. (1962). *J. Phys. Chem.,* **66,** 947
35. Melton, C. E. and Hamill, W. H. (1964). *J. Chem. Phys.,* **41,** 546, 698, 1469, 3464
36. Delwiche, J., (1970). Reference 6, p.71
37. Marmet, P., (1964). *Can. J. Phys.,* **42,** 2102
38. Winters, R. E., Collins, J. H. and Courchene, W. L., (1966). *J. Chem. Phys.,* **45,** 1931
39. Morrison, J. D., (1963). *J. Chem. Phys.,* **39,** 200
40. Harland, P. W., MacNeil, K. A. G., and Thynne, J. C. J. (1970). Reference 6, p.105
41. Studier, M. H., (1963). *Rev. Sci. Instr.,* **34,** 1367
42. Studier, M. H., (1967). U.S. Pat. 3,296,434. *(Nucl. Sci. Abstr,* **21,** N12372)
43. Miller, C. D., Tiernan, T. O. and Futrell, J. H., (1969). *Rev. Sci. Instr.,* **40,** 503
44. Dietz, L. A., (1965). *Rev. Sci. Instr.,* **36,** 1763
45. Bitner, E. D., Rohwedder, W. K. and Selke, E., (1964). *Appl. Spectr.,* **18,** 134
46. Haumann, J. R. and Studier, M. H., (1968). *Rev. Sci. Instr.,* **39,** 169
47. Studier, M. H. and Hayatsu, R., (1968). *Anal. Chem.,* **40,** 1011
48. Meyer, R. T., (1969). Reference 5, p.61
49. Moulton, D. McL. and Michael, J. V. (1965). *Rev. Sci. Instr.,* **36,** 226
50. Dove, J. E. and Moulton, D. McL., (1965). *Proc. Roy. Soc. (London),* **A283,** 216
51. Kistiakowsky, G. B. and Kydd, P. H. (1957). *J. Amer. Chem. Soc.,* **79,** 4825
52. Kende, P., (1968). *Rev. Sci. Instr.,* **39,** 270
53. Greene, F. T., (1968). *16th Annual Conf. of Mass Spect., ASTM-E-14.* (Pittsburgh, USA)
54. Meyer, R. T. and Freese, J. M. (1965). *13th Annual Conf. of Mass Spect. and Allied Topics* (St. Louis, USA)

55. D'Oyly-Watkins, C. and Gaythorpe, S. N., (1970). Reference 6, p.215
56. Meyer, R. T., Olson, C. E., and Berlint, R. R. (1966). *Report number SC-R-66-928,* (Sandia Corporation).
57. Fallgatter, M. B. and Hanrahan, R. J. (1970), *Report number OR 0-3106-33,* (University of Florida, Gainesville, U.S.A.)
58. Carrico, J. P., Ferguson, L. D. and Mueller, R. K. (1971). *Advan. Mass Spec.,* **5,** in the press
59. Carrico, J. P., Ferguson, L. D. and Mueller, R. K. (1970). *Appl. Phys. Lett.,* **17,** 146
60. French, J. B. and Locke, J. W. (1967). *Rarified Gas Dynamics,* Ed. by Brundin, C. L., **2,** (New York: Academic Press)
61. Crosby, D. A. and Zorn, J. C., (1969), *J. Vac. Sci. Tech.,* **6,** 82
62. Clampitt, R., (1969). *Entropie,* **30,** 36
63. Studier, M. H. and Huston, J. L., (1967). *J. Phys. Chem.,* **71,** 457
64. Huston, J. L., (1966). *Chem. and Eng. News,* Sept. 18, 1966, p.58
65. Huston, J. L. (1968). *Inorg. Nucl. Chem. Lett.,* **4,** 29
66. Appelman, E. H., Sloth, E. N. and Studier, M. H. (1966). *Inorg. Chem.,* **5,** 766
67. Breisacher, P. and Siegel, B., (1965). *J. Amer. Chem. Soc.,* **87,** 4255
68. Timms, P. L., Kent, R. A., Ehlert, T. C. and Margrave, J. L. (1965). *J. Amer. Chem. Soc.,* **87,** 2824
69. Margrave, J. L. and Thompson, J. C. (1967). *Science,* **155,** 669
70. Timms, P. L., Ehlert, T. C., Margrave, J. L., Brinkmann, F. E., Farrar, T. C. and Coyle, T. D. (1965). *J. Amer. Chem. Soc.,* **87,** 3819
71. Timms, P. L., Strump, D. D., Kent, R. A. and Margrave, J. L. (1966). *J. Amer. Chem. Soc.,* **88,** 940
71a. Margrave, J. L., Sharp, K. G. and Wilson, P. W. (1970). *J. Inorg. Nucl. Chem.,* **32,** 1813
71b. Margrave, J. L., Sharp, K. G. and Wilson, P. W., (1970). *J. Inorg. Nucl. Chem.,* **32,** 1817
72. Timms, P. L. (1967). *J. Amer. Chem. Soc.,* **89,** 1629
73. Kirk, R. W. and Timms, P. L. (1967). *Chem. Commun.,* 18
74. Wong, E. L. and Potter, E., *OTS document NASA TND-2648.* (Lewis Research Centre, Cleveland, U.S.A.)
75. Kohout, F. C. and Lampe, F. W. (1965). *J. Amer. Chem. Soc.,* **87,** 5795
76. Maycock, J. N., Pai Verneker, V. R. and Jacobs, P. W. M., (1967). *J. Chem. Phys.,* **46,** 2857
77. Meyer, R. T. and Nelson, L. S. (1967). Paper presented at *15th Annual Conf. on Mass Spectrometry and Allied Topics,* (Denver, Colo., U.S.A.)
78. McKinley, J. D. (1965). *J. Chem. Phys.,* **42,** 615
79. McKinley, J. D. (1964). *J. Chem. Phys.,* **41,** 2814
80. McKinley, J. D. (1964). *J. Chem. Phys.,* **40,** 120
81. Wessler, M. A., (1967). *Dissert. Abstr.,* **27,** 3955
82. Hanrahan, R. J., Hagopian, A. K. E. and Davis, D. D. (1967). *Chem. and Eng. News.,* **Ap.24,** 46
83. Brown, R. A., Kay, M. I., Kelliher, J. M. and Dietz, W. A., (1967). *Anal. Chem.,* **39,** 1805
84. Hughes, B. M. and Hanrahan, R. J., (1965). *J. Phys. Chem.,* **69,** 2707
85. Griesbaum, J., Naegele, W. and Wanless, G. G. (1965). *J. Amer. Chem. Soc.,* **87,** 3151
86. Rohwedder, W. K., Mabrouk, A. F. and Selke, E., (1965). *J. Phys. Chem.,* **69,** 1711
87. Ruth, J. M. and Philippe, R. J., (1966). *Anal. Chem.,* **38,** 720
88. Grigsby, R. D., Hamming, M. C., Eisenbraun, E. J. Hertzler, D. V. and Bradley, N., (1966). Paper presented at *14th Annual Conf. on Mass Spect. and Allied Topics, ASTM E-14.* (Dallas, U.S.A.)
89. McFadden, W. H., Stevens, J. L., Karabatsos, G. J., Meyerson, S. and Orzech, C. E. (1965). *J. Phys. Chem.,* **69,** 1742
90. Quin, L. D. and Shelburne, F. A. (1965). *J. Org. Chem.,* **30,** 3135
91. Kropp, P. J., (1965). *J. Amer. Chem. Soc.,* **87,** 3914
92. Studier, M. H., Hayatsu, R. and Fuse, K. (1968). *Anal. Biochem.,* **26,** 320
93. Denning, R. G. and Wentworth, R. A. D. (1966). *J. Amer. Chem. Soc.,* **88,** 4619
94. Winters, R. E. and Kiser, R. W. (1967). *J. Organometal. Chem.,* **10,** 7
95. Winters, R. E. and Kiser, R. W. (1966). *J. Phys. Chem.,* **70,** 1680

96. Angelini, P., Forss, D. A., Bazinet, M. L. and Merritt, C. (1967). *J. Amer. Oil. Chem. Soc.*, **44,** 26
97. Wick, E. L., Underriner, E. and Paneras, E. (1967). *J. Food Sci.*, **32,** 365
98. Schultz, T. H., Black, D. R., Bomben, T. L., Mon, T. R. and Teranishi, R. (1967). *J. Food Sci.*, **32,** 698
99. Stern, D. J., Lee, A., McFadden, W. H. and Stevens, K. L. (1967). *J. Agr. Food Chem.*, **15,** 1100
100. Buttery, R. G., Black, D. R., Lewis, M. J. and Ling, L. (1967). *J. Food Sci.*, **32,** 414
101. Lee, W. L. (1961). Paper presented at *Aerospace medical panel of NATO (AGARD)* (July 1961, Oslo, Norway)
102. Felic, P. (1965). *Aerospace medicine*, **36,** 858
103. Kupel, R. E. and Scheel, L. D., (1968). *Amer. Ind. Hygiene Assoc. J.*, **29**
104. Waritz, R. S. and Kwon, B. K., (1968). *Amer. Ind. Hygiene Assoc. J.*, **29,** 19
105. Coleman, W. E., Scheel, L. D., Kupel, R. E. and Larkin, R. L. (1968). *Amer. Ind. Hygiene Assoc.* J., **29,** 54
106. Scheel, L. D., Lane, W. C. and Coleman, E. W. (1968). *Amer. Ind. Hygiene Assoc. J.*, **29,** 41
107. Scheel, L. D., McMillan, L. and Phipps, F. C. (1968). *Amer. Ind. Hygiene Assoc. J.*, **29,** 49
108. Euler-Chelpin, H. von. (1970). *The Swedish Cancer Society, Yearbook 4,* Chap. XIII, No. 116, p.507
109. Saalfield, F. E., McDowell, M. V., Gondal, S. and MacDiarmid, A. G. (1967). Paper presented at *15th Ann. Conf. on Mass Spect. and Allied Topics* (Denver, Colo. U.S.A., 1967)
110. Winters, R. E. and Kiser, R. W. (1965). *J. Phys. Chem.*, **69,** 1618, 3198
111. McDowell, M. V. and Saalfeld, F. E. (1968). Paper presented at *16th Conf. on Mass Spect., ASTM E-14,* (Pittsburgh, Pa, U.S.A., 1968)
112. Saalfeld, F. E. (1966). Paper presented at *14th Conf. on Mass Spect. & Allied Topics,* (Dallas, U.S.A., 1966)
113. Hess, G. G., Lampe, F. W. and Sommer, L. H. (1965). *J. Amer. Chem. Soc.*, **87,** 5327
114. Martin, R. M., Lampe, F. W. and Taft, R. W. (1966). *J. Amer. Chem. Soc.*, **88,** 1353
115. Taft, R. W., Martin, R. H. and Lampe, F. W. (1965). *J. Amer. Chem. Soc.*, **87,** 2490
116. Ames, L. L., Walsh, P. N. and White, D. (1967). *J. Phys. Chem.*, **71,** 2707
117. Dillard, J. G. and Kiser, R. W. (1965). *J. Phys. Chem.*, **69,** 3893
118. Dyson, K. O. (1970). Reference 6, p.23
119. Dyson, K. O. (1971). *Advan. Mass. Spect.*, **5,** in the press
120. Shadoff, L. A. (1967). *Nucl. Sci. Abstr.*, **21,** 14989; *Dissert. Abs.*, **27,** 14458
121. Stuber, F. A. (1965). *J. Chem. Phys.*, **43,** 2639
122. Kahng, S. K., Whitehead, W. D. and Landes, H. S. (1967). *Bull. Amer. Phys. Soc.*, **12,** HG9
123. McCulloh, K. E., Sharp, T. E. and Rosenstock, H. M. (1965). *J. Chem. Phys.*, **42,** 3501
124. Ferguson, R. E., McCulloh, K. E. and Rosenstock, H. M. (1965). *J. Chem. Phys.*, **42,** 100
125. Hunt, W. W., Huffmann, R. E., and McGee, K. E. (1964) *Rev. Sci. Instr,* **35,** 82. (see also ibid. p.88)
126. Hunt, W. W. and McGee, K. E. (1964). *J. Chem. Phys.*, **41,** 2709
127. Johnsen, R. H. and Mooberry, F. (1968). *Nucl. Sci. Abstr.*, **22,** 846 (No. 8408)
128. Dugger, D. L. and Kiser, R. W. (1967). *J. Chem. Phys.*, **47,** 5054
129. Haddon, W. F. (1969). *Dissert. Abst.*, **29,** 4070B
130. Fromont, M. J. and Johnsen, R. H. (1970). *Int. J. Mass. Spectrom. Ion. Phys.*, **4,** 235
131. McLafferty, F. W., Gohlke, R. S. and Golesworth, R. C. (1964). Paper presented at *12th Ann. Conf. on Mass Spect., ASTM E-14,* Montreal, 1964
132. Grayson, M. A. and Conrads, R. J. (1970). *Anal. Chem.*, **42,** 456
133. Becker, P. and Lampe, F. W. (1965). *J. Chem. Phys.*, **42,** 3857
134. DeCorpo, J. J. and Lampe, F. W. (1968). Paper presented at *15th Conf. on Mass Spect., ASTM E-14*
135. Winters, R. E. and Kiser, R. W. (1966). *J. Phys. Chem.*, **70,** 1680
136. Thynne, J. C. J., MacNeil, K. A. G., and Caldwell, K. J. (1969). Reference 5, p.117
137. Edelson, D., Griffiths, J. E. and McAfee, K. B. (1962). *J. Chem. Phys.*, **37,** 917
138. Compton, R. N., Christophorou, L. G., Hurst, G. S. and Reinhardt, P. W. (1966). *J. Chem. Phys.*, **45,** 4634

139. Winters, R. E. and Kiser, R. W. (1966). *J. Chem. Phys.,* **44,** 1964
140. Bouby, L., Compton, R. N. and Souleyrol, A. (1968). *Compt. Rend. Ser. C.,* **266,** 1250
141. MacNeil, K. A. G. and Thynne, J. C. J. (1969). *Intern. J. Mass Spectrom. Ion Phys.,* **2,** 1
142. Dillard, J. C. and Franklyn, J. L. (1968). *J. Chem. Phys.,* **48,** 2349
143. Lampe, F. W. and Hess, G. G. (1964). *J. Amer. Chem. Soc.,* **86,** 2952
144. Becker, P. M. and Lampe, F. W. (1965). *J. Chem. Phys.,* **42,** 3857
145. Cress, M. C., Becker, P. M. and Lampe, F. W. (1966). *J. Chem. Phys.,* **44,** 2212
146. Sieck, L. W., Abramson, F. P. and Futrell, J. H. (1966). *J. Chem. Phys.,* **45,** 2859
147. Abramson, F. P. and Futrell, J. H. (1967). *J. Phys. Chem.,* **71,** 3791
148. McKnight, L. G., McAfee, K. B. and Sipler, D. P. (1968). *Phys. Rev.,* **164,** 62
149. Fluegge, R. A. (1969. *J. Chem. Phys.,* **50,** 4373
150. McAffee, K. B., Sipler, D. and Edelson, D. (1967). *Phys. Rev.,* **160,** 130
151. Thynne, J. C. J., MacNeil, K. A. G. and Caldwell, K. J. (1969). Reference 5, p.140
152. Stockdale, J. A. D., Compton, R. N. and Reinhardt, P. W. (1968). *Phys. Rev. Lett.,* **21,** 664
153. Ferguson, R. E., McCulloh, K. E. and Rosenstock, H. M. (1965). *J. Chem. Phys.,* **42,** 100
154. Pike, G. W. F. (1970). Reference 6, p.139
155. Tiernan, T. O., Futrell, J. H., Abramson, F. P. and Miller, C. D. (1967). Presented at *15th Ann. Conf. on Mass Spect., ASTM E-14* (Denver, Colo, U.S.A.)
156. Hughes, B. M. and Tiernan, T. O. (1969). *J. Chem. Phys.,* **51,** 4373
156a. Potzinger, P. and Lampe, F. W. (1971). *J. Phys. Chem.,* **75,** 13
157. Beggs, D. and Lampe, F. W. (1968). Presented at *16th Ann. Conf. on Mass Spect., ASTM E-14* (Pittsburgh, Pa., U.S.A.)
158. Beggs, D. P. and Lampe, F. W. (1969). *J. Phys. Chem.,* **73,** 3315
159. Beggs, D. P. and Lampe, F. W. (1969). *J. Phys. Chem.,* **73,** 3307
160. Beggs, D. P. and Lampe, F. W. (1969). *J. Phys. Chem.,* **73,** 4194
161. Herman, Z., Hierl, P., Lee, A. and Wolfgang, R. (1969). *J. Chem. Phys.,* **51,** 454
162. Franklin, J. L., Wada, Y., Natalis, P. and Hierl, P. M. (1966). *J. Phys. Chem.,* **70,** 2353
163. Hand, C. W. and Weyssenhoff, H. von. (1964). *Can. J. Chem.,* **42,** 2385
164. Conway, D. C. (1968). Presented at *Amer. Chem. Soc. 156th Nat. Meeting; Abs. No. Phys. 42.* (Atlantic City, N.J., U.S.A.)
165. Young, C. E., Edelson, D. and Falconer, W. E. (1971). *Advan. Mass Spectrom,* **5,** in the press
166. Lehrle, R. S., Robb, J. C. and Thomas, D. W. (1962). *J. Sci. Instr.,* **39,** 458
167. Homer, J. B., Lehrle, R. S., Robb, J. C., Takahasi, M. and Thomas, D. W. (1963). *Advan. Mass Spectrom.,* **2,** 503
168. Homer, J. B., Lehrle, R. S., Robb, J. C. and Thomas, D. W. (1964). *Advan. Mass Spectrom.,* **3,** 415
169. Thomas, D. W. (1969). Reference 4, p.171
170. Hughes, B. M. and Tiernan, T. O. (1968). Presented at *16th Conf. on Mass Spect., ASTM E-14* (Pittsburgh, Pa., U.S.A.)
171. Knewstubb, P. F. and Field, D. (1971). *Int. J. Mass Spectrom. Ion Phys.,* **6,** 45
172. Homer, J. B., Lehrle, R. S., Robb, J. C. and Thomas, D. W. (1964). *Nature (London),* **202,** 795
173. Siek, L. W., Abramson, F. P. and Futrell, J. H. (1967). Presented at *14th Ann. Conf. on Mass Spect. and Allied Topics* (Dallas, Texas, U.S.A.)
174. Hyatt, D. and Lacmann, K. (1968). *Z. Naturforsch.,* **23A,** 2019
175. Homer, J. B., Lehrle, R. S., Robb, J. C. and Thomas, D. W. (1966). *Trans. Faraday Soc.,* **62,** 619
176. Lehrle, R. S., Parker, J. E., Robb, J. C. and Scarborough, J. (1968). *Int. J. Mass Spectrom. Ion Phys.,* **1,** 455
177. Zvenigorosky, A. (1967). *Thesis* (University of Toulouse, France, 1967), *Vacuum index, Index Biblio du Vide,* **3,** 102 (1968)
178. Lehrle, R. S., Robb, J. C., Scarborough, J. and Thomas, D. W. (1968). *Advan. Mass Spectrom,* **4,** 687
179. Lehrle, R. S., Parker, J. E. and Robb, J. C. (1970). Presented at the *Triennial International Conference on MS,* (Brussels, 1970). To be published in 1971
180. McKinley, J. D. (1964). *J. Chem. Phys.,* **40,** 576

181. McKinley, J. D. (1964). *J. Chem. Phys.*, **40,** 120
182. McKinley, J. D. (1966). *J. Chem. Phys.*, **45,** 1690
183. Ptushinskii, Yu. G., and Chuikov, B. A. (1964). *Ukr. Fiz. Zh.*, **9,** 1035 (*Chem. Abstr.*, **62,** (1965) 52f)
184. Kaposi, O., Rideal, M. and Toro, F. (1969). *Chem. Abstr.*, **71,** 96186e
185. Ageev, V. N. and Ionov, N. I. (1970). *Sov. Phys. Tech. Phys.*, **14,** 1142
186. Thomas, D. W. (1969). Reference 5, p.171
187. Dillon, A. F. (1968). *Ph.D. Thesis,* University of Birmingham
188. Dillon, A. F., Lehrle, R. S., Robb, J. C. and Thomas, D. W. (1968). *Advan. Mass Spectrom.*, **4,** 477
189. Robb, J. C., Terrell, D. R. and Thomas, D. W. (1970). Reference 6, p.87
190. Woolley, R. L., Warner, A. G. and Poole, D. H. (1970). *Central Electricity Generating Board,* RD/B/N 1320, Research and Development Department, Berkeley Nuclear Laboratory
191. Muller, E. W. and Panitz, J. (1967). *14th Field Emission Symposium,* Washington D.C.
192. Muller, E. W., McLane, S. B. and Panitz, J. A. (1969). *Surface Science,* **17,** 430
193. Muller, E. W., Krishnaswamy, S. V. and McLane, S. B. (1970). *Surface Science,* **23,** 112
194. Muller, E. W. (1970). *Naturwissenschaften,* **57,** 222
195. Muller, E. W. (1971). *Advan. Mass Spectrom.*, **5,** in the press
196. Brenner, S. S. and McKinney, J. T. (1970). *Surface Science,* **23,** 88
197. Turner, P. J. and Southon, M. J. (1970). Reference 6, p.147
198. Rauh, E. G., Sadler, R. C. and Thorn, R. J. (1962). *Argonne National Labs. ANL Report 6536,* Argonne, Ill.
199. Bowles, R. (1969). Reference 5, p.211
200. McKinley, J. D. (1965). *J. Chem. Phys.*, **42,** 2245
201. Hildebrand, D. L. and Murad, E. (1966). *J. Chem. Phys.*, **44,** 1524
202. Cater, E. D., Lee, T. E., Johnson, E. W., Rauh, E. G. and Eick, H. A. (1965). *J. Phys.* Chem., **69,** 2684
203. Cater, E. D. and Steiger, R. P. (1968). *J. Phys. Chem.*, **72,** 2231
204. Muenow, D. W. and Margrave, J. L. (1970). *J. Phys. Chem.*, **74,** 2577
205. Suzuki, A. and Wahlbeck, P. G. (1966). *J. Phys. Chem.*, **70,** 1914
206. Kant, A. (1966). *J. Chem. Phys.*, **44,** 2451
207. Kant, A. and Strauss, B. H. (1966). *J. Chem. Phys.*, **45,** 822
208. Kant, A. and Strauss, B. (1966). *J. Chem. Phys.*, **45,** 3161
209. Kent, R. A., McDonald, J. D. and Margrave, J. L. (1966). *J. Phys. Chem.*, **70,** 874
210. Zavitsanos, P. D. (1966). *G. E. Space Sci. Lab., Tech. Info. Ser. No. R66 SD31*
211. Zavitsanos, P. D. (1970). Reference 6, p.1
212. Cater, E. D., Rauh, E. G. and Thorn, R. J. (1968). *J. Chem. Phys.*, **49,** 5244
213. Kant, A. (1968). *J. Chem. Phys.*, **49,** 5144
214. Kant, A. (1968). *J. Chem. Phys.*, **48,** 523
215. Lyubimov, A. P., Zobers, L. and Rakhovski, V. (1958). *Zh. Fiz. Khim.*, **32,** 1804
216. Belton, G. R. and Fruehan, R. J. (1967). *J. Phys. Chem.*, **71,** 1403
217. Belton, G. R. (1971). *Advan. Mass Spectrom.*, **5,** in the press
218. Alcock, C. B., Sridhar, R. and Svedberg, R. C. (1969). *Acta Met.*, **17,** 839
219. Hager, J. P., Howard, S. M. and Jones, J. H. (1970). *Met. Trans.*, **1,** 415
220. Rolinski, E. J., Oblinger, C. J. and Hoch, M. (1971). *Advan. Mass Spectrom.*, **5,** in the press
221. Zitomer, F. (1968). *Anal. Chem.*, **40,** 1091
222. Langer, H. G. and Gohlke, R. S. (1966). *14th Annual Conf. on Mass Spec. and Allied Topics,* Dallas, Texas
223. Gohlke, R. S. (1963). *Chem. Ind. (London)*, 946
224. Langer, H. G. and Gohlke, R. S. (1963). *Anal. Chem.*, **35,** 1301
225. Harless, H. R. and Anderson, R. L. (1970). *Text. Res. J.*, **40,** 448
226. D'Oyly-Watkins, C. and Winsor, D. E. (1970). Reference 6, p.175
227. Schmidt, G. A. and Gaulin, C. A. (1967). *J. Appl. Polymer Sci.*, **11,** 357
228. Pai Verneker, V. R. and Maycock, J. N. (1967). *J. Chem. Phys.*, **47,** 3618
229. Maycock, J. N., Pai Verneker, V. R. and Jacobs, P. W. M. (1967). *J. Chem. Phys.*, **46,** 2857
230. Maycock, J. N. and Pai Verneker, V. R. (1967). *J. Phys. Chem.*, **71,** 4077
231. Pottie, R. F. (1965). *J. Chem. Phys.*, **42,** 2607

232. Wacks, M. E. (1964). *J. Phys. Chem.*, **68**, 2725
233. Goldstein, H. W. (1971). *Advan. Mass Spectrom.*, **5**, in the press
234. Friedman, H. L., Goldstein, H. W. and Griffiths, G. A. (1967). *15th Annual Conf. on Mass Spectrom. and Allied Topics,* Denver, Colorado
235. Korobeinichev, O. P., Aleksandrov, V. V. and Lyakhov, N. Z. (1970). *Izv. Akad. Nauk SSSR, Ser Khim.*, **3**, 562
236. Meyer, R. T. and Ames, L. L. (1967). *Applications of Mass Spectrom. in Inorg. Chem.* Ed. by Margrave, T. L., Adv. Chem. Series (Washington: American Chemical Society)
237. Koch, V., van Heek, K. H. and Juntgen, H. (1970). Reference 5, p.15
238. Lincoln, K. A. (1965). *Anal. Chem.*, **37**, 541
239. Lincoln, K. A. (1964). *Rev. Sci. Instr.*, **35**, 1688
240. Lincoln, K. A. and Werner, D. (1967). *15th Annual Conf. on Mass Spectrom. and Allied Topics,* Denver, Colorado
241. Lincoln, K. A. (1965). *Pyrodynamics,* **2**, 133
242. Friedman, H. L. (1965). *J. Appl. Polymer Sci.*, **9**, 651
243. Joy, W. K. (1970). Reference 6, p. 183
244. Joy, W. K. (1969). Reference 5, p.45
245. Bowles, R. (1969). Reference 5, p.211
246. Bernal, E., Ready, J. F. and Levine, L. P. (1966). *Phys. Lett.*, **19**, 645
247. Vastola, F. J., Pirone, A. J. and Knox, B. E. (1966). *14th Annual Conf. on Mass Spectrom. and Allied Topics,* Dallas, Texas
248. Vastola, F. J. and Pirone, A. J. (1967). *Advan. Mass Spectrom.*, **4**, 107
249. Vastola, F. J., Pirone, A. J. and Mumma, R. O. (1968). *16th Annual Conf. Mass Spectrom. and Allied Topics,* Pittsburgh, Pa.
250. Bernel, E., Levine, L. P. and Ready, J. F. (1966). *Rev. Sci. Instr.*, **27**, 938
251. Knox, B. E. (1968). *16th Annual Conf. Mass Spectrom. and Allied Topics,* Pittsburgh, Pa.
252. Knox, B. E. and Ban, V. S. (1968). *16th Annual Conf. Mass Spectrom. and Allied Topics,* Pittsburgh, Pa.
253. Knox, B. E. (1967). *Intl. Mass Spectrom. Conf.*, Berlin
254. Zavitsanos, P. D. (1968). *Carbon,* **6**, 731
255. Joy, W. K., Ladner, W. R. and Pritchard, E. (1970). *Fuel,* **49**, 26
256. Joy, W. K., Ladner, W. R. and Pritchard, E. (1968). *Nature (London),* **217**, 640
257. Lincoln, K. A. (1969). *Int. J. Mass Spectrom. Ion Phys.*, **2**, 75
258. Bykovskii, Yu. A., Dorofeer, V. I., Dymovich, V. I., Nikolaer, B. I., Ryzhikh, S. V. and Silnov, S. M. (1970). *Sov. Phys. Tech. Phys.*, **14**, 955
259. Apollonov, V. V., Bykovskii, Yu. A., Degtyarenko, N. M., Elesin, V. F., Kozyrev, Yu. P. and Silnov, S. M. (1970). *J.E.T.P. Lett.*, **11**, 252
260. Agnew, W. G. and Agnew, J. T. (1964). *10th Intern. Combust. Symp.*, Cambridge, England, and (1965). *Chem. Abst.*, **63**, 16204f
261. Milne, T. A. and Greene, F. T. (1966). *J. Chem. Phys.*, **44**, 2444
262. Greene, F. T. and Milne, T. A. (1963). *J. Chem. Phys.*, **39**, 3150
263. King, I. R. and Scheurich, J. T. (1966). *Rev. Sci. Instr.*, **37**, 1219
264. Carabetta, R. and Kaskan, W. E. (1968). *J. Phys. Chem.*, **72**, 2483
265. Milne, T. A. and Greene, F. T. (1968). *U.S. Govt. Res. and Dev. Rep.*, **68**, No. 7, 149
266. Dymshits, B. M., Koretski, Y. P., Ler, V. A., Turkena, M. Y. and Dobachin, S. L. (1969). *Zh. Prikl. Khim. (Leningrad),* **42**, 2470
267. Brinkman, F. E. and Gordon, G. (1969). *J. Res. Nat. Bur. Stand.*, **73A**, 383
268. Ashby, D. E. T. F., Burcham, J. N. and Chambers, R. G. (1969). *CLM R102,* Culham Laboratory, Abington, Berkshire
269. Dyson, K. O. (1969). Reference 5, p.89
270. Robinson, P. A. (1969). Reference 4, p.53
271. Bradley, J. N. and Kistiakowsky, G. B. (1961). *J. Chem. Phys.*, **35**, 256
272. Bradley, J. N. and Kistiakowsky, G. B. (1961). *J. Chem. Phys.*, **35**, 264
273. Bradley, J. N. (1961). *Trans. Faraday Soc.*, **57**, 1750
274. Hand, C. W. and Kistiakowsky, G. B. (1962). *J. Chem. Phys.*, **37**, 1239
275. Kistiakowsky, G. B. and Michael, J. V. (1964). *J. Chem. Phys.*, **40**, 1447
276. Glass, G. P. and Kistiakowsky, G. B. (1964). *J. Chem. Phys.*, **40**, 1448
277. Glass, G. P., Kistiakowsky, G. B., Michael, J. V. and Niki, H. (1965). *J. Chem. Phys.*, **42**, 608

278. Gay, I. D., Glass, G. P., Kern, R. D. and Kistiakowsky, G. B. (1967). *J. Chem. Phys.*, **47,** 313
279. Gay, I. D., Glass, G. P., Kistiakowsky, G. B. and Niki, H. (1965). *J. Chem. Phys.*, **43,** 4017
280. Gay, I. D., Kern, R. D., Kistiakowsky, G. B. and Niki, H. (1966). *J. Chem. Phys.*, **45,** 2371
281. Gay, I. D., Kistiakowsky, G. B., Michael, J. V. and Niki, H. (1965). *J. Chem. Phys.*, **43,** 1720
282. Modica, A. P. (1965). *J. Phys. Chem.*, **69,** 2111
283. Modica, A. P. and La Graff, J. E. (1965). *J. Chem. Phys.*, **43,** 3383
284. Modica, A. P. (1967). *J. Chem. Phys.*, **46,** 3663
285. Modica, A. P. and La Graff, J. E. (1966). *J. Chem. Phys.*, **44,** 3375
286. Diessen, R. W. (1966). *J. Chem. Phys.*, **44,** 3662
287. Diesen, R. W. and Felmlee, W. J. (1963). *J. Chem. Phys.*, **39,** 2115
288. Diesen, R. W. (1963). *J. Chem. Phys.*, **39,** 2121
289. Diesen, R. W. (1964. *J. Chem. Phys.*, **41,** 3256
290. Diesen, R. W. (1966). *J. Chem. Phys.*, **45,** 759
291. Modica, A. P. and Honig, D. F. (1966). *J. Chem. Phys.*, **45,** 760
292. Diesen, R. W. (1968). *J. Phys. Chem.*, **72,** 108
293. Appleby, S. E., Howarth, S. B., Heald, P., Lippiatt, J. H., Madeley, J. D., Myers, P., Orville-Thomas, W. J., Price, D. and Ward, G. B. (1971). *Advan. Mass Spectrom.*, **5,** in the press
294. Garnett, S. H., Kistiakowsky, G. B. and O'Grady, B. V. (1969). *J. Chem. Phys.*, **51,** 84
295. Skinner, G. T. and Moyzis, J. (1965). *Phys. Fluids,* **8,** 452
296. Becker, E. W., Bier, K. and Burghoff, H. (1955). *Z. Naturforsch,* **10a,** 565
297. Andres, R. P., Anderson, J. B. and Fenn, J. B. (1966). *Advan. Chem. Phys.*, **10,** 275
298. Ausloos, P. (1966). *Ann. Rev. Phys. Chem.*, **17,** 205
299. Meyer, R. T. (1968). *Nucl. Sci. Abst.*, **22,** 4384
300. Meyer, R. T. and Price, D. (1969). Reference 5, p.23
301. Meyer, R. T. (1967). *Rev. Sci. Instr.*, **44,** 422
302. Meyer, R. T. (1967). *J. Chem. Phys.*, **46,** 967
303. Meyer, R. T. (1967). *J. Chem. Phys.*, **46,** 4146
304. Meyer, R. T. (1968). *J. Phys. Chem.*, **72,** 1583
305. Berry, R. S., Clardy, J. and Schafer, M. E. (1964). *J. Amer. Chem. Soc.*, **86,** 2738
306. Schäfer, M. E. and Berry, R. S. (1965). *J. Amer. Chem. Soc.*, **87,** 4497
307. Berry, R. S., Clardy, J. and Schafer, M. E. (1965). *Tetrahedron Lett.*, **15,** 1003 and 1011
308. Appleby, S. E., Howarth, S. B., Jones, A. T., Lippiatt, J. A., Orville-Thomas, W. J., Price, D. and Heald, P. (1970). Reference 6, p.37
309. Knewstubb, P. F. and Reid, N. W. (1970). Reference 6, p.59
310. Knewstubb, P. F. and Reid, N. W. (1970). *Int. J. Mass Spectrom. Ion Phys.*, **5,** 361
311. Harvey, A., Monteiro, M.-De-L. and Reed, R. I. (1970). *Int. J. Mass Spectrom. Ion Phys.*, **4,** 365
312. Kamaratos, E. and Lampe, F. W. (1970). *J. Phys. Chem.*, **74,** 2267
313. Gohlke, R. S. (1959). *Anal. Chem.*, **31,** 535
314. Ebert, A. A. (1961). *Anal. Chem.*, **33,** 1865
315. Dutton, H. J. (1961). *J. Amer. Oil. Chem. Soc.*, **38,** 660
316. McFadden, W. H. and Teranishi, R. (1963). *Nature (London)*, **200,** 329, 435
317. Hunter, G. L. K. and Brogden, W. B. (1964). *Anal. Chem.*, **36,** 1122
318. Corse, J. W., Kilpatrick, P. W., McFadden, W. H., Schultz, T. H. and Teranishi, R. (1964). *J. Food Sci.*, **29,** 790
319. Grigsby, R. D., Eisenbraun, E. J., Hertzler, D. V. and Piel, K. M. (1967). Paper presented at *15th Ann. Conf. on Mass Spect., ASTM E-14,* (Denver, Colo. U.S.A., 1967)
320. McFadden, W. H. (1966). *Separation Sci.*, **1,** 723
321. D'Oyly-Watkins, C., Hillman, D. E., Winsor, D. E. and Ardrey, R. E. (1970). Reference 6, p.163
322. Ryhage, R. (1964). *Anal. Chem.*, **36,** 759
323. Watson, J. T. and Bieman, K. (1965). *Anal. Chem.*, **37,** 1135, 8446
324. Lipsky, S., Horvath, C. G. and McMurray, W. J. (1966). *Anal. Chem.*, **38,** 1585
325. Blume, M. (1968). *Anal. Chem.*, **40,** 1590
326. Llewellyn, P. M. and Littlejohn, D. P. (1966). *Pittsburgh Conf. on Anal. Chem.*, Feb. 1966

327. Luchte, A. J. and Damoth, D. C. (1970). Reference 6, p.219
328. Agnew, W. G. and Agnew, J. T. (1964). Paper presented at *10th Internat. Symp. on Combustion,* (Cambridge, England, 1964)
329. Fallgalter, M. B. and Hanrahan, R. T. (1970). *J. Phys. Chem.,* **74,** 2806
330. Bufalini, M. and Todd, J. E. (1969). *J. Phys. Chem.,* **72,** 3367
331. Tewarson, A. and Lampe, F. W. (1968). *J. Phys. Chem.,* **72,** 3261
332. Greene, F. T. and Milne, T. A. (1966). Paper presented at *14th Ann. Conf. on Mass Spect. and Allied Topics, ASTM E-14,* (Dallas, Texas, U.S.A., 1966)
333. Thomas, D. W. (1969). Reference 5, p.196
334. Malone, T. J. and McGee, H. A. (1965). *J. Phys. Chem.,* **69,** 4338
335. McGee, H. A., Malone, T. J. and Martin, W. J. (1966). *Rev. Sci. Instr.,* **37,** 561
336. McGee, H. A., Bivins, D. B., Malone, T. J. and Wilson, J. H. (1966). Paper presented at *14th Ann. Conf. on Mass Spect. and Allied Topics, ASTM E-14,* (Dallas, Texas, U.S.A., 1966)
337. Bivins, D. B. (1966). *Ph.D. Thesis,* (Georgia Inst. of Tech., Atlanta, Ga., U.S.A.)

8
Metastable Ions in Mass Spectrometry

J. L. HOLMES and F. M. BENOIT
University of Ottawa, Ontario

8.1 HISTORICAL INTRODUCTION AND EARLY EXPERIMENTAL WORK

The subject of this review article began in 1945 when Hipple and Condon[1] first explained the origin of the weak diffuse peaks observed in mass spectra. In a further publication with Fox[2] they showed that metastable ions have lifetimes $\geqslant 10^{-6}$ s and that they are produced by ions which, after full acceleration, dissociate in the field-free region between the ion source and analyser (single-focusing magnetic instrument). It was considered possible that these dissociations proceeded with some conversion of internal energy into translational kinetic energy, but the familiar equation

$$m^* = m_2^2/m_1 \tag{8.1}$$

relating the apparent mass of the metastable ion peak, m^*, to the masses of the daughter (m_2) and precursor (m_1) ions was derived assuming that such an energy conversion is negligible. They proposed that the observed peak broadening was caused by both energy release and focusing properties of the instrument.

An important problem in early experiments was the need to distinguish between collision-induced metastable ions and truly unimolecular dissociations. Hipple, Condon and Fox[2] studied the metastable ion peaks in the mass spectrum of n-butane as a function of pressure; they were able to show that some processes could indeed be attributed to unimolecular metastable dissociations, by virtue of the non-disappearance of some metastable ions at low pressure and that their abundances were proportional to the first power of the pressure. Furthermore, the abundance of these unimolecular metastables was strongly dependent on the ion source draw-out potential, i.e. with large draw-out potentials, the shorter the residence time of ions within the source and the greater the relative abundance of metastable ions (compared with the rest of the mass spectrum). By use of an electrostatic energy filter (retarding potential) they confirmed the identity of precursor and daughter ions giving rise to a particular metastable.

In a third paper, Hipple[3] described experiments designed to measure the half-lives of metastable ions. He showed that for the dissociations

$$C_4H_{10}^{+\cdot} \rightarrow C_3H_7^+ + CH_3^{\cdot} \tag{8.2}$$

$$C_4H_{10}^{+\cdot} \rightarrow C_3H_6^{+\cdot} + CH_4 \tag{8.3}$$

the half-lives for the metastably dissociating molecular ions were $\sim 2 \times 10^{-6}$ s.

Fox and Langer[4] measured appearance potentials of metastable ions in the mass spectrum of n-butane and other C_4 hydrocarbons; they found that the appearance potentials of the metastable ions were *c.* 2 eV higher than those of their precursor ions. They equated this excess energy with that required for the precursor ion to dissociate but it was later shown by Hertel and Ottinger[5] that this simple interpretation is incorrect in terms of the quasi-equilibrium theory. Note that the appearance potential of a *daughter* ion is frequently found to be higher than that of its generating metastable ion. This energy difference is known as the 'kinetic shift' and will be referred to later[6].

An important, controversial issue which arose early in the study of metastable ions concerned their origin. According to the quasi-equilibrium theory of mass spectra[7] (Q.E.T.), metastable ions arise from unimolecular dissociations among a large assemblage of electronic and vibrational energy states, too numerous (in the case of polyatomic molecules) to be formally individually identifiable by kinetic studies; this theory has proved capable of explaining all the major phenomena in mass spectrometry. An alternative view, namely that metastable transitions occur from few (or single), perhaps identifiable, excited states of the precursor ion, was advanced by Momigny[8]. While this hypothesis is not generally accepted, there is good evidence that *some* metastable dissociations can be related to particular ion energies. The quasi-equilibrium theory was first thoroughly formulated in 1952 by Rosenstock *et al.*[7]. The major assumption in Q.E.T. can be stated as follows; molecular ions are formed with a certain amount of excess electronic and vibrational energy above the minimum of their lowest electronic state. The majority of molecular ions do not dissociate immediately but undergo at least several vibrations. During this period there is a high probability that radiationless transitions take place among the many potential energy surfaces for the molecular ions, resulting in random distribution of the excitation energy. A given molecular ion only dissociates when its nuclei are in the proper configuration and sufficient vibrational energy has concentrated in the necessary internal degrees of freedom. It must be stressed that the Q.E.T. is to be expected to hold only for large polyatomic molecules. For small molecules discrete energy states will apply, e.g. the work of Newton and Sciamanna on the mass spectra of N_2O and NO_2 [9].

Insofar as the existence of metastable ions is concerned, their chief significance for the Q.E.T. is their confirmation of the theory's energy distribution function which predicts that a small fraction of ions will have internal energies resulting in lifetimes of the order of microseconds. Rosenstock, Wahrhaftig and Eyring[10] stated that ions having half-lives from 10^{-12} s upward should be observable in mass spectrometers of suitable versatility and they re-iterated their belief that any discussion in terms of isolated individual or sets of electronic states must be precluded by the high probability of radiationless transitions among the closely-spaced electronic states of a large polyatomic ion. They further pointed out that since the rate constant for unimolecular dissociations is a function both of an energy term and a frequency factor, no simple relationship could be predicted for the relative abundances of precursor, daughter and metastable ion peaks.

From the experimental point of view, the improvements in pumping equipment and the greater sensitivity of ion detectors have made unimolecular metastable ion identification much easier. The most important experimental development in recent years has been the introduction of methods for separating metastable ions from the ions making up the 'normal' mass spectrum – the so-called 'metastable defocusing' techniques; these will be described in detail later.

8.2 METASTABLE ION CHARACTERISTICS

8.2.1 The shape of metastable ion peaks

This short section will deal with theories proposed to account for the shapes of metastable ion peaks as observed in conventional single- or double-focusing mass spectrometers.

Hipple and Condon[1] concluded that the Gaussian shape of the diffuse peaks was due to an instrument focusing effect associated with metastable dissociations in all regions of the mass spectrometer. It was soon proposed[2], however, that the relative broadness of metastable ions could arise from the conversion of a small amount of internal energy into translational kinetic energy when dissociation takes place. This postulate was denied by Coggeshall[11] who concluded from his experiments with a 180 degree, single-focusing instrument, that metastable peak broadening was essentially an ion-focusing phenomenon and did not constitute evidence for energy release. According to Coggeshall's predictions, metastable ions observed with a Dempster-type instrument should exhibit a sharp discontinuity at the calculated apparent mass m^* and this peak should tail in the direction of higher mass. This particular controversy was resolved by Newton[12] who convincingly showed that the shape of the metastable ion for equation (8.2) could best be correlated with a small kinetic energy release of 0.015 ± 0.005 eV. The effect of small energy releases on metastable peak shapes was also discussed by Ottinger[13] with particular reference to the problem of disparities in relative metastable peak abundances depending on the instrument employed. It should be pointed out that the idea of fragmentation with energy release was not in itself new but had largely been associated with processes involving doubly charged ions[14]. Energy release was evidenced by the appearance of daughter ions possessing excess kinetic energy and had not been related to metastable ion phenomena.

8.2.1.1 'Flat-top' metastable ion peaks

An important development came in 1965 with the first of two papers by Beynon *et al*[15]. In their first report, the phenomenon of very broad (so-called 'flat-top') metastable ion peaks was considered in detail. It had been observed[16] that the width of the 'flat-top' metastable associated with the process

$$CO_2^{2+} \rightarrow CO^+ + O^+ \tag{8.4}$$

varied approximately inversely with acceleration potential: the equation

$$d = \frac{4m_2^2}{m_1}\left(\frac{\mu T}{eV}\right)^{\frac{1}{2}} \tag{8.5}$$

derived by Beynon *et al.*[15] was in accord with this observation. In the above equation d is the width of the flat section of the metastable peak (see Figure 8.1(a)) measured in atomic mass units (as would be observed in a conventional mass spectrometer record), μ is the mass ratio $(m_1 - m_2)/m_2$, eV is the ion acceleration potential in electron volts and T is the energy released (also in electron volts). The energy releases in the metastable processes (m^* centred at $m/e = 85.5$) for the three nitrophenol isomers were measured (i) by finding

$$[\text{NO}_2\text{-C}_6\text{H}_4\text{-OH}]^{+\cdot} \xrightarrow{*} C_6H_5O_2^{+} + NO^{\cdot} \tag{8.6}$$

the retarding potential over which the metastable ion disappeared, i.e. an electrode was placed in front of the ion collector in a single-focusing Nier-type (90 degree) instrument; at a certain potential the low-mass end of the metastable peak began to disappear. As the retarding potential was increased more of the metastable peak was removed until, at a limiting upper potential,

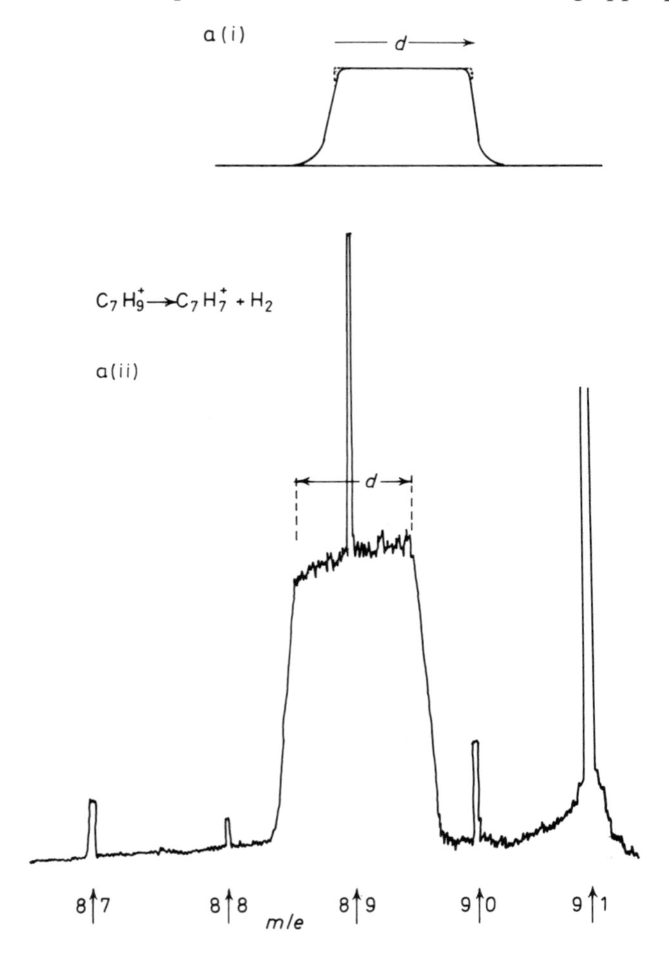

Figure 8.1 (a) (i) Flat-top metastable ion (schematic) and (ii) flat-top metastable ion observed with a single-focusing instrument

the peak finally disappeared; (ii) by measuring the width of the flat-top metastable in the conventional mass spectrum and substituting the appropriate values in equation (8.5). The *meta*-isomer displayed a normal Gaussian metastable peak while the widths of the flat-top metastables observed with both the *ortho*- and *para*-isomers gave T(i) = 0.74 ± 0.16 eV and T(ii) = 0.76 ± 0.02 eV. This difference in behaviour between the *meta* and the other two isomers has been explained in terms of the potential stability of the *ortho*- and *para*-daughter ions in their quinonoid forms[17]:

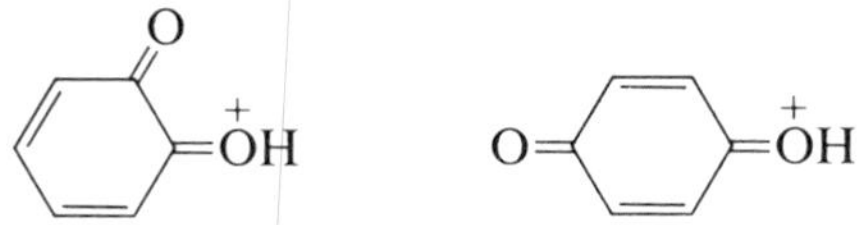

At the same time that Beynon's results appeared, Higgins and Jennings[18] reported their observation of two flat-top metastable peaks in the mass spectrum of benzene, corresponding to the dissociation of the doubly charged molecular ion into two charge-bearing fragments

$$C_6H_6^{2+} \rightarrow CH_3^+ + C_5H_3^+ \tag{8.7}$$

$$m^*(CH_3^+) = 5.77; m^*(C_5H_3^+) = 101.8$$

(That such processes in the mass spectrum of benzene involved kinetic energy release had been postulated much earlier[19].) Using equation (8.5) it was found that the energy release associated with the methyl ion (2.16 ± 0.08 eV) was in excellent agreement with the value obtained[14] from a direct study of the energy content of the CH_3^+ daughter ions, 2.27 eV. Further results of energy-release determinations and conclusions drawn therefrom concerning ion structures will be discussed later. Also in 1965, Flowers[20] commented on the metastable peaks reported by Higgins and Jennings[18] which were somewhat dish-shaped (see Figure 8.1(b)); Flowers proposed that the shape was the consequence of a preferred energy release, the distance between the wings (d' in Figure 8.1(b)) representing a most-probable energy release. Shannon *et al.*[21] were, however, able to induce a concave top in the metastable peak corresponding to H_2 loss from the CH_2OH^+ ion (from methanol) by merely reducing the acceleration potential. This result indicated that instrument parameters were important in determining metastable peak shapes. In addition they showed that the energy release derived from the metastable peak width in the *p*-nitrophenol dissociation (equation (8.6)) agreed well with Beynon's result[15] when a 90 degree, single–focusing Hitachi RMU-6A instrument was used (0.72 eV) but was very different for their CEC 21-110B double-focusing, Mattauch–Herzog instrument (1.77 eV). This latter effect they ascribed to the very narrow source exit slit of the Hitachi instrument producing a better defined image at the collector compared with that obtained with the wide β-stop (between electrostatic and magnetic sectors) of the double-focusing machine.

In the second paper, Beynon and Fontaine[22] showed that concavity in flat-top metastables resulted from ion-collection discrimination due to the combined effect of a finite collector slit width and the release of energy in a direction perpendicular to the long axis of the slit. A computation based on the geometry of an AEI MS-9 mass spectrometer led to the prediction that collection of *all* fragment ions would yield a metastable peak such as Figure 8.1(a) while discrimination by a narrow collector slit would yield a peak such as Figure 8.1(b).

In the energy-release equation (8.5) it had been assumed that T was single-valued and the peaks are thereby predicted to have sharply rising well-defined edges. The effect of a range of released energies was considered by Smyth and

Shannon[23] who showed that this can give rise to the diffuse and gently-sloping sides commonly observed (Figure 8.1(b)). Smyth and Shannon discussed the problem of energy release in normal (Gaussian) metastable peaks and showed that a Boltzmann distribution of kinetic energy releases yields a Gaussian-shaped metastable peak when the energy release is *small*. As an example it was shown that for the loss of N_2 from the molecular ion

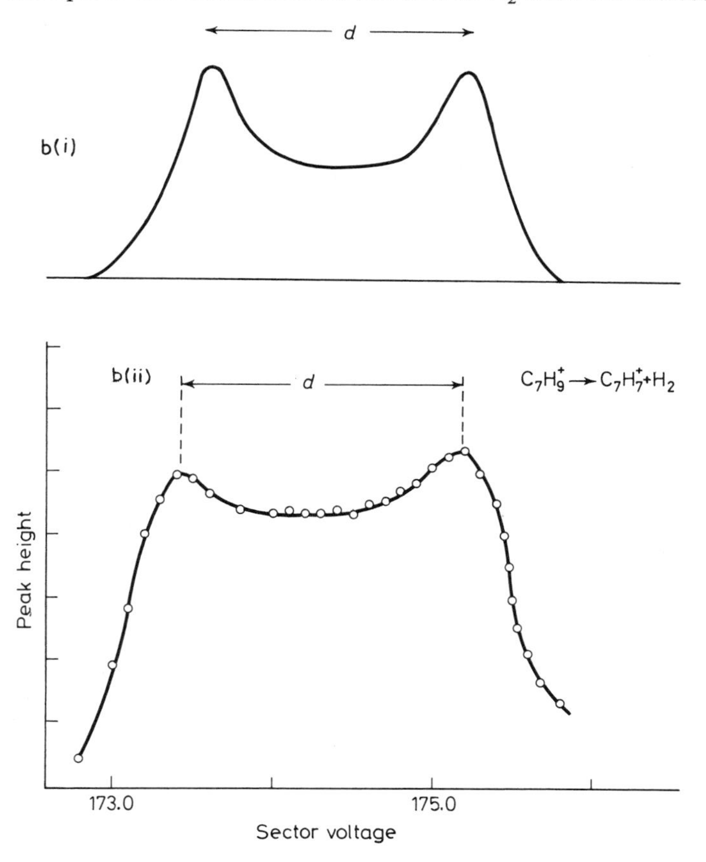

Figure 8.1(b) (i) Dished flat-top metastable (schematic) and (ii) metastable ion, 8.1(a)(ii), defocused by Kiser's method[56]

of benzocinnoline ($C_{12}H_8N_2$), the half-height width of the (Gaussian) metastable peak for this dissociation corresponded to a translational 'temperature' of 250 °C, only slightly greater than ion source temperature and indicating that the loss of N_2 probably occurs from the ground state of the molecular ion. Flat-top metastable ions can be accommodated by a similar scheme using a reversed Boltzmann distribution of energy releases (i.e. where the average energy is *less* than the most probable energy). Calculations [23] performed for the process

$$CH_2^{+}OH \longrightarrow CHO^{+} + H_2 \tag{8.8}$$

in the methanol mass spectrum yielded a metastable profile in excellent agreement with experiment.

The range of energies released in metastable decompositions was discussed by Beynon *et al.*[24] with respect to the lifetime of the decomposing molecular ion of pyrazine

$$C_4H_4N_2^{+\cdot} \longrightarrow C_3H_3N^{+\cdot} + HCN \tag{8.9}$$

This metastable ion peak is narrow and of Gaussian shape; its true half-height width, which was taken as a measure of the energy release, was determined by plotting the metastable peak width against that of a normal ion peak at several instrument resolving powers. Extrapolation to infinite resolution (i.e. to a normal peak width of zero) yielded the 'true' metastable peak half-height width. The ion acceleration potential was varied to provide a range of source residence and time-of-flight times over which the energy release, determined as above, was evaluated. The longer ion lifetimes were associated with the smaller energy release values; in the lifetime range $10–26 \times 10^{-6}$ s the energy release fell from 0.0222 to 0.0187 eV. Using these results and the simple harmonic oscillator form for the Q.E.T. equation, conclusions were drawn as to the possible effective number of oscillators participating in the dissociation.

Reichert *et al.*[25] have shown that Kiser's electric-sector method of metastable 'defocusing'[56] (see later) is particularly suited to measuring metastable peak widths.

Rowland[26] has discussed the same benzocinnoline fragmentation referred to above; provided that the following inequality obtains:

$$\left(d + \frac{1}{2}\pi R + d_1\right)\left(\frac{\mu T}{eV}\right)^{\frac{1}{2}} < L/2$$

(where d is the length of the field-free region in front of the magnetic field; R is the radius of the magnetic analyser, d_1 is the distance between magnetic field and collector slit, L is the collector slit length) all ions will be collected and with single-valued T a flat-top metastable will be observed centred about $m^* = m_2^2(1 + \mu T/eV)/m_1$. (When T is small $m^* = m_2^2/m_1$.) A centre-of-mass kinetic energy distribution function can be derived from a study of the (experimental) first derivative of metastable peak height with respect to mass number, plotted on an energy scale. Examination of the results for the benzocinnoline fragmentation yields a slightly non-Boltzmann distribution function.

The simple equation (8.2) of Beynon *et al.*[15,22] relating kinetic energy release with the width of the metastable peak and apparatus parameters was tested by Barber *et al.*[27]; they employed a double-focusing (MS-9) instrument and compared the energy release calculated (by equation (8.2)) for the flat-top metastable ion peak for the process

$$C_3H_4F^+ \longrightarrow C_3H_3^+ + HF \tag{8.10}$$

observed in the normal mass spectrum, with those obtained by the following method.

At a given electrostatic sector potential, E, the daughter ion (m_2) was brought into focus in the normal manner. The acceleration potential (initially V_0) was then increased until the (metastable) m_2 ions were again collected; for this 'flat-top' metastable peak a range of acceleration potentials was required ($V_L \rightarrow V_H$) to sweep over its profile, the width being proportional to ($V_H - V_L$). The appropriate kinematic equation for the energy release, T, is given by

$$T = \frac{V}{4\mu}\left(\frac{m_1 V_0}{m_2} \cdot \frac{1}{V} - 1\right)^2 \qquad (8.11)$$

It was observed that T was a function of E; a linear plot was obtained when one parameter is plotted against the other; the intercept value of T at $E = 0$ gave an energy release in excellent agreement with that obtained from measuring the width of the metastable peak in the normal mass spectrum. The procedure was justified theoretically by consideration of the fraction of ions which reach the collector as a function of the acceleration potential.

To summarise, metastable peak shapes observed in conventional mass spectra depend on the magnitude and distribution of the kinetic energy released upon fragmentation. Small energy releases (<2 kcal) having an essentially Boltzmann energy distribution yield the 'normal' Gaussian-shaped peaks: 'flat-top' and 'dished' metastable peaks are associated with larger energy release and, for these dissociations, collector-efficiency problems may give rise to incorrect values for the energy release.

8.2.2 Dissociations in non-field-free regions

Phenomena arising from (slow) dissociation of ions in non-field-free regions of single- and double-focusing mass spectrometers were thoroughly discussed by Beynon and Fontaine[22]. Such ions are either lost by collision with the instruments' walls or are collected at apparent masses other than $m^*(= m_2^2/m_1)$. In a double-focusing instrument, first field-free region metastable ions (dissociations between source and electrostatic sector) will not be transmitted (unless the instrument is operated in the 'defocused' mode—see later). It is second field-free region metastable ions (dissociations between electrostatic and magnetic sectors) which appear at m^* in the conventional mass spectrum obtained with a double-focusing spectrometer.

(i) Ions which dissociate *during acceleration* appear as tails on either the low mass side of m_2 or the high mass side of m^* with a continuum between them (Figure 8.2(a)).

(ii) Ions which dissociate *within the magnetic field* (single- or double-focusing instruments) produce a continuum between m^* and m_1 with slight intensity increases at the limits (Figure 8.2(b)).

(iii) Ions which dissociate in the *electrostatic sector* of a double-focusing instrument have been discussed by Beynon *et al.*[28] with particular respect to dissociations involving small mass change (e.g. loss of $H^{\bullet}$ from $C_6H_5CH_3^{+\bullet}$). Such daughter ions will be transmitted by the instrument and appear between

m_1 and m_2. Schulze and Burlingame[29] have discussed the matter in greater detail (see also under metastable defocusing).

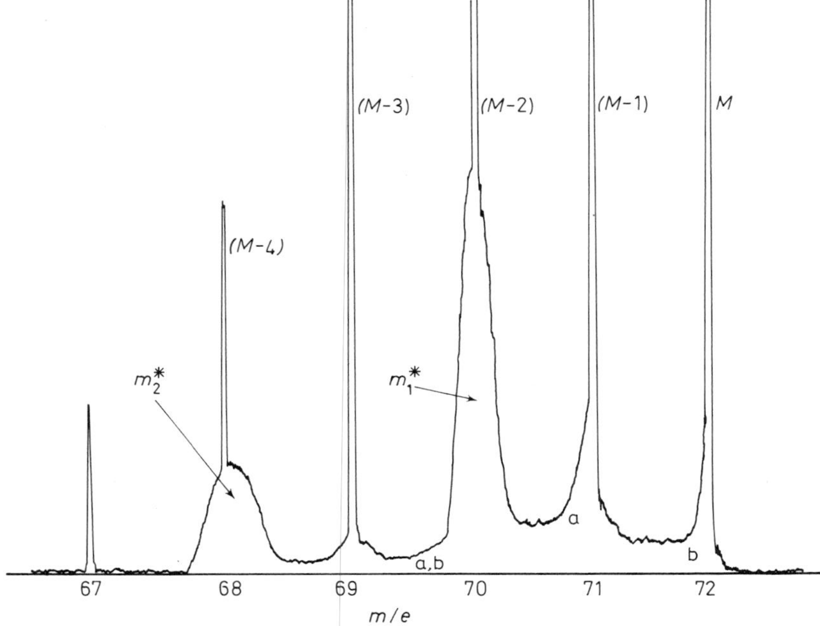

Figure 8.2 Dissociations in non-field-free regions. Conventional Gaussian metastable ions m_1^* ($72 \rightarrow 71+1$), m_2^* ($72 \rightarrow 70+2$) in a normal mass spectrum. Also showing tails and continua for ions (a) dissociating during acceleration and (b) within the magnetic field. $m/e = 72$ ($C_5H_4D_4$), $m/e = 71$ ($C_5H_3D_4$), $m/e = 70$ ($C_5H_4D_3$)

(iv) Dissociations within the field-free region *between magnetic analyser and detector* will cause tailing on each side of the precursor ion peak and the spread of the tails can be related to the kinetic energy released.

8.2.3 Metastable ion lifetimes

For the majority of experimentalists using a conventional single- or double-focusing instrument the range of ion lifetimes which can be examined is limited to *c.* 10^{-5}–10^{-7} s. Although the time of flight of the ions between source and collector can be readily estimated, the residence time of charged species within the source (from their genesis to full acceleration) is uncertain because the potential gradient distribution within the source is generally not known; however, in many instruments this residence time lies in the range 1 to 10 μs.

Using an experimental technique similar to that of Hipple[3], in which the ion repeller potential was accurately varied, Momigny[8] observed a non-linear dependence of the logarithm of the relative metastable ion abundance with respect to source residence time (first-order plot). The effect of ionising-electron energy variation on the benzonitrile dissociation

$$C_6H_5CN^{+\cdot} \rightarrow C_6H_4^{+\cdot} + HCN \tag{8.12}$$

indicated the existence of a second type of metastable ion which had a very short lifetime and only appeared at electron energies in excess of 15 eV. He proposed that his results cast some doubt on the major assumption of the Q.E.T., namely that dissociations from discrete excited states will be non-observable. Momigny's viewpoint was further supported by the results of Coggeshall[11] from experiments using a 180 degree single-focusing mass spectrometer. By varying the ion-repeller potential he could change the source residence time of ions from $0.3–3 \times 10^{-6}$ s. He found that his first-order plots for the n-butane dissociation (equation (8.2)) could be well represented by three intersecting linear sections each purporting to show dissociation of a different excited state of the n-butane molecular ion. Muccini *et al.*[30] studied the first-order decay of $C_3H_7^+$ ions

$$C_3H_7^+ \rightarrow C_3H_5^+ + H_2 \tag{8.13}$$

produced from a variety of n-alkanes, C_3–C_{10}, by a similar experimental method. They observed $k_{8.13} = 0.46 \pm 0.24–0.72 \pm 0.23 \times 10^6\ s^{-1}$ with no apparent trend with parent hydrocarbon. Although they were unable to relate their results to participation of individual excited states they felt that the essential invariance of $k_{8.13}$ with both hydrocarbon precursor and electron energy was inconsistent with the Q.E.T. In 1964, Schug[31] repeated the experiments of Momigny and Coggeshall and could find no evidence for breaks in the first-order plots for the benzonitrile dissociation (equation 8.12)). He declared that the introduction of discrete rate constants arising from separate energy states was unnecessary to explain the curvature of the first-order plots and the latter could be well explained within the framework of the statistical theory. (The effect of electron energy on metastable ion abundances would be absent only in a case where a *single* dissociating state were involved[43].) A careful test of the Q.E.T. with respect to metastable ions was made by Rosenstock *et al.*[32], also in 1964. They compared the relative abundances of metastable ions for the two processes

$$C_6H_{13}^+ \rightarrow C_4H_9^+ + C_2H_4\ (m^* = 38.2) \tag{8.14}$$

$$C_6H_{13}^+ \rightarrow C_3H_7^+ + C_3H_6\ (m^* = 21.8) \tag{8.15}$$

as a function of precursor compound and ionising-electron energy. A wide range of precursor compounds was selected – eight n-alkanes, n-hexyl bromide and di-n-hexyl ether. The metastable abundance ratios were carefully measured using peak areas rather than peak heights. The results were statistically analysed in order to determine whether there was any dependence on molecular weight of precursor compound or on ionising-electron energy. The results were fitted to straight lines of the form

$$R = R_0 + b(M - 154.3) \tag{8.16}$$

where R is the metastable abundance ratio and M is the molecular weight. R_0 and b are shown in Table 8.1 with their probable errors (95% confidence limits).

The invariance of the 30 eV ratios indicated that in the rate constant range for these dissociations (10^5–10^7 s^{-1}) there are no non-equilibrium effects due to the mode of preparation of the hexyl ion. At higher electron energies there was an unexplained, slight inverse effect on the ratio with increasing molecular weight. Implicit in this work is the important corollary proposed

Table 8.1 Dependence of metastable ratio on molecular weight of precursor for two dissociations of the $C_6H_{13}^+$ ion[31]

Electron energy/eV	R_0	b
30	1.569 ± 0.062	−0.00018 ± 0.00177
50	1.556 ± 0.065	−0.00252 ± 0.00187
70	1.607 ± 0.037	−0.00260 ± 0.00118

by Shannon and McLafferty[33] namely that if the metastable abundance ratios for several dissociations of a given ion produced from different precursors are the same, then the given dissociating ion is either a unique species or consists of the same assemblage of species both with respect to their structure and energy. This observation has application in the identification of ion structures and its use will be discussed in a separate section later (see Sections 8.3.2. and 8.3.5.1).

Further work in 1965 by Osberghaus and Ottinger[34] involved use of a modified ion source wherein the ion residence time could be varied from 2×10^{-8} to 3×10^{-6} s; they studied the two n-butane dissociations (equations (8.2) and (8.3)) and could find no evidence for the participation of discrete excited states. Benz and Brown[35] determined half-lives of some metastable ions in a cycloidal mass spectrometer. Their mean half-life for the n-butane dissociation (equation (8.2)), 1.45×10^{-6} s, was similar to previous values; their experiments also yielded no results indicative of discrete lifetimes.

A possible test of Q.E.T. lies in isotope effects where it had been predicted[36] that metastable peaks arising from perdeuteriated molecular ions should be more intense than those of the unlabelled molecules. This is because owing to the relatively smaller vibrational frequencies and larger moments of inertia in the deuteriated molecule, the density of vibrational–rotational states of the activated ion is larger for the heavy ion; thus the rate constant for dissociation should be *smaller* for the deuteriated ion and will therefore have a greater probability of falling in the range of dissociations observable as metastable processes. Lifshitz and Shapiro[37] studied the metastable ions for methane loss from the molecular ions of C_3H_8 and C_3D_8. Vestal's[36] prediction was correct, $m_D^*:m_H^* \approx 7.0$ for 70 eV electrons. However, at a low electron energy, 13 eV, and using a pulsed ion source to evaluate residence times they found $k(C_3H_8) = 3.2 \times 10^4$ s^{-1}, $k(C_3D_8) = 8.3 \times 10^4$ s^{-1} in apparent contradiction of Q.E.T. Furthermore, the first-order plot for C_3D_8 showed a pronounced break possibly indicative of two discrete rate constants. Theory and experiment could, however, be reconciled by proposing that allowance be made for the presence and difference in endothermicities of both a 1,2 and 1,3 elimination reaction, i.e.

$$CH_3CH_2CH_3^{+\bullet} \rightarrow CH_2CH_2^{+\bullet} + CH_4 \quad (8.17)$$

$$CH_3CH_2CH_3^{+\bullet} \rightarrow CH_3CH^{+\bullet} + CH_4 \quad (8.18)$$

Reaction (8.18) is more endothermic than (8.17) and therefore has a higher activation energy. On this basis they proposed that equation (8.17) contributes strongly to the metastable ion peak for C_3D_8 (at low electron energy) but not for C_3H_8. It followed from the above observations that a study of the molecules $CH_3CD_2CH_3$ and $CD_3CH_2CD_3$ should be rewarding and the appropriate experiments were carried out by Lifshitz and Shapiro[38] in the following year. They used the same apparatus as before and determined ion lifetimes with their pulsed ion source. The normal mass spectra confirmed earlier results indicating that methane elimination is mainly a 1,3 reaction. Unfortunately only a metastable peak for loss of CH_4 from $CH_3CD_2CH_3$ was observed, those for CD_4 or CD_3H from $CD_3CH_2CD_3$ and CH_3D from $CH_3CD_2CH_3$ were surprisingly absent.

$$CH_3CD_2CH_3^{+\bullet} \rightarrow CH_4^{+} + C_2H_2D_2^{+\bullet} \quad (8.19)$$

An additional embarrassment was the presence of an apparent break in the first-order plot for the metastable ion in which 1,2 and 1,3 eliminations cannot both yield CH_4 unless there is appreciable scrambling within the molecular ion before dissociation! This latter possibility was discounted because no metastable ion was observed for CH_3D. Ottinger[39] independently studied metastable ions for hydrogen and deuterium losses from the molecular ions of $CH_3CD_2CH_3$ and CD_3CHDCD_3. He showed that there was no evidence for hydrogen scrambling in the molecular ions and that only s-propyl ions are produced in the metastable decay. Where loss of a hydrogen atom competed directly with loss of a deuterium atom, as in the processes

$$CD_3CHDCH_3^{+\bullet} \begin{array}{l} \rightarrow C_3D_7^{+} + H^{\bullet} \\ \rightarrow CD_3CHCD_3 + D^{\bullet} \end{array} \quad (8.20)$$

a very large isotope effect was observed in favour of hydrogen ($m_H^* : m_D^* \approx 300$).

The problem of the observation of discrete lifetimes from individual excited states was further discussed by Lifshitz[40] with respect to the observations of Meyer and Harrison[41] on the behaviour of the toluene molecular ion in a pulsed ion source; the latter authors examined precursor and daughter ion intensities and found that most of the loss of a hydrogen atom from the molecular ion occurred from a state (or states) with a half-life of less than 10^{-7} s but a significant fraction of the molecular ions dissociated by H loss from a state with a greater half-life of the order of 5×10^{-6} s. Lifshitz[40] accordingly studied the *metastable* ion intensities as a function of delay time in the range $1\text{–}10 \times 10^{-6}$ s; she observed non-linear first-order plots without any sign of discontinuities, indicating that more than one discrete lifetime was involved. The curves were approximately parallel from 10–16 eV ionising electron energies. These results are in keeping with Q.E.T. In 1967, Hertel and Ottinger[42] described an improved apparatus in which the available time scale for the study of ion dissociations had been extended from 5×10^{-9} to 5×10^{-6} s. From their results they were able to determine the distribution of rate constants with energy content of the ions dissociating within the above time interval. The molecules studied were n-butane, benzonitrile, n-heptane and benzene. More recently, Yeo *et al.*[43] repeated some previously erroneous[44] experiments on 2,4,6-d_3-benzonitrile and showed, by comparing the appropriate metastable peaks, that with 20–70 eV electrons hydrogen and deuterium are randomly lost in the dissociation

$$C_6H_2D_3CN^{+\cdot} \rightarrow HCN + C_6HD_3^{+\cdot} \quad \rightarrow DCN + C_6H_2D_2^{+\cdot} \quad (m^*_{HCN}/m^*_{DCN} = 0.73 \pm 0.04) \tag{8.21}$$

The ratio of daughter ion peak heights lay in the range 0.67–0.82 under similar experimental conditions.

However, the ratio of the metastable ions was observed to fall markedly at low electron energy (to 0.45 ± 0.1 at 16 eV) and thus the dissociation is inconsistent with a single-state model[31]. The authors felt unable to decide in favour of either Momigny's[8] several-state model or the continuous distribution of lifetimes proposed by Schug[31] and Ottinger[45]. For 2-d_1-benzothiazole and 2,4,6-d_3-benzoic acid larger changes in competing metastable ratios were observed with variation of electron energy near the threshold. While the large variation of $m^*(M-\mathrm{DCN})/m^*(M-\mathrm{HCN})$ in the former molecule (by a factor of 3.7 from 70 eV to 14 eV) could be explained by the generation of metastable ions from a wide energy band (i.e. a relatively large kinetic shift for threshold appearance potentials of metastable and daughter ions) it was entertained that contributions to the two processes from isolated electronic states should be seriously considered. In support of this proposal it was shown that the metastable ion ratio was also a function of ion-repeller potential (at low electron energy) and of accelerating potential (at 70 eV), both these adjustments facilitating a change of time spent in the ion source and analyser sections of the mass spectrometer.

The unresolved propane/deuterated propanes problem was tackled again in 1970 by Vestal and Futrell[46] who studied defocused metastable ions in the mass spectra of C_3H_8, C_3D_8, $CH_3CD_2CH_3$ and $CD_3CH_2CD_3$ and made detailed calculations of the mass spectra of these molecules based on developments in the Q.E.T.[47]. Their calculated results were in good qualitative agreement with earlier experiments[37–39] and the observed isotope effects were effectively accounted for.

Further support for the Q.E.T. has come from field ionisation mass spectra; in a report by Beckey *et al.*[48] the rate constant for the unimolecular dissociation

$$\text{n-}C_7H_{16}^{+\cdot} \rightarrow C_5H_{11}^{+} + C_2H_5^{\cdot} \tag{8.22}$$

was evaluated over a time range 5×10^{-11} to 10^{-6} s and a quasi-continuous distribution of rate constants was observed in striking qualitative agreement with the statistical theory.

8.3 THE APPLICATION OF OBSERVATIONS ON METASTABLE IONS TO SOME CURRENT PROBLEMS IN MASS SPECTROMETRY

The last five years have seen a great upsurge of interest in metastable ions, indeed 80% of the references cited in this review were published post 1965.

One major reason for this increase of activity lies in the introduction of a variety of methods whereby metastable ions can be observed free from the precursor and daughter ion peaks which make up a 'normal' mass spectrum. The next section will describe the techniques by which metastable ions can be (i) enhanced in a single-focusing mass spectrometer; (ii) separately observed in double-focusing instruments of various geometries.

8.3.1 Instrumentation

8.3.1.1 Single-focusing instruments

Until 1964 (when Barber and Elliott[49] described their method for 'defocusing' metastable ion peaks), studies of metastable ions had been confined to observing them in a normal mass spectrum. The presence, at (almost) every integral mass, of the multitudinous daughter ion peaks makes difficult the determination of the exact apparent mass, relative abundance and width of metastable peaks. Metastable peaks occurring at or very near to integral masses are often obscured by intense fragment ion peaks and may be missed altogether if they are narrow. The 180 degree magnetic instrument

Table 8.2 Relative abundance of the metastable ion peak $C_6H_6^{+\cdot} \rightarrow C_4H_4^{+\cdot} + C_2H_2$ ($m^* = m/e$ 34.7) in the mass spectrum of benzene as a function of electron energy, resolution and repeller potential. Hitachi–Perkin-Elmer RMU 6D single-focusing instrument

Energy/eV (*nominal*)	*Relative abundance* % *base peak*	m^*(× 100) % *total ionisation*
70	10	4.5
20	9.7	6.1
15	6.8	5.2
12	3.8	3.2
10	2.4	2.1
Repeller potential (*electron energy* 70 eV)		
1	3.5	1.5
12	10	4.5
28	15.5	6.4
Repeller potential (*electron energy* 20 eV)		
2	2.1	1.6
12	9.7	6.1
28	20	9.4
Resolution (*electron energy* 70 eV)		
400	10	4.6
800	8.0	3.6
1600	7.1	3.2
2500	6.5	2.9

(such as was used by Coggeshall)[11] is not very suitable for metastable ion studies owing to the small region of the ion path from which metastable transitions can be observed; this results in low metastable ion peak intensities.

A single focusing 90 degree sector magnetic instrument of modest resolving power (~1 : 5000) can, however, be used to study metastable ions under favourable operating conditions.

Metastable ion peak intensities are quite sensitive to the following instrument parameters: (i) resolving power can be increased by narrowing the collector slit but while this operation narrows and better defines the profile of normal ion peaks, the intensity of metastable ion peaks is sharply reduced; (ii) the intensities of normal ion peaks can be reduced by operating at low electron energies, say in the range 15–20 eV; metastable ion peaks derived from major fragmentations will be easier to observe against the much reduced normal peak 'background'; (iii) source residence time can be moderately well controlled by varying the repeller potential. At high repeller potential, ions are more quickly ejected from the source and an increase in metastable ion peak intensities is observed. Caution must be exercised with this adjustment because ion beam focusing is affected also.

For a given instrument, optimum conditions for the above variables can be determined. As an example some observations with a Hitachi RMU 6D single-focusing instrument are shown in Table 8.2.

An interesting new ion-detector system pioneered by Daly *et al.*[50] will soon become commercially available. The detector consists of a glass scintillator placed in front of a photomultiplier and the instrument is claimed to be as sensitive as the electron multipliers commonly found in mass spectrometers. The important feature of this new detector insofar as metastable ions are concerned is the presence of a retarding grid which allows selection of ions with respect to their energy before they reach the detector. Enhancement of metastable ion peak intensity over that for normal ion peaks of up to 1000 is anticipated. The same authors have reported results obtained on the metastable ion spectra of *cis*- and *trans*-but-2-ene[51].

The observation of metastable transitions in a time-of-flight mass spectrometer has been described by Dugger and Kiser[52].

8.3.1.2 Double-focusing instruments

The commonly encountered double-focusing mass spectrometers have either Nier–Johnson or Mattauch–Herzog geometry. These are schematically illustrated in Figure 8.3. The above remarks pertaining to metastable ion studies in the normal mass spectrum apply similarly to double-focusing instruments where the metastable dissociations will have taken place in the second field-free region (see Figure 8.3).

Two principal methods of metastable separation have been described with particular reference to Nier–Johnson type instruments. The first, (i), involves varying the acceleration potential V and the second, (ii), involves varying the electrostatic sector potential E.

(i) This is the method described by Barber and Elliot[49]. If the acceleration potential is changed from V_0 to V where

$$V = V_0 \times m_1/m_2 \tag{8.23}$$

then only daughter ions formed in the first field-free region by the process

$$m_1^+ \rightarrow m_2^+ + \text{neutral fragment}$$

now have the requisite energy to pass through the electrostatic sector at its normal (main-beam) potential E_0. They will be observed at a (magnetic field strength) mass corresponding to m_2. By such a technique many precursor–daughter ion relationships can be unequivocally determined[26]. A disadvantage of this method is that the source and hence ion-focusing conditions are changed when the acceleration potential is varied. A semi-automated method for identifying metastable transitions was described by Barber *et al.*[53].

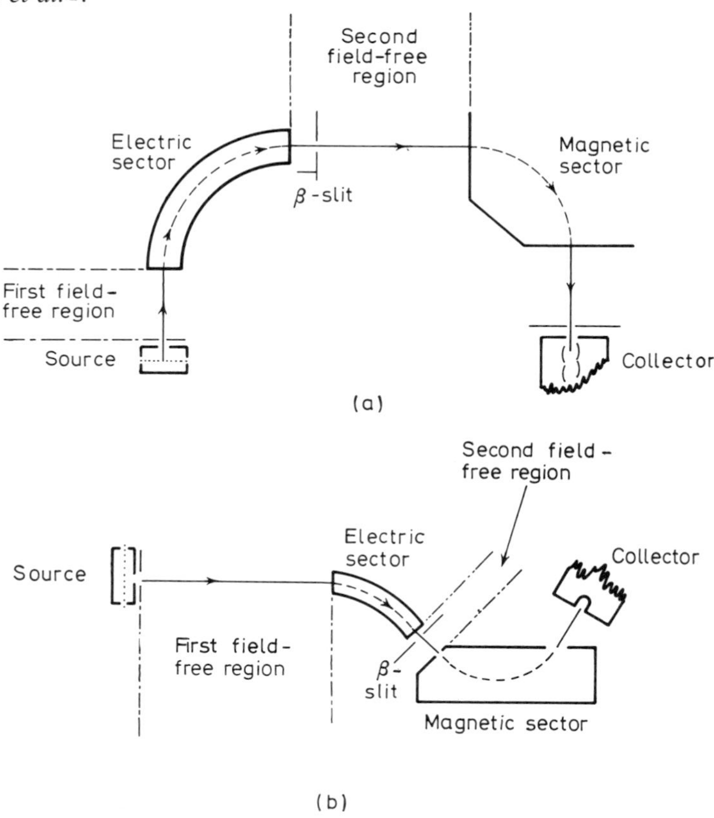

Figure 8.3 Schematic drawings of Nier–Johnson (a) and Mattauch–Herzog (b) double-focusing mass spectrometers

(ii) Variation of the electrostatic sector potential, keeping source conditions constant, is the basis of the method of Beynon *et al.*[54] who have coined the name 'ion-kinetic energy spectra' (I.K.E.S.) for the observed results. In this technique ions are detected at the β-slit position, i.e. the slit between the electrostatic and magnetic sectors. In a voltage sweep from zero to E_0, ions resulting from metastable processes will pass through the sector and be detected thereafter whenever this condition applies

$$E_{m^*} = E_0(m_2/m_1) \tag{8.24}$$

(Note that for doubly charged ions dissociating into two singly charged

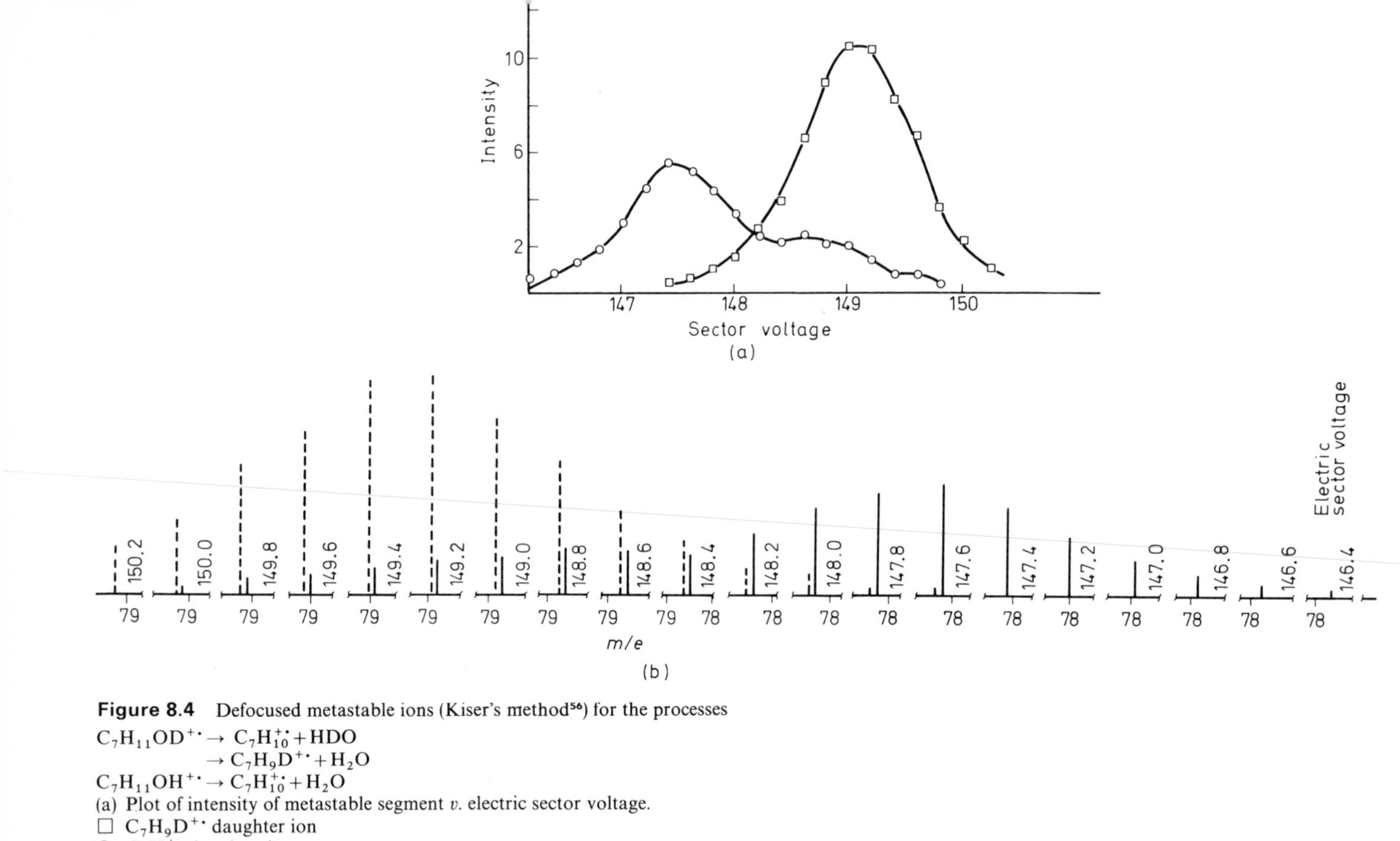

Figure 8.4 Defocused metastable ions (Kiser's method[56]) for the processes

$C_7H_{11}OD^{+\cdot} \rightarrow C_7H_{10}^{+\cdot} + HDO$

$\rightarrow C_7H_9D^{+\cdot} + H_2O$

$C_7H_{11}OH^{+\cdot} \rightarrow C_7H_{10}^{+\cdot} + H_2O$

(a) Plot of intensity of metastable segment *v.* electric sector voltage.

□ $C_7H_9D^{+\cdot}$ daughter ion

○ $C_7H_{10}^{+\cdot}$ daughter ion

(b) Sector voltage increased on 0.2 V steps, magnetic field scanned in the region m^*. Solid lines are metastable segments for $C_7H_{10}^{+\cdot}$ daughter ions; broken line for $C_7H_9D^{+\cdot}$. N.B. Maxima occur at $E = E_0 \times m_2/m_1$ and $m/e = m^{*\cdot}$

species E_{m*} may *exceed* E_0.) Following a brief report by McLafferty *et al.*[55], Kiser *et al.*[56] pointed out that metastable ion peaks can be studied in greater detail if the electric sector potential is varied stepwise in small increments of $\sim$0.1 V above and below E_{m*} and the magnetic field scanned in the mass region close to m^*. The mass scale remains unchanged but the observed metastable ion peak appears to move with each successive value of E. Such a series of scans for a deuterium-labelled compound are shown in Figure 8.4. From these data a graph can be constructed of the metastable intensities as a function of sector voltage; the maxima occur at $(m_2/m_1) \times E_0$. This method has also been discussed by Tajima and Seibl[57] with particular reference to the observation of multistep processes.

For Mattauch–Herzog instruments similar experiments have been described[58–60]. An interesting variation is the use of a reverse-geometry double-focusing mass spectrometer in which mass selection precedes energy selection. Such an instrument is now commercially available from M.A.T. and has been described by Maurer *et al.*[61].

The separation of metastable ions from the normal spectrum obtained in a time-of-flight mass spectrometer has been described by Haddon and McLafferty[62].

8.3.2 Energy release

The relationship between the shape of a metastable peak and the energy released in the fragmentation has been described above but the uses to which such information can be put will be described hereunder. Although a large number of energy releases have been determined by observations of metastable ions relatively little attention has been directed to relating this energy to properties of the precursor ion. Collected below are energy release data

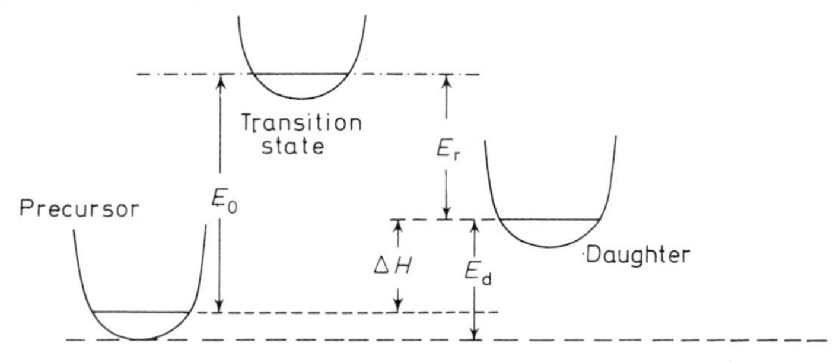

Figure 8.5 Energy diagram showing the relationship between thermochemical properties and energy release

with appropriate discussion sections where attempts have been made to explain the observations.

The relationship between thermochemical properties and energy release (T) derived from metastable ion widths has been discussed by Beynon *et al.*[63] and can be represented by the following equation (see also Figure 8.5):

$$T = E_0 + \Sigma\tfrac{1}{2}h\nu_p - E_d - \Sigma\tfrac{1}{2}h\nu_d + (\text{kinetic shift} - \text{vibrational energy of products})$$

(where E_0 = true activation energy, ν_p and ν_d = vibrational frequencies in precursor and daughter ions respectively, E_d = the difference in electronic energy of daughter and precursor ions) where E_r is the reverse activation energy.

8.3.2.1 *Aromatic systems*

The benzene fragmentation

$$C_6H_6^{2+} \rightarrow CH_3^+ + C_5H_3^+ \qquad (8.25)$$

is accompanied by a pair of prominent 'flat-top' metastable peaks at m/e = 5.77 and 101.8 corresponding to the methyl cation and $C_5H_3^+$ species respectively[18]. The energy release associated with the former process was evaluated as 2.16 ± 0.08 eV (2.27 eV)[14]; the average kinetic energy release was 2.67 ± 0.1 eV[18, 64, 65]. Beynon and Fontaine[66] proposed that this energy could reasonably be wholly due to repulsion between the two positive charges and thus should be equal to the work required to bring two charges from infinity to a distance apart equal to the charge separation in the doubly charged benzene molecular ion. This will be true provided that no energy is derived from any particular stability of the two fragment ions. For this simple proposal an energy release of 2.8 eV corresponds to an intercharge distance of 5.14 Å (in the latter case above this distance will be too small). Beynon and Fontaine suggested several possible acyclic structures which could arise directly from the benzene molecular ion without hydrogen migration, only ring opening and bond rotation are allowed.

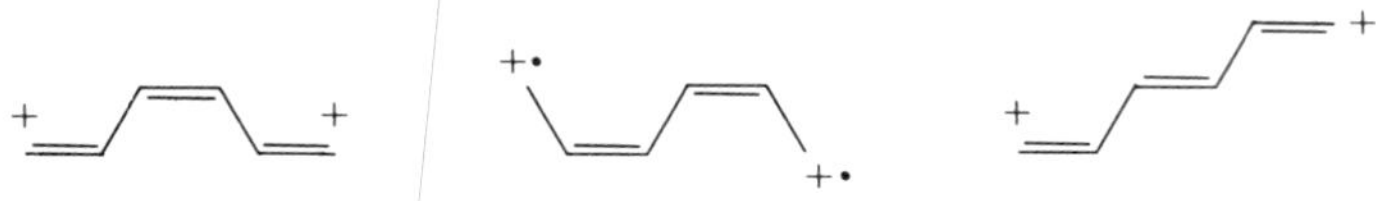

d = 5.58 Å $\qquad$ d = 5.13 Å $\qquad$ d = 6.13 Å

One limitation of this proposal is that it does not go far towards explaining this particular fragmentation which involves loss of a methyl cation. A possibly more appropriate structure with a charge separation of 5.04 Å is shown below.

$$CH_3-\overset{+}{C}H-C{\equiv}C-CH{=}\overset{+}{C}H$$

$$\downarrow$$

$$CH_3^+ + CH{\equiv}C-\overset{+}{C}H-C{\equiv}CH$$

Using a metastable ion defocusing technique, Beynon *et al.*[67] have shown that the processes

$$C_6H_4^{2+} \rightarrow C_5H^+ + CH_3^+ \qquad (8.26)$$

and

$$C_6H^{2+} \rightarrow C_5H^+ + C^+ \qquad (8.27)$$

take place with energy releases of 2.33 and 2.25 eV respectively; they state that these energies, if solely due to charge repulsion, are consistent with a charge separation of *c.* 6.2 Å.

Energy releases associated with dissociations of a wide range of doubly charged aromatic hydrocarbon molecular ions have also been reported[64, 65, 68] but not discussed in detail. The triply charged molecular ion of 9,10-diphenylanthracene fragments by loss of $C_6H_6^{+\bullet}$ together with an energy release of 2.4 ± 0.3 eV [69]. Nounou[70] observed a large energy release of 5.63 eV in the process (equation (8.28))

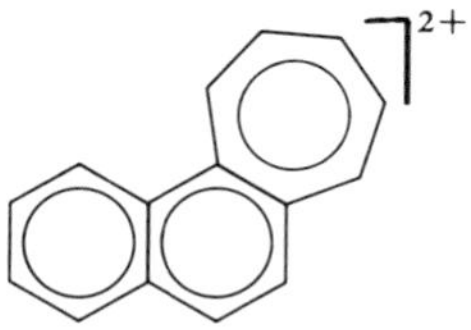

$$\longrightarrow C_{13}H_9^{+} + C_2H_2^{+} \qquad (8.28)$$

and he concluded that the pseudo-tropylic $(M-H)^{2+}$ ion of methylphenanthrene probably retains its cyclic structure (in contrast to the benzene fragmentations) since a charge separation of only 2.6 Å would suffice to produce the observed energy release. A decomposition of the triply charged molecular ion of biphenyl has also been observed.

$$C_{12}H_{10}^{3+} \longrightarrow C_{11}H_7^{2+} + CH_3^{+} + 4.5\,\text{eV} \qquad (8.29)$$

and the charge distribution in the precursor ion was discussed[71]. The different energy releases associated with loss of NO from the molecular ions of *meta*- and *para*-substituted nitrobenzenes have already been referred to (Reference 17) with respect to the earlier observations of Beynon *et al.*[15]. In the later paper[17] Bursey and McLafferty measured energy releases for the *p*-NMe_2, *p*-NH_2, *p*-OH, *p*-OMe and *p*-Ph nitrobenzenes; the observed values were 0.84 ± 0.02, 0.85 ± 0.04, 0.74 ± 0.06, 0.56 ± 0.06 and 0.38 ± 0.06 eV respectively. Only these five compounds among 13 *para*-substituted nitrobenzenes were found to display a 'flat-top' metastable ion. This observation correlated well with the criterion that release of NO with kinetic energy requires stabilisation of the product ion by the substituent; all the above ions are capable of participating in resonance hybrids of the type

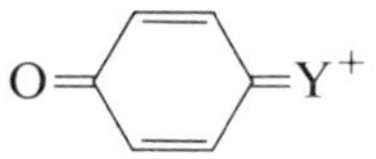

O= ⟨ring⟩ =Y⁺

Bursey and Hoffmann[72] studied the effect of steric hindrance on the above process by examining the mass spectra of the corresponding dimethyl compounds; the methyl groups are placed *ortho* to the resonance donors. In terms of the 'flat-top' metastables for the $[M-NO]$ reaction, the methyl substituents had little effect in the cases of the *p*-NH_2, *p*-OH, *p*-OMe compounds (0.76 ± 0.05, 0.65 ± 0.08, 0.58 ± 0.07 eV respectively) but for the *p*-NMe_2 compound the 'flat-top' metastable ion was transformed into an intense narrow Gaussian metastable ion (energy release $\not> 0.05$ eV); this was interpreted as resulting from extreme steric hindrance in the attainment of the planar quinonoid resonance form of the daughter ion

$$O{=}C_6H_2(CH_3)_2{=}\overset{+}{N}(CH_3)_2$$

8.3.2.2 *Small molecules*

For N_2O and NO_2 [9], some success has been achieved in relating energy release to specific excited states of the molecular ions. Similar observations have been made for the dissociation[73]

$$CO_2^{2+} \longrightarrow CO^+ + O^+$$

The mass spectra of methanol and its deuterium-labelled analogues have been thoroughly investigated by Beynon *et al.*[63]. Just before this work appeared the same authors reported the observation of an anomalous metastable peak in the methanol mass spectrum[74]; this peak appeared adjacent to the metastable for $M^{+\cdot} \rightarrow M^+ + H^\cdot$, $m^* = 30.042$, m^*(unknown) $= 30.122$. (The promised explanation for this phenomenon has not yet appeared.)

Energy releases were measured for the exothermic processes

$$H{-}\dot{C}{=}\overset{+}{O}{-}H' \longrightarrow HCO^+ + H'^\cdot \quad (8.30)$$

$$CH_2\overset{+}{O}H' \longrightarrow HCO^+ + HH' \quad (8.31)$$

	(8.30) eV	(8.31) eV
CH_3OH	0.19	1.42
CH_3OD	0.39	1.64
CD_3OH	0.18	1.27
CD_3OD	0.38	1.48

Thermochemical calculations on the above data for reaction (8.30) yielded the apparent result that $D(\overset{+}{O}{-}H)$ is greater than $D(\overset{+}{O}{-}D)$ by 0.20 ± 0.02 eV opposite to that expected from zero-point energy differences. Similar calculations yield $D(C{-}D) - D(C{-}H) = 0.20$ eV if H_2 is produced or ~ 0.15 eV if two H atoms are formed in reaction (8.31). The large exothermicity of reaction (8.31) was attributed to the reverse activation energy and it was considered that this dissociation takes place from an electronic excited state of the precursor ion rather than a vibrationally excited ground state thereof. These observations were discussed in detail with respect to a possible quantum mechanical tunnelling process*.

---

*Lifshitz, Shapiro and Sternberg[171] have also examined metastable transitions in the mass spectra of deuterium-labelled methanols. Their results are essentially in agreement with those of Beynon *et al.*[63]; they also observed the anomalous metastable peak and found that its intensity was partly pressure dependent and displayed a strong isotope effect.

In addition to their metastable abundance ratio hypothesis (see Section 8.2.3 p. 270) Shannon and McLafferty[33] pointed out that energy release should also be independent of the origin of the precursor ion; this was observed to be approximately true for the $C_2H_5O^+$ ions produced from three series of compounds $HOCH_2CH_2Y$, $CH_3CH(OH)Y$, CH_3CH_2OY (where Y indicates a wide variety of substituents), the 'flat-top' metastable peak energy releases for the processes

$$C_2H_5O^+ \longrightarrow CHO^+ + CH_4 \tag{8.32}$$

being 0.50 ± 0.04, 0.44 ± 0.02 and 0.52 ± 0.05 eV respectively. (In sharp contrast, the ion $CH_2\overset{+}{O}CH_3$ which had different metastable abundance ratios gave an energy release of <0.1 eV for this dissociation.) In terms of the Q.E.T. this result implies that for ions dissociating with lifetimes in the metastable range, the internal energy is rapidly randomised among the internal degrees of freedom to form a common electronic state. There was, however, one noteworthy anomaly in a system which could be expected to conform to the above behaviour. Ottinger[45] reported that the energy release in the process

$$C_4H_9^+ \longrightarrow C_3H_5^+ + CH_4 \tag{8.33}$$

was 0.16 ± 0.02 eV when the ion originates from n-heptane but 0.30 ± 0.13 eV in n-butane; this result appeared to be due to non-equilibrium behaviour. Khodadi, Botter and Rosenstock[75] carefully re-examined this dissociation using both double- and single-focusing mass spectrometers and concluded that within experimental error the energy releases are the same (0.09 ± 0.02 eV). Other processes examined in this study are listed below

$$C_2H_4^{+\cdot} \longrightarrow C_2H_2^{+\cdot} + H_2 + 0.016 \pm 0.006 \text{ eV (5 precursors)} \tag{8.34}$$

$$C_2H_5^+ \longrightarrow C_2H_3^+ + H_2 + 0.019 \pm 0.008 \text{ eV (5 precursors)} \tag{8.35}$$

$$C_3H_5^+ \longrightarrow C_3H_3^+ + H_2 + 0.042 \pm 0.011 \text{ eV (5 precursors)} \tag{8.36}$$

The limits are generous and it may prove that more detailed studies will reveal small energy differences which lie outside the range of experimental error and which are indicative of different kinetic energy distributions among the internal degrees of freedom of the common ion.

An interesting deuterium isotope effect on energy release has been reported by Briggs *et al.*[76] who observed the following processes in the mass spectrum of s-triazole.

$$[\text{s-triazole (N–H)}]^{+\cdot} \longrightarrow N_2 + C_2NH_3^{+\cdot} + 1.41 \pm 0.02 \text{ eV} \tag{8.37}$$

$$[\text{s-triazole (N–D)}]^{+\cdot} \longrightarrow N_2 + C_2NH_2D^{+\cdot} + 1.57 \pm 0.02 \text{ eV} \tag{8.38}$$

They relate this energy difference to a rate-controlling hydrogen shift with which the N_2 elimination is concerted.

Bursey and Dusold[77] described the effect of substituents on the energy release associated with the elimination of CO from the molecular ion of a large number of pyrone derivatives. They found the energy released, 0.17 ± 0.07 eV, to be independent of position and nature of the substituent, and they conclude that the substituents do not give any resonance stabilisation to the daughter $(M-CO)^{+\cdot}$ ions, thus ruling out some possible structures for the daughter ions. Kinetic energy release data for a number of metastable transitions involving doubly charged ions of alkanes, alkenes and cycloalkenes have been reported by Barber and Jennings[78]. They measured metastable ion and satellite peak widths for the process

$$M^{2+} \longrightarrow X^{+} + CH_3^{+} \tag{8.39}$$

There was no correlation between energy release and number of carbon atoms in the hydrocarbon in the range C_3–C_8; thus the prediction that energy release would fall with increase of chain length (assuming the molecular ions to be linear with charges located on terminal carbon atoms) was not supported.

Jennings[79] performed a detailed study of metastable transitions in the mass spectra of six fluoroethylenes. Of particular interest here was the comparison between energy release evaluated from metastable widths and the ionisation and appearance potential results of Lifshitz and Long[80, 81]. For the process

$$CH_2CHF^{+\cdot} \rightarrow C_2H_2^{+\cdot} + HF \tag{8.40}$$

the endothermicity, $\Delta H_{calc} = 1.75$ eV (calculated from the known heats of formation of the three species). The appearance potential results[81] give $\Delta H_{app} = 3.28$ eV while Jennings'[79] results show an energy *release* of 0.68 eV. Thus about half of the excess energy $(\Delta H_{app} - \Delta H_{calc})$ can be accounted for as kinetic energy released. It is important to note here that the kinetic energy released appears to be *less* than the reverse activation energy whose minimum value is 1.5 eV $(\Delta H_{app} - \Delta H_{calc})$. Two alternative explanations were given; either the unaccounted for excess energy (~ 0.8 eV) is partitioned as vibrational energy of the products, or the reverse activation energy is 0.68 eV and the excess measured by Lifshitz and Long is due to a kinetic shift of about 0.8 eV. Similar results were observed for the process

$$CH_2CF_2^{+\cdot} \rightarrow C_2HF^{+} + HF \tag{8.41}$$

$\Delta H_{calc} = 2.9$ eV, $\Delta H_{app} = 4.0$ eV, m^* energy release $= 0.47$ eV. Lifshitz and Sternberg[82] recently reported observations on the metastable peaks associated with the processes

$$C_2H_6^{+\cdot} \rightarrow C_2H_4^{+\cdot} + H_2 \ (+0.22 \pm 0.03 \text{ eV}) \tag{8.42}$$

$$CD_3CH_3^{+} \rightarrow CD_2CH_2^{+\cdot} + HD \ (+0.12 \pm 0.02 \text{ eV}) \tag{8.43}$$

Their kinetic energy release values are in excellent agreement with the discrepancy between appearance potential for the daughter ion, accurately determined by Chupka and Berkowitz[170] by photoionisation (12.08 ± 0.03 eV), and the calculated value (11.86 eV). The latter authors attribute the difference to a small excess activation energy and/or kinetic shift. Lifshitz and Sternberg's result shows that in this example the excess (activation) energy is entirely converted to translational kinetic energy. From thermochemical calculations similar to those of Beynon *et al.*[63], a value of 0.14 eV for $D(C-D) - D(C—H)$ was derived.

8.3.3 Methylene elimination

The loss of neutral methylene, CH_2, from large molecular and fragment ions of organic compounds has yet to be substantiated. In one of the earliest collections of data on metastable transitions[83] in the mass spectra of hydrocarbons it was noted that metastable ion peaks resulting from CH_2 loss were conspicuously absent and that if such fragmentations ever took place they must necessarily be very rapid reactions. In spite of several reports in the literature[84–86] recent careful repeat experiments[87, 88] have been unable to substantiate the earlier observation of a metastable ion peak corresponding to methylene loss (from chromanes[84, 88] or thioxanthene[85–87]). The claim of Brittain *et al.*[89] to have observed metastable ion peaks corresponding to the process

$$C_7H_9^+ \rightarrow C_6H_7^+ + CH_2 \; m/e(m^*) = 67.1 \qquad (8.44)$$

in the mass spectra of several $C_{10}H_{16}$ isomers deserves re-investigation.

8.3.4 Metastable ion peaks for consecutive fragmentations

A number of reports have been made for the observation of metastable ion peaks corresponding to two-step fragmentations taking place in flight between the ion source and collector.

$$M_1^+ \rightarrow (M_2^+ + A) \rightarrow M_3^+ + B \; m^{**} = m_3^2/m_1 \qquad (8.45)$$

For example, loss of 2CO from the molecular ions of some substituted cyclobutane-1,3-diones[90], loss of $2H_2O$ in some steroidal systems[91] and loss of $H_3O_2^{\cdot}$ from the molecular ions of some ketone 2,4-dinitrophenylhydrazones[92]. These and other similar processes[93, 94] were identified by the presence of a metastable peak at the appropriate mass m^{**}. Jennings[95] described the identification of the consecutive fragmentations

$$C_7H_7^+ \rightarrow C_5H_5^+ \; (+C_2H_2) \rightarrow C_3H_3^+ \quad (+C_2H_2) \qquad (8.46)$$

by use of the acceleration-potential variation method of metastable-peak defocusing, wherein the magnet current (in a Nier–Johnson type MS 9 instrument) was set to the apparent mass m_3^2/m_2 so as to observe the second dissociation ($m_2 \rightarrow m_3$) in the field-free region between the electric and magnetic sectors. By increasing the acceleration potential in the ratio $m_1 : m_2$ and keeping all other instrument parameters constant, ions of mass m_2 formed in the field-free region between source and electrostatic sector will be transmitted through the sector. Jennings still observed a signal at an apparent mass m_3^2/m_2 and therefore concluded that the two-step dissociation $m_1^+ \rightarrow m_2^+ \rightarrow m_3^+$ was taking place.

This method for identifying two-step fragmentations, however, has been criticised by Beynon *et al.*[96] who point out that ions of mass m_3 formed by decomposition of ions of mass m_1 will always possess the same fraction m_3/m_1 of the kinetic energy of the m_1 ions *irrespective* of the number of dissociation steps which yield m_3 from m_1. Furthermore, in the MS 9 instrument ions of mass m_3 formed in a single step from m_1 within the electric sector can also pass through the energy resolving (β) slit and be recorded at an apparent mass m_3^2/m_1. Ions formed thus throughout the electric sector will form a continuum in the plane of the β slit. These authors then undertook to examine the origin of the ions giving rise to the peak at m_3^2/m_1 by studying the effect of energy-resolving slit width on the height of the metastable peak. In their experiments the electric-sector potential was changed by a ratio $m_2 : m_1$ at fixed acceleration potential (with the advantage that this technique does not affect source tuning conditions). Examination of the metastable processes

$$91^+ \rightarrow 65^+$$
$$65^+ \rightarrow 39^+$$

in the mass spectrum of toluene showed that the intensities of the metastable ion peaks $65^2/91$, $39^2/65$ were (after normalisation) identical functions of β-slit width, each rising to a constant value thus showing that the decomposition

$$91^+ \rightarrow 39^+$$

(for which only a weak metastable could be observed) does at least in part take place as a two-step metastable process. This technique has been discussed further by Tajima and Seibl[57] who have identified the presence of two 3-step metastable dissociations in the mass spectrum of 3-(2,4,6-trimethoxyphenyl)-butan-2-one, namely

$$([M-43]-30-15-28) \text{ and } ([M-43]-30-30-30).$$

The theoretical aspects of such dissociations have been discussed by Hills *et. al.*[97].

Generally, ions which have sufficiently long lifetimes to dissociate metastably are believed to contain little excess energy above the threshold required for dissociation; thus they are not expected to undergo a second dissociation. Hills *et al.*[97] calculated the relative abundances of the metastable ion peaks as a function of the internal energy of the tropylium ion (undergoing the above dissociations observed by Jennings[95]). Activation energies and structures for the activated complexes were assumed and the ion-breakdown curves were calculated using the Q.E.T. equations. The agreement between calculated metastable abundances (for the consecutive reactions in the first and second field-free regions) and experimental observation was very satisfactory and can be taken as additional support for the validity of the Q.E.T. Tou[98] has discussed the dissociations

$$C_6H_5OX^{+\cdot} \rightarrow C_6H_5O^+ + X^\cdot \rightarrow C_5H_5^+ + X^\cdot + CO \qquad (8.47)$$

$$C_6H_5OX^{+\cdot} \rightarrow C_5H_5X^{+\cdot} + CO \rightarrow C_5H_5^+ + X^\cdot + CO \qquad (8.48)$$

$$X = \text{Br and Cl}$$

also from the point of view of Q.E.T.

8.3.5 Determination of fragmentation mechanisms

The original use of metastable ion peaks was simply to confirm by their presence that a particular dissociation had indeed taken place; this was a powerful aid to the elucidation of fragmentation sequences. However, more sophisticated experiments involving the use of 'defocused' metastable ions, isotopic labelling, energy release measurements, relative metastable ion abundances and the 'degrees-of-freedom' effect have contributed greatly to our understanding of fragmentation mechanisms and, to a lesser extent, to ideas concerning the structures of molecular and fragment ions.

8.3.5.1 Common ion structures

The observations of Rosenstock[32] and McLafferty[33] led to the important hypothesis that if a given ion observed in the mass spectra of a variety of molecules displays the same metastable ion peaks with the same relative abundances, then the said given ion has a common structure and/or energy irrespective of the structure of its precursor ion. Failure of a given ion to meet this criterion need not, however, constitute proof that it is structurally different from its fellows. Ions having the same structure but different internal energies can give rise to different metastable ion abundance ratios[99–101]. Similarly (as has been described in the section on energy release) the observation that a given ion dissociates yielding a 'flat-top' metastable ion corresponding to a constant energy release, irrespective of its precursor, is again evidence for a common intermediate. Many examples of the use and limitations of these criteria are in the recent literature and some of the more striking cases will be discussed below.

(a) *Hydrocarbon ions*—An interesting example of ions displaying unexpectedly dissimilar behaviour was recently reported by Bursey *et al.*[101]. They observed that the benzene molecular ions produced in the fragmentation of benzene-chromium tricarbonyl show metastable ion abundance ratios appreciably different from those from benzene itself and benzenetungsten tricarbonyl; the latter pair behave with close similarity.

The alkyl ions $C_4H_9^+$, generated from n-, iso-, s- and t-butyl bromides all showed metastable ion peaks corresponding to the processes

$$C_4H_9^+ \rightarrow CH_4 + C_3H_5^+ \tag{8.49}$$

$$\rightarrow C_2H_4 + C_2H_5^+ \tag{8.50}$$

First field-free metastable ion abundance ratios were not exactly the same (n:i:s:t = 17:34:36:50) but more important was the observation that irrespective of the C_4 precursor, 94% or more of the ions decomposing in the first field-free region of the mass spectrometer did so by loss of methane yielding metastable peaks of the same shape[102]. The observations on atom scrambling (discussed later) led to the conclusion that the $C_4H_9^+$ ions have isomerised completely (or almost so) to the same structure (or mixture of structures) before they dissociate by loss of methane or ethylene. Similar work was described for the ions $C_6H_{13}^+$ and $C_8H_{17}^+$ generated from primary, secondary and tertiary bromides or iodides[103]. The competitive (first field-free region) metastable transitions were

$$C_6H_{13}^+ \rightarrow C_4H_9^+ + C_2H_4 \quad (8.51)$$

$$\rightarrow C_3H_7^+ + C_3H_6 \quad (8.52)$$

$m^*_{8.51}/m^*_{8.52} = 1.58 \pm 0.05$ for primary and secondary compounds studied and 1.6 ± 0.3 for the tertiary halides.

$$C_8H_{17}^+ \rightarrow C_5H_{11}^+ + C_3H_6 \quad (8.53)$$

$$\rightarrow C_4H_9^+ + C_4H_8 \quad (8.54)$$

$m^*_{8.53}/m^*_{8.54} = 1.48 \pm 0.05$ and 1.5 ± 0.3 (as above).

The metastable behaviour of the ions $C_4H_7^+$, $C_5H_9^+$, $C_6H_{11}^+$, $C_5H_7^+$ and $C_6H_9^+$ has recently been described[104]. The $C_4H_7^+$ ions were generated from but-2-enyl bromide (I), cyclopropylmethyl bromide (II) and 2-methylprop-2-enyl bromide (III); the (first field-free region) metastable ion abundance ratios are shown below for the dissociations

$$C_4H_7^+ \rightarrow C_4H_5^+ + H_2 \quad (8.55)$$

$$\rightarrow C_3H_3^+ + CH_4 \quad (8.56)$$

$$\rightarrow C_2H_5^+ + C_2H_2 \quad (8.57)$$

Generating compound	$m^*_{8.55}$	$m^*_{8.56}$	$m^*_{8.57}$
I	1	1.46	2.51
II	1	1.36	2.08
III	1	1.28	1.42

Lossing[105] has recently accurately measured the heat of formation of $C_4H_7^+$ ions generated by mono-energetic electrons[106] from five isomeric C_4H_8 molecules (by H loss) and eight isomeric C_5H_{10} molecules (by CH_3 loss). The heats of formation of all $C_4H_7^+$ ions studied lie in the narrow range 204.4–207.7 kcal mol^{-1}; this lies close to values for ΔH_f for the two isomeric methylallyl cations $CH_3\overset{+}{C}HCH{=}CH_2$, $\Delta H_f = 204 \pm 3$ kcal mol^{-1} and $CH_2{=}C(CH_3){-}\overset{+}{C}H_2$, $\Delta H_f = 211 \pm 5$ kcal mol^{-1}. The heat of formation of other $C_4H_7^+$ ions is greater than 225 kcal mol^{-1}. A combined metastable and thermochemical approach of this kind would be of immense value in properly identifying the common ion in a series of $C_xH_y^+$ ions generated from different precursors.

The metastable abundance ratios for the $C_5H_9^+$ ions are shown below: the loss of C_2H_2 from these ions only produced a very weak metastable ion in each case.

$$C_5H_9^+ \rightarrow C_5H_7^+ + H_2 \quad (8.58)$$

$$\rightarrow C_3H_5^+ + C_2H_4 \quad (8.59)$$

Generating compound	$m^*_{8.58}$	$m^*_{8.59}$
Cyclopentyl bromide	1	10.8
3-Methylbut-2-enyl bromide	1	5.5
2-Methylbut-2-enyl bromide	1	13.3

1-Methylbut-2-enyl bromide	1	9.5
1,2-Dimethylprop-1-enyl bromide	1	11.5
Cyclobutylmethyl bromide	1	12.5
Pent-2-enyl bromide	1	17.2
Pentamethylene dibromide	1	19.0

For $C_6H_{11}^+$ ions the only intense metastable observed was for C_2H_4 loss but this was the predominant metastable irrespective of the precursor compound. Data for $C_5H_7^+$ and $C_6H_9^+$ are listed below.

$$C_5H_7^+ \rightarrow C_5H_5^+ + H_2 \tag{8.60}$$

$$\rightarrow C_3H_5^+ + C_2H_2 \tag{8.61}$$

$$\rightarrow C_3H_3^+ + C_2H_4 \tag{8.62}$$

Generating compound	$m^*_{8.60}$	$m^*_{8.61}$	$m^*_{8.62}$
Cyclopentyl bromide	25.2	17.3	1
3-Methylbut-2-enyl bromide	45.3	12.5	1
2-Methylbut-2-enyl bromide	50.7	28.3	1
1-Methylbut-2-enyl bromide	35.5	13.5	1
Cyclobutylmethyl bromide	27.2	21.7	1
Pent-2-enyl bromide	27.5	21.5	1
1,2-Dimethylprop-1-enyl bromide	35.6	30.0	1

$$C_6H_9^+ \rightarrow C_6H_7^+ + H_2 \tag{8.63}$$

$$\rightarrow C_5H_5^+ + CH_4 \tag{8.64}$$

$$\rightarrow C_4H_5^+ + C_2H_4 \tag{8.65}$$

Generating compound	$m^*_{8.63}$	$m^*_{8.64}$	$m^*_{8.65}$
Cyclohexyl bromide	25	1	3.3
Cyclopentylmethyl bromide	25	1	4.3
Hex-2-enyl bromide	19	1	4.3
1,1-Dimethylbut-3-enyl bromide	26	1	3.7

Several remarks should be made concerning the above data; it is noticeable that the metastable ion abundance ratios are frequently only approximately constant. Just how large a deviation from constancy is required to invalidate the 'common-ion' hypothesis is still a matter for speculation. It has been suggested[104] that variations up to a factor of 15 are tolerable but Yeo and Williams[100] revised this generous figure to a factor of not greater than five. In this later work[100] the authors discussed the dependence of metastable ion abundance ratios for two competing reactions as a function of the internal energy of the dissociating ions. Calculations were performed using the simple Q.E.T. equation relating the dissociation rate constant with the internal energy and the number of effective oscillators in the fragmenting ion. The width of the internal energy band which yields ion dissociations observable as metastable ions, depends on the frequency factor associated with the process. For the more commonly encountered dissociations which have high frequency factors (10^{10}–10^{13}) the energy band is relatively narrow.

The problem of the change of metastable abundance ratios with origin and/or energy content of an ion has also been discussed by Occolowitz[107] who concludes that variations of up to a factor of three in metastable ion abundance ratios may well be expected for ions of the same structure but different energy contents.

A careful comparison of defocused metastable ions in the fragmentations of the isomeric $C_{13}H_{10}$ hydrocarbons, fluorene and phenalene, was made by Bowie and Bradshaw[108]. The primary aim of this study was to compare the behaviour of the $C_{13}H_9^+$ ions derived from fluorene and phenalene with those produced by a range of 17 other aromatic molecules. Metastable ion abundance ratios for the losses of $C_2H_2(m_1^*)$ and $C_4H_2(m_2^*)$ were measured. m_2^*/m_1^* (fluorene) = 1.15 and so those compounds whose $C_{13}H_9^+$ fragment ion had $m_2^*/m_1^* = 1.13 \pm 0.02$ were considered to yield the fluorene $(M-H)^+$ ion. The value m_2^*/m_1^* for phenalene was 0.81 and a second group of compounds was similarly identified. Although very reasonable correlations with postulated fragmentation mechanisms were made on the basis of these results, the difference in metastable ion abundance ratios for the fluorene and phenalene groups is small and no useful conclusions as to whether the difference originates in ion structure and/or energy content can be drawn at this time.

McLafferty and Bryce[109] described the first field-free-region metastable ions observed in the mass spectra of five isomeric hexanes (C_6H_{14}), two octanes (C_8H_{18}) and two n-hexanes. For these compounds the metastable abundance ratios for molecular ion dissociations were very different (i.e. by factors as large as 150) and thus with confidence can be ascribed to ions of different structure. This result is not unexpected but is in striking contrast with both the behaviour of the alkyl and alkenyl cations described above and that of aromatic hydrocarbons[110]. The widely differing metastable peak ratios observed in hydrocarbon spectra have also been described by Wanless and Glock[111] with respect to their use in hydrocarbon identification and analysis using a dual source (field-ionisation, electron impact) mass spectrometer.

A detailed study of the defocused metastable ions in the mass spectra of the n-alkanes, C_3–C_7, has recently been reported[112]; this valuable work not only supplements the much older data[83, 113] but provides evidence for 'fine structure' in metastable ion peaks. In this work a variable-width monitor slit was used[114], thus increasing the energy resolving power of the MS 9 instrument. Thus, for example, metastable ion peaks normally observed as Gaussian in shape were found to be flat-topped or dish-shaped allowing accurate evaluation of the energy release associated with the generating fragmentation. Also observed was evidence for two superimposed metastable peaks for the process

$$C_3H_5^+ \longrightarrow C_3H_3^+ + H_2 \tag{8.66}$$

centred at the same point on the energy scale; the different widths of the peaks indicate that the daughter ions possess different energy distributions although arising from the same precursor ion. Two general explanations for this effect were proposed: (i) that two different stable excited states of either the precursor ion, the daughter ion or the neutral fragment are involved; (ii) that two different structures exist of either the precursor ion, daughter ion or neutral fragment and that both are present in the system. After detailed consideration, the second alternative was considered to be the more attractive. Some support for this particular proposal can be found in observations[115] on the metastable ion peak associated with loss of water from the $[M-CO_2]^{+\cdot}$ fragment ion in the mass spectrum of oxalic acid; the metastable ion appears to be made up of two superimposed peaks arising from dissociation of the two isomeric ions

$$HO\text{—}\dot{C}\text{=}\overset{+}{O}H \quad \text{and} \quad O\text{=}\dot{C}\text{—}\overset{+}{O}H_2$$

the former representing a species of lower energy than the latter.

(b) *Heteroatom-containing ions* – The identification of $C_2H_5O^+$ ions from different precursors by comparing metastable ion peak abundances[33] has been described above, this note being the first of many such experiments. The problem of the isomeric $(M-NO)^+$ ions in the mass spectra of substituted nitrobenzenes[15, 17] has also been discussed in an earlier section.

8.3.5.2 The vibrational-degrees-of-freedom (V.D.F.) effect

In the 1964 paper by Rosenstock *et al.*[32] it had been observed that the metastable abundance ratio for two competing hexyl ion dissociations was weakly dependent on the molecular weight of the precursor. The ratio fell with increasing molecular weight of the precursor. In 1966 McLafferty and Shannon[33] briefly reported a similar observation by Pike and, in 1967, McLafferty and Pike[116] showed that plots of the logarithm of the ratio metastable ion abundance:abundance of precursor ion *v.* the reciprocal of the number of vibrational degrees of freedom (V.D.F.) in the original molecule were straight lines having a positive slope. This was demonstrated for the dissociations of $C_2H_5O^+$ ions

$$M \longrightarrow C_2H_5O^+ + X \tag{8.67}$$

$$C_2H_5O^+ \longrightarrow H_3O^+ + C_2H_2 \tag{8.68}$$

$$\longrightarrow CHO^+ + CH_4 \tag{8.69}$$

The $C_2H_5O^+$ ions were derived from homologous primary and secondary alcohols and terminal diols. Also shown were plots for dissociations in homologous alkanones. The qualitative explanation of this phenomenon which is in agreement with the basic assumption of the Q.E.T., is simply that when the neutral fragment X is large it can carry off more of the internal energy of the ion as vibrational energy, leaving after dissociation a relatively energy-poor $C_2H_5O^+$ daughter ion; more of the latter will be detected as $C_2H_5O^+$ ions rather than as metastable ions and hence the positive slope of the plots. Before discussing examples of the use of the V.D.F. effect, reference should be made to a theoretical treatment by Lin and Rabinovitch[117]. They applied a random statistical theory to the internal energy partitioning that accompanies the molecular ion dissociation for the C_3–C_7 primary alkanols (equation (8.67)). It was felt that the necessary assumptions regarding energetics and ion structure would not be too critical in this test case. The agreement between their calculation and McLafferty and Pike's observations was qualitatively good; the theory, however, did not yield strictly linear plots and the experimental fall of abundance ratio with increasing molecular size was less than predicted; this could arise from incomplete energy randomisation in the larger molecules. The original, related observation of Rosenstock *et al.*[32] probably reflects the effect of energy content of the precursor ion on the relative rate of the competing metastable dissociations therefrom. Cooks and Williams[118] examined the relative rates of the benzoyl ion fragmentation

$$C_6H_5CO^+ \longrightarrow C_6H_5^+ + CO \qquad (8.70)$$

as a function of precursor molecule. They observed that the ratios ((m/e = 105)/$m^*_{8.70}$) and (m/e = 105)/(m/e = 77) both increased with increase in the number of degrees of freedom in the precursor molecule, C_6H_5COX. It is noteworthy that in this system the V.D.F. effect outweighs the effect of excess energy of the benzoyl ions at threshold. An unanswered problem in this study was the observation that the (m/e = 105)/(m/e = 77) ratios were greater for X = s-propyl or isobutyl than for their respective n-alkyl analogues, indicating a *lower* rate of dissociation of benzoyl cations produced from the branched alkyl phenyl ketones.

8.3.5.3 Labelling experiments and atom scrambling

Much of our knowledge concerning the mechanisms of ion fragmentations has come from isotopic labelling experiments in which specific atoms in the molecule being studied are replaced with heavier isotopes (e.g. deuterium, ^{13}C, ^{18}O); their subsequent presence or absence in fragment ions has allowed the identification of many ion-dissociation mechanisms. The method and elegant examples of its use are too numerous and well known to bear repetition in this article.

However, the use of defocused metastable ion peaks to study the fate of an isotopic label is particularly valuable because the behaviour of precursor ions may thus be studied free from interference of all other ions recorded in a normal mass spectrum. As shown in Figure 8.4 and explained in Section (8.3.1.2(ii)), with this technique the specific labelling in the molecule being studied does not have to be complete. It is becoming increasingly evident that in many fragmenting ions some or all of the atoms appear to have lost their positional identity and are lost in a random or statistically predictable manner. This phenomenon, generally called atom 'scrambling', has now been so widely observed that a better understanding of its origin and the implications for ion structures and energies are essential to the development of the subject. As in the preceding section the phenomenon will be described for classes of compounds and general discussion will be included wherever relevant.

(a) *Molecular and fragment ions in hydrocarbons*—The dissociations of methane and labelled methanes have been studied by several workers. A question of some interest arose when Dibeler and Rosenstock[119] reported that only one truly unimolecular metastable ion could be observed in the mass spectrum of methane and methane-d_4

$$CH_4^{+\cdot} \xrightarrow{*} CH_3^+ + H^\cdot \quad (8.71)$$

$$CD_4^{+\cdot} \xrightarrow{*} CD_3^+ + D^\cdot \quad (8.72)$$

Vestal[120] reconciled these findings with the Q.E.T. and the problem was re-investigated by Ottinger[121] who not only observed more metastable ion peaks (whose abundances did not tend to zero with decreasing pressure) but with intensities different from those previously observed[120]. Ottinger also showed that there was a preference for H loss over D in the mixed deuterio-methanes by a factor of at least 30 for CH_3D, >80 for CH_2D_2 and >70 for CHD_3. Beynon *et al.*[122] have described yet more metastable ions in the methane fragmentation observed in I.K.E. spectra; this technique allows detection of metastable ion peaks corresponding to ~1 p.p.m. of base-peak intensity[123]. The large isotope effect in methane also appears in labelled ethanes[124]. There is no evidence for any hydrogen scrambling in $CH_3CD_3^{+\cdot}$ prior to its dissociation by hydrogen loss[124, 82]. Lifshitz and Sternberg[82] studied only the loss of molecular hydrogen from the molecular ions of C_2H_6 and CH_3CD_3 and from the fragment ions $C_2H_5^+$, $C_2H_4^+$, and their deuterated analogues; they observed a small isotope effect on the intensity of the molecular ion dissociations and on the kinetic energy released therein (see Section 8.3.2.2). The metastable ion peak abundances for the losses of H_2, HD and D_2 from the labelled ethyl ion $C_2H_2D_3^+$ were observed in the ratio 4.9 : 6.9 : 1.0.

Similar losses from the $C_2H_3D_2^+$ ion were not easily estimated since this is a relatively much less abundant fragment ion than $C_2H_2D_3^+$. The study by Löhle and Ottinger[124] contained considerably greater detail but their results are not in good agreement with those of Lifshitz and Sternberg[82]. For example, for the hydrogen losses quoted above, their metastable ion ratio was 55 : 61 : 1 respectively. The former result weakly approaches that for random loss (a scrambled ethyl ion), 1 : 6 : 3 while the latter values are far from statistical. It is difficult to reconcile these two sets of data. Both results refer to first field-free-region metastable ions; in Lifshitz's experiments a single-focusing (Atlas CH4) instrument was used and Ottinger employed a reverse-geometry mass spectrometer (magnetic field before electric sector). It is possible that the discrepancy lies in the presence of collision-induced metastables in the CH4 instrument, but the authors were careful to measure metastable ion abundances by extrapolations to zero pressure. Löhle and Ottinger's experiments were performed at low source pressures, $\sim 2 \times 10^{-7}$ torr, and the unimolecular nature of the dissociations was confirmed by admitting krypton gas into the first field-free region. The explanation for their observations was that the hydrogen atoms *are* scrambled but the deviation from statistical loss lies in a strong isotope effect. They based their assumption that scrambling was complete on their observation that in the losses of CH_2, CHD and CD_2 from the same ion a metastable ratio 1 : 5.1 : 2.3 was observed – certainly close to the statistical values (1 : 6 : 3)*. It was

*As noted in Section 8.3.3, methylene elimination has never been substantiated; the metastable ion for loss of methylene from the $M-1$ ion can easily be mistaken for the isotopic contribution to the loss of CH_3 from the molecular ion.

realised that this assumption may prove to be invalid because the activation energy for C—C rupture in the ethyl ion is ~2.5 eV higher than that for hydrogen loss. These systems certainly deserve further study. (See also the continuing discussion below.)

The labelled propane system has been referred to in Section 8.2.3 but atom scrambling was not discussed therein.

In the paper by Ottinger[36] metastable ion abundances were recorded for the (analogous) competing reactions of the $C_3H_5D_2^+$ ion derived from $CH_3CD_2CH_3$.

$$C_3H_5D_2^+ \rightarrow C_3H_3D_2^+ + H_2 \quad m^* = 92 \tag{8.73}$$

$$C_3H_5D_2^+ \rightarrow C_3H_4D^+ + HD \quad m^* = 15 \tag{8.74}$$

$$C_3H_5D_2^+ \rightarrow C_3H_5^+ + D_2 \quad m^* < 3 \tag{8.75}$$

These ratios are again far from the statistical random numbers 10 : 10 : 1. In this case, Ottinger chose to regard the result as lack of scrambling rather than complete scrambling plus a strong isotope effect. Again, in contrast, the results of Vestal and Futrell[46] show a metastable ion abundance ratio of 26 : 12 for the first two dissociations of $C_3H_5D_2^+$ shown above. The latter authors do not discuss the possibility of scrambling in these secondary fragmentations.

The fragmentation behaviour of benzene and toluene has been studied in considerable detail by labelling experiments. Jennings[125] studied metastable ion peaks in the mass spectra of 1,4-dideuteriobenzene and 1,3,5-trideuterio-benzene and observed that the relative abundances for all major fragmentations involved random selection of hydrogen and deuterium atoms*. This scrambling was tentatively explained by the participation of Dewar benzenes, prismanes and benzvalenes – behaviour analogous to photochemical transformations – which would effectively transpose the carbon atoms without necessitating C—H bond breaking. Two definitive experiments have recently been attempted; Horman *et al.*[126] prepared 1,3,5[$^{13}C_3$] benzene and studied the defocused first field-free region metastable ion peaks corresponding to the losses of C_2H_2, $^{13}CCH_2$ and $^{13}C_2H_2$ from the molecular ion. After correction had been made for overlapping metastable peaks arising from incomplete labelling and dissociations of the $(M-1)^+$ ion, a ratio of $21 \pm 4 : 61 \pm 4 : 18 \pm 2$ was derived for the three acetylene losses; this is in excellent agreement with the random-loss ratio 1:3:1 and thus the carbon atoms in the molecular ion completely lose their positional identity before the above fragmentation takes place. (Cooks and Bernasek[127] similarly described complete carbon scrambling in benzthiophene.) The second study of multi-labelled benzene by an international group of workers[128] appeared later in the same year. This ambitious project involved the synthesis of 1,2[$^{13}C_2$] – 3,4,5,6[$^2H_4$] benzene; unfortunately the measurements were confined to a high-resolution study of the isotopic composition of the $C_4H_4^+$ and $C_3H_3^+$ fragment ions. Within the limits of uncertainty of the experiment

*McDonald, C. G. and Shannon, J. (1962). *Austral. J. Chem.*, **15**, 771, had reached similar conclusions by a study of fragment-ion peak intensities (*not* metastable ion peaks) in the mass spectrum of 1,2,3-trideuteriobenzene.

it was concluded that scrambling of the carbon atoms without breaking C—H bonds, scrambling of hydrogens with no loss of positional identity of carbon atoms and independent scrambling of carbons and hydrogens are probably *all* taking part in the molecular ion fragmentations which produce $C_4H_4^{+}$ and $C_3H_3^{+}$. It should be borne in mind that this later study was for short-lived ions which dissociate within the ion source whereas the previous study by Horman *et al.* involved the slower metastable species. A more complete study of the doubly labelled benzene system has recently been reported by Dickinson and Williams[172]. Complete scrambling of the carbon and hydrogen atoms was identified for both the benzene molecular ion and the $C_6H_5^{+}$ ion; the molecular ion of 1-^{13}C phenol was shown to dissociate by ^{13}CO loss only, i.e. *without* scrambling. The phenyl and phenoxy cations have also been independently examined with the same result[173].

Toluene and the tropyl cation have received considerable attention. The early labelling study reported by Grubb and Meyerson[129] showed that the hydrogen atoms in toluene lose their positional identity prior to the formation of $C_7H_7^{+}$ while the results of Rinehart *et al.*[130] on α,1-[$^{13}C_2$] toluene showed a similar result for carbon atoms. Two studies involving deuterium labelling and metastable ion measurements recently have been reported[131, 132]. Beynon *et al.*[131] studied the I.K.E. spectra of toluene-α-d_3 and toluene-2,3,4,5,6-d_5 with the aim of determining the manner in which a hydrogen or deuterium atom is lost from the molecular ion. They concluded that for the slowly dissociating ions examined, the probability that hydrogen is lost from the side chain was *equal* to the probability that hydrogen be lost from the ring ('preference' factor = 1); however, with regard to the probability for loss of a deuterium atom from a specific site relative to the loss of hydrogen from *the same site* an isotope factor of 3.50 was deduced. The major assumption in this work was that the preference factor and the isotope factor operate *independently* and that the isotope factor is the same for ring and α-deuterium atoms. The value of unity for the preference factor indicates that for metastably dissociating molecular ions complete 'scrambling' has taken place and the non-observation of random loss is due to an isotope effect.

The work of Howe and McLafferty[132] confirms that for the low-energy molecular ions of toluene and cycloheptatriene scrambling is complete but the degree of scrambling decreases with increasing internal energy. The isotope effect also decreases with increasing internal energy. Their isotope effect, 2.80, is in reasonable agreement with that of Beynon *et al.*[131]. At low ionising electron energies, near to threshold, the isotope effect increases (3.6 at 14 eV, 4.5 at 13 eV for labelled toluenes). The behaviour of higher-energy precursor ions was studied by collisionally inducing first field-free-region dissociations with argon; the isotope effect fell to $\sim$2.2. Ions of high energy which dissociate within the source yielded daughter ions which showed that preferential loss of α-hydrogens occurred, and the isotope effect was $\sim$1.65.

A very important lesson to be learned from these results is that the analysis of data from labelling studies, particularly where atom scrambling is involved, may be rendered very difficult by the presence of primary isotope effects. It is noteworthy that Bowie and Bradshaw in their study of fluorene and phenalene[108] found random loss of hydrogen and deuterium from their dideuteriated molecular ions with *no* isotope effect. This type of behaviour is somewhat rarely observed; for example the molecular ion $C_6D_5C{\equiv}C{-}C_6H_4CH_3^{+\bullet}$ dissociates exclusively by loss of H presumably to yield a tropyl phenyl acetylene[133]. This ion, which then dissociates by elimination of acetylene, does so with partial hydrogen scrambling; whether this involves skeletal rearrangement and/or hydrogen–deuterium migration across the acetylene bridge is not known. The molecular ion and the $(M-H)$ ion of the diphenyl methanes $C_6H_5CH_2C_6D_5$ and $C_6H_5{}^{13}CH_2C_6H_5$ fragment by loss of methyl in which both the carbon and hydrogen atoms are randomly selected[134]; this disproved an earlier claim[135] that the methyl radical was specifically generated in the $(M-H)$ fragment, by the central CH unit departing together with two ortho hydrogen atoms. Much work on atom scrambling in aromatic systems has been completed without recourse to metastable ion abundance measurements, the relatively simple fragmentations in many of these molecules making it possible to draw valid conclusions from only daughter ion peak abundances.

For labelled alkanes, alkenes, alkyl ions, alkenyl ions and related species where daughter ions frequently have several precursors, metastable ion defocusing is necessary to separate the many overlapping processes and measure relative abundances. The labelled isomeric butyl cations were studied by Davis *et al.*[102], who concluded that not only did they isomerise to a common intermediate (see Section 8.3.5.1(a)) but that complete hydrogen and carbon scrambling takes place in these ions before they metastably dissociate by loss of methane or ethylene. A discussion of the energy barrier for hydrogen and deuterium scrambling in the t-butyl cation was presented by Yeo and Williams[136]; their upper limit 44 ± 7 kcal mol^{-1} may be compared with the estimated minimum energy requirement, 28 kcal mol^{-1}, from solution studies of the labelled t-butyl cation[137]. Further experiments have been described for ions of the $C_5H_9^+$ family[104], labelled ions being generated from 2,2,5,5-d_4-cyclopentyl bromide; both the first and second field-free region metastable ions showed hydrogen randomisation preceding fragmentation by ethylene loss. It seems probable that many even-electron hydrocarbon ions will undergo complete atom scrambling in low-energy dissociations observable by metastable ion peaks; even the (bicyclic) norbornyl cation displays complete hydrogen randomisation before dissociating by ethylene loss[138]. The molecular ions of the isomeric butenes, methylpropene, methylcyclopropane and cyclobutane are believed to isomerise to a common structure before dissociation (appearance potential measurements[139] and fragment ion peak studies on labelled compounds[140, 141]).

(b) *Heteroatom-containing molecules* – Williams and Ronayne[142] studied metastable peaks in the mass spectra of 2,6-d_2 and 2-d_1-pyridine. They found that the losses of HCN and DCN from the molecular ion involved random selection of H and D. Similar experiments[143] showed that both benzonitrile and its isomer, phenyl isocyanide, randomly lose hydrogen in HCN loss from their molecular ions. The same is true for HCN expulsion from the molecular ions of 2-methylpyridine, 4-methylpyridine and quinoline[144]: however it is noteworthy that the degree of hydrogen scrambling in the methylpyridines before CH_3 loss is dependent on ionising electron energy. The high-energy ions of the trideuteriomethyl pyridines fragment with almost complete retention of the label in the expelled methyl group.

That hydrogen scrambling may be dependent upon ionising electron energy was clearly illustrated by a study of the olefin elimination reaction of alkyl ketones (McLafferty rearrangement)[145]. In the same work, the time dependence of hydrogen scrambling was demonstrated; 2,2,4,4-d_4-heptane-3-one showed mainly loss of C_3H_6 in the normal (70 eV) spectrum (daughter ion peaks), mostly C_3H_6 and C_3H_5D loss in first field-free region metastable ions and approaching random values for the second field-free region metastable ions. This reaction is usually represented as the specific McLafferty rearrangement shown below,

$$CH_3CD_2C(=O)CD_2CH_2CH_2CH_3^{+\cdot} \longrightarrow C_3H_6 + C_4H_4D_4O^{+\cdot} \quad (8.76)$$

and this example serves well to illustrate the *apparent* complexity of the fragmentation resulting from partial hydrogen scrambling. In this example the alkyl groups were separated by a keto group; complete hydrogen scrambling was observed among aromatic groups separated by a heteroatom (or heteroatom-bearing carbon) in the compounds diphenyl ether and diphenyl carbonate[146]; partial hydrogen scrambling was found in diphenylmethanol and diphenylmethyl chloride[147]. In contrast with these observations, McLafferty and Schiff[148] showed that $C_6H_5OC_2D_5$ dissociates by loss of C_2D_4 exclusively. In the same work they compared metastable ion abundances for the processes

$$YC_6H_5O^{+\cdot} \rightarrow YC_5H_5^{+\cdot} + CO \quad (8.77)$$

$$\rightarrow YC_5H_4^{+} + CHO^{\cdot} \quad (8.78)$$

as a function of Y in the two precursor molecules $YC_6H_4OC_2H_5$ and YC_6H_4OH. Their conclusion was that the ions $YC_6H_5O^{+\cdot}$ observed in the mass spectra of alkyl aryl ethers are formed mainly in the phenolic form without loss of ring position identity[149]. Scrambling in benzyl phenyl ketone derivatives has been described by Bowie *et al.*[150].

Yeo[151] recently provided evidence which distinguishes between intra- and inter-alkyl hydrogen scrambling in the mass spectra of labelled ethyl acetates. He showed that water loss from the molecular ion can be described as follows. Decomposition in the source involves complete scrambling of the ethyl hydrogens with no isotope effect; first and second field-free region metastable ion dissociations, however, are best explained in terms of hydrogen randomisation throughout the whole molecular ion; scrambling is more extensive in the longest-lived ions.

Although extensive or complete hydrogen scrambling was found in fragmentations of phthalic anhydride, biphenyl, cyanonaphthalenes and benzothiophene, the loss of HCN from thiazole involved the specific loss of the C-2 hydrogen while HCN loss from benzothiazole was $>90\%$ specific for ion source dissociations, tending towards random loss for long-lived ions[152].

The distinction between energy content and ion structure was emphasised by Harrison and Keyes[153] who concluded from a study of propan-2-ol-2-^{13}C that the $C_2H_5O^+$ ions derived from alkan-2-ols have the structure $CH_3CH=$

$\overset{+}{O}H$ at threshold but that rearrangement to a symmetrical structure (such as that shown below) takes place in ions with excess energy.

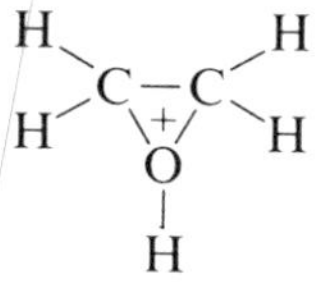

These conclusions are in agreement with the earlier proposals of Shannon and McLafferty[33].

Partial hydrogen scrambling has been reported in some phenyl substituted cyclopentadienols; the unusual fragment ion $(C_6H_5)_4C_4^{+\cdot}$ is suggested to have the tetraphenyltetrahedrane structure[154].

A fragmentation that has attracted considerable attention is hydroxyl loss from the molecular ion of benzoic acid. Beynon *et al.*[155] and Meyerson and Corbin[156] showed that the slow (metastable) dissociation of the molecular ion involved scrambling of the hydroxyl and *ortho* ring hydrogens prior to hydroxyl loss, while the rapid dissociation essentially involved direct C_6H_5CO—OH cleavage without hydrogen mixing. The mechanism of the *ortho*-exchange reaction has been considered by Shapiro *et al.*[157, 158] as a process involving carboxyl group rotation.

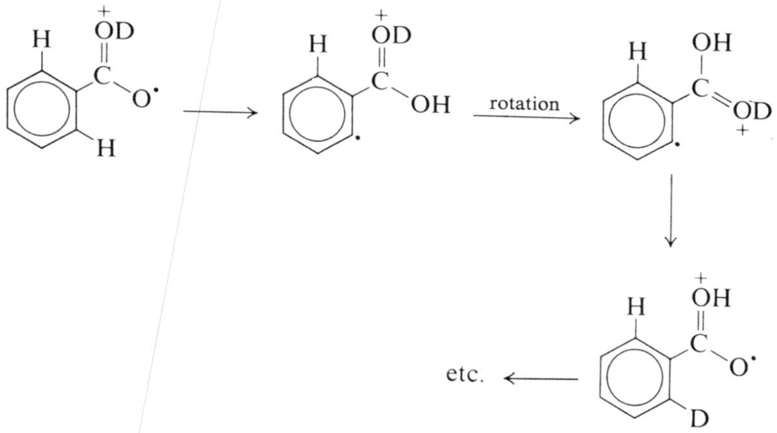

Equivalence of the oxygen atoms has been demonstrated by Shapiro and Tomer[159].

Holmes and Benoit[160] have discussed the structures of the $(M-OH)$ and $(M-OD)$ daughter ions on the basis of their subsequent fragmentation behaviour compared with that of isomeric $C_7H_5O^+$ ions produced from different precursors. Howe and McLafferty[161] propose that the *ortho*-hydrogen exchanging ions are low-energy species which completely equilibrate the three hydrogens prior to dissociation and thereafter do not dissociate further. (In the C_6H_5COOD acid *only* the $M-OH$ and $M-CO_2$ fragment ions retain deuterium.) This proposal necessitates no new structural formulation for the (benzoyl) daughter ions.

The difficulties associated with the interpretation of labelling experiments in which there are hydrogen–deuterium rearrangements and competing specific fragmentation pathways, each having time and energy dependence, are well exemplified by the studies of water loss from labelled cyclohexanols[162–165]. The only established conclusion to date is that some specifically *cis* 1,4 elimination of water does take place. 1,2 and 1,3 eliminations apparently occur but hydrogen rearrangements preclude their positive identification. Water loss from the molecular ion of cyclohexane-1,2-diol has been identified as consisting of a specific (rapid, no metastable ion) 1,2 elimination plus a process in which the two hydroxyl and 4- and 5-methylene hydrogens scramble prior to water loss; the latter process is also rapid, there being no evidence for incomplete scrambling either from metastable or daughter ion peaks[166].

A possible future development lies in the use of collisionally-induced metastable decompositions. Collisionally induced mass spectra of aromatic molecular ions[167] are found to be closely similar to their normal (70 eV) mass spectra. The relative metastable ion abundance test for a common ion structure can be satisfactorily applied to collision-induced metastable ions[168] and, more important, there is good evidence that some isomerisation and scrambling processes are greatly reduced in this method of ion excitation[169].

References

1. Hipple, J. A. and Condon, E. U. (1945). *Phys. Rev.*, **68,** 54
2. Hipple, J. A., Fox, R. E. and Condon, E. U. (1946). *Phys. Rev.*, **69,** 347
3. Hipple, J. A. (1947). *Phys. Rev.*, **71,** 594
4. Fox, R. E. and Langer, A. (1950). *J. Chem. Phys.* **18,** 460
5. Hertel, I. and Ottinger, Ch. (1967). *Z. Naturforsch.*, **22a,** 40
6. Chupka, W. A. (1959). *J. Chem. Phys.*, **30,** 191
7. Rosenstock, H. M., Wallenstein, M. B., Wahrhaftig, A. L. and Eyring, H. (1952). *Proc. Nat. Acad. Sci.*, **38,** 667
8. Momigny, J. (1961). *Bull. Soc. Chim. Belge.*, **70,** 291
9. Newton, A. S. and Sciamanna, A. F. (1966). *J. Chem. Phys.*, **44,** 4327; (1969). *J. Chem. Phys.*, **50,** 4868; (1970). *J. Chem. Phys.*, **52,** 327
10. Rosenstock, H. M. Wahrhaftig, A. L. and Eyring, H. (1955). *J. Chem. Phys.*, **23,** 2200
11. Coggeshall, N. D. (1962). *J. Chem. Phys.*, **37,** 2167
12. Newton, A. S. (1966). *J. Chem. Phys.*, **44,** 4015
13. Ottinger, Ch. (1967). *Z. Naturforsch.*, **22a,** 1157
14. Olmsted, J., Street, K., Jr. and Newton, A. S. (1964). *J. Chem. Phys.*, **40,** 2114
15. Beynon, J. H., Saunders, R. A. and Williams, A. E. (1965). *Z. Naturforsch.*, **20a,** 180
16. Newton, A. S. and Sciamanna, A. F. (1964). *J. Chem. Phys.*, **40,** 718
17. Bursey, M. M. and McLafferty, F. W. (1966). *J. Amer. Chem. Soc.*, **88,** 5023
18. Higgins, W. and Jennings, K. R. (1965). *Chem. Commun.*, 99
19. Hustrulid, A., Kusch, P. and Tate, J. T. (1938). *Phys. Rev.*, **54,** 1037
20. Flowers, M. C. (1965). *Chem. Commun.*, 235
21. Shannon, T. W., McLafferty, F. W. and McKinney, C. R. (1966). *Chem. Commun.*, 478
22. Beynon, J. H. and Fontaine, A. E. (1967). *Z. Naturforsch.*, **22a,** 334
23. Smyth, K. C. and Shannon, T. W. (1969). *J. Chem. Phys.*, **51,** 4633
24. Beynon, J. H., Hopkinson, J. A. and Lester, G. R. (1968). *Int. J. Mass Spectrom. Ion Phys.*, **1,** 343
25. Reichert, C., Fraas, R. E. and Kiser, R. W. (1970). *Int. J. Mass Spectrom. Ion Phys.*, **5,** 457
26. Rowland, C. G. (1971). *Int. J. Mass Spectrom. Ion Phys.*, **6,** 155
27. Barber, M., Jennings, K. R. and Rhodes, R. (1967). *Z. Naturforsch.*, **22a,** 15

28. Beynon, J. H., Saunders, R. A. and Williams, A. E. (1964). *Nature (London)*, **204**, 67
29. Schulze, P. and Burlingame, A. L. (1968). *J. Chem. Phys.*, **49**, 4870
30. Muccini, G. A., Hamill, W. H. and Barker, R. (1964). *J. Phys. Chem.*, **68**, 261
31. Schug, J. C. (1964). *J. Chem. Phys.* **40**, 1283
32. Rosenstock, H. M., Dibeler, V. H. and Harllee, F. N. (1964). *J. Chem. Phys.*, **40**, 591
33. Shannon, T. S. and McLafferty, F. W. (1966). *J. Amer. Chem. Soc.*, **88**, 5021
34. Osberghaus, O. and Ottinger, Ch. (1965). *Phys. Lett.*, **16**, 121
35. Benz, H. and Brown, H. W. (1968). *J. Chem. Phys.*, **48**, 4308
36. Vestal, M. L. (1965). *J. Chem. Phys.*, **43**, 1356
37. Lifshitz, C. and Shapiro, M. (1966). *J. Chem. Phys.*, **45**, 4242
38. Lifshitz, C. and Shapiro, M. (1967). *J. Chem. Phys.*, **46**, 4912
39. Ottinger, Ch. (1967). *J. Chem. Phys.*, **47**, 1452
40. Lifshitz, C. (1967). *J. Chem. Phys.*, **47**, 1870
41. Meyer, F. and Harrison, A. G. (1965). *J. Chem. Phys.*, **43**, 1778
42. Hertel, I. and Ottinger, Ch. (1967). *Z. Naturforsch.*, **22a**, 1141
43. Yeo, A. N. H., Cooks, R. G. and Williams, D. H. (1969). *J. Chem. Soc. B*, 149
44. Cooks, R. G., Ward, R. S. and Williams, D. H. (1967). *Chem. Commun.*, 850
45. Ottinger, Ch. (1967). *Z. Naturforsch.*, **22a**, 20
46. Vestal, M. and Futrell, J. H. (1970). *J. Chem. Phys.*, **52**, 978
47. Rosenstock, H. M. (1968). *Advan. Mass Spec.*, **4**, 523
48. Beckey, H. D., Hey, H., Levson, K. and Tenschert, G. (1969). *Int. J. Mass Spectrom. Ion Phys.*, **2**, 101
49. Barber, M. and Elliot, R. M. (1964). *'Proceedings of the 12th Annual Conference on Mass Spectrometry and Allied Topics'*. Montreal, A.S.T.M. E14 Committee. (Also presented as A.E.I. Technical Publication no. 5).
50. Daly, N. R., McCormick, A. and Powell, R. E. (1968). *Rev. Sci. Instr.*, **39**, 1163
51. Daly, N. R., McCormick, A. and Powell, R. E. (1968). *Org. Mass Spectrom.*, **1**, 167
52. Dugger, D. L. and Kiser, R. W. (1967). *J. Chem. Phys.*, **47**, 5054
53. Barber, M., Wolstenholme, W. A. and Jennings, K. R. (1967). *Nature (London)*, **214**, 664
54. Beynon, J. H., Amy, J. W. and Baitinger, W. E. (1969). *Chem. Commun.*, 723
55. McLafferty, F. W., Okamoto, J., Tsuyama, H., Nakajima, Y., Noda, T. and Major, H. W. (1969). *Org. Mass Spectrom.*, **2**, 751
56. Kiser, R. W., Sullivan, R. E. and Lupin, M. S. (1969). *Anal. Chem.*, **41**, 1958
57. Tajima, E. and Seibl, J. (1969). *Int. J. Mass Spectrom. Ion Phys.*, **3**, 245
58. Futrell, J. H., Ryan, K. R. and Sieck, L. W. (1965). *J. Chem. Phys.*, **43**, 1833
59. Sasaki, S., Watanabe, E., Itagaki, Y., Aoyama, T. and Yamauchi, E. (1968). *Anal. Chem.*, **40**, 1000
60. Shannon, T. W., Mead, T. E., Warner, C. G. and McLafferty, F. W. (1967). *Anal. Chem.*, **39**, 1748
61. Maurer, K. H., Brunnée, C., Kappus, G., Habfast, K. E., Schulze, P. and Schräder, U. (1971). *'Proceedings of the 19th Annual Conference on Mass Spectrometry and Allied Topics'*, Atlanta, A.S.T.M. E14 Committee
62. Haddon, W. F. and McLafferty, F. W. (1969). *Anal. Chem.*, **41**, 31
63. Beynon, J. H., Fontaine, A. E. and Lester, G. R. (1968). *Int. J. Mass Spectrom. Ion Phys.*, **1**, 1
64. Higgins, W. and Jennings, K. R. (1966). *Trans. Faraday Soc.*, **62**, 97
65. Ottinger, Ch. (1965). *Z. Naturforsch.*, **20a**, 1229
66. Beynon, J. H. and Fontaine, A. E. (1966). *Chem. Commun.*, 717
67. Beynon, J. H., Caprioli, R. M., Baitinger, W. E. and Amy, J. W. (1970). *Org. Mass Spectrom*, **3**, 963
68. Beynon, J. H., Caprioli, R. M., Baitinger, W. E. and Amy, J. W. (1970). *Org. Mass Spectrom.*, **3**, 455
69. Jennings, K. R. and Whiting, A. F. (1967). *Chem. Commun.*, 820
70. Nounou, P. (1970). *Int. J. Mass Spectrom. Ion Phys.* **4**, 219
71. Beynon, J. H., Caprioli, R. M., Baitinger, W. E. and Amy, J. W. (1970). *Org. Mass Spectrom.*, **3**, 661
72. Bursey, M. M. and Hoffman, M. K. (1969). *J. Amer. Chem. Soc.*, **91**, 5023
73. Fuchs, R. and Taubert, R. (1965). *Z. Naturforsch.*, **20a**, 823
74. Beynon, J. H., Fontaine, A. E. and Lester, G. R. (1968). *Chem. Commun.*, 265
75. Khodadi, G., Botter, R. and Rosenstock, H. M. (1969). *Int. J. Mass Spectrom. Ion Phys.*, **3**, 397

76. Briggs, P. R., Parker, W. L. and Shannon, T. W. (1968). *Chem. Commun.*, 727
77. Bursey, M. M. and Dusold, L. R. (1967). *Chem. Commun.*, 712
78. Barber, M. and Jennings, K. R. (1969). *Z. Naturforsch.*, **24a,** 134
79. Jennings, K. R. (1970). *Org. Mass Spectrom.*, **3,** 85
80. Lifshitz, C. A. and Long, F. A. (1963). *J. Phys. Chem.*, **67,** 2463
81. Lifshitz, C. A. and Long, F. A. (1965). *J. Phys. Chem.*, **69,** 3731
82. Lifshitz, C. A. and Sternberg, R. (1969). *Int. J. Mass Spectrom. Ion Phys.*, **2,** 303
83. Bloom, E. G., Mohler, F. L., Lengel, J. H. and Wise, C. E. (1948). *J. Res. Nat. Bur. Stds.*, **40,** 437
84. Willholm, B., Thomas, A. F. and Gautschi, F. (1964). *Tetrahedron.*, **20,** 1185
85. Tilak, B. D., Das, K. G. and El-Namaky, H. M. (1967). *Experientia.*, **23,** 609
86. Cooks, R. G. (1969). *Org. Mass Spectrom.*, **2,** 481
87. Heiss, J. and Zeller, K. P. (1969). *Org. Mass Spectrom.*, **2,** 1039
88. Nibbering, N. M. M. and De Boer, Th. J. (1970). *Org. Mass Spectrom.*, **3,** 409
89. Brittain, E. F. H., Wells, C. H. J. and Paisley, H. M. (1968). *J. Chem. Soc. B*, 304
90. Turro, N. J., Neckers, D. C., Leermakers, P. A., Seldner, D. and D'Angelo, P. (1965). *J. Amer. Chem. Soc.*, **87,** 4097
91. Caspi, E., Wicha, J. and Mandelbaum, A. (1967). *Chem. Commun.*, 1161
92. Sample, S. D. and Djerassi, C. (1965). *Nature (London)*, **208,** 1314
93. Seibl, J. (1967). *Helv. Chim. Acta*, **50,** 263
94. Letcher, R. M. and Eggers, S. H. (1967). *Tetrahedron Lett.*, 3541
95. Jennings, K. R. (1966). *Chem. Commun.*, 283
96. Beynon, J. H., Baitinger, W. E., Amy, J. W. and Caprioli, R. M. (1969). *Int. J. Mass Spectrom. Ion Phys.*, **3,** 309
97. Hills, L. P., Futrell, J. H. and Wahrhaftig, A. L. (1969). *J. Chem. Phys.*, **51,** 5255
98. Tou, J. C. (1970). *J. Phys. Chem.*, **74,** 3076
99. Williams, D. H., Cooks, R. G. and Howe, I. (1968). *J. Amer. Chem. Soc.*, **90,** 6759
100. Yeo, A. N. H., and Williams, D. H. (1971). *J. Amer. Chem. Soc.*, **93,** 395
101. Bursey, M. M., Tibbetts, F. E. III and Little, W. F. (1969). *Tetrahedron Lett.*, **40,** 3469
102. Davis, B., Williams, D. H. and Yeo, A. N. H. (1970). *J. Chem. Soc. B*, 81
103. Cole, W. G. and Williams, D. H. (1969). *Chem. Commun.*, 784
104. Shaw, M. A., Westwood, R. and Williams, D. H. (1970). *J. Chem. Soc. B*, 1773
105. Lossing, F. A. Private communication
106. Maeda, K., Semeluk, G. P. and Lossing, F. P. (1968). *Int. J. Mass Spectrom. Ion Phys.*, **1,** 395
107. Occolowitz, J. L., (1969). *J. Amer. Chem. Soc.*, **91,** 5202
108. Bowie, J. H. and Bradshaw, T. K. (1970). *Aust. J. Chem.*, **23,** 1431
109. McLafferty, F. W. and Bryce, T. A. (1967). *Chem. Commun.*, 1215
110. Grubb, H. M. and Meyerson, S. (1963). *Mass Spectrometry of Organic Ions*, Ed. by F. W. McLafferty. Ch. 10 (New York: Academic Press).
111. Wanless, G. G. and Glock, G. A. Jr. (1967). *Anal. Chem.*, **39,** 2
112. Goldberg, P., Hopkinson, J. A., Mathias, A. and Williams, A. E. (1970). *Org. Mass Spectrom.*, **3,** 1009
113. Bloom, E. G., Mohler, F. L., Wise, C. E. and Wells, E. J. (1949). *J. Res. Nat. Bur. Std.*, **43,** 65
114. Beynon, J. H., Fontaine, A. E., Hopkinson, J. A. and Williams, A. E. (1969). *Int. J. Mass Spectrom. Ion Phys.*, **3,** 143
115. Holmes, J. Unpublished results
116. McLafferty, F. W. and Pike, W. T. (1967). *J. Amer. Chem. Soc.*, **89,** 5951
117. Lin, Y. N. and Rabinovitch, B. S. (1970). *J. Phys. Chem.*, **74,** 1769
118. Cooks, R. G. and Williams, D. H. (1968). *Chem. Commun.*, 627
119. Dibeler, V. H. and Rosenstock, H. M. (1963). *J. Chem. Phys.*, **39,** 1326
120. Vestal, M. L. (1964). *J. Chem. Phys.*, **41,** 3997
121. Ottinger, Ch. (1965). *Z. Naturforsch.*, **20a,** 1232
122. Beynon, J. H., Caprioli, R. M., Baitinger, W. E. and Amy, J. W. (1970). *Org. Mass Spectrom.*, **3,** 479
123. Beynon, J. H., Baitinger, W. E. and Amy, J. W. (1969). *Int. J. Mass Spectrom. Ion Phys.*, **3,** 55
124. Löhle, U. and Ottinger, Ch. (1969). *J. Chem. Phys.*, **51,** 3097
125. Jennings, K. R. (1967). *Z. Naturforsch.*, **22a,** 454

126. Horman, I., Yeo, A. N. H. and Williams, D. H. (1970). *J. Amer. Chem. Soc.*, **92,** 2131
127. Cooks, R. G. and Bernasek, S. L. (1970). *J. Amer. Chem. Soc.*, **92,** 2129
128. Perry, W. O., Beynon, J. H., Baitinger, W. E., Amy, J. W., Caprioli, R. M., Renaud, R. N., Leitch, L. C. and Meyerson, S. (1970). *J. Amer. Chem. Soc.*, **92,** 7236
129. Grubb, H. M. and Meyerson, S. (1963). *Mass Spectrometry of Organic Ions,* Ed. by McLafferty, F. W., Chapter 10 (New York; Academic Press)
130. Rinehart, K. L., Buchholz, A. C., Van Lear, G. E. and Cantrell, L. C. (1968). *J. Amer. Chem. Soc.*, **90,** 2983
131. Beynon, J. H., Corn, J. E., Baitinger, W. E., Caprioli, R. M. and Benkeser, R. A. (1970). *Org. Mass Spectrom.*, **3,** 1371
132. Howe, I. and McLafferty, F. W. (1971). *J. Amer. Chem. Soc.*, **93,** 99
133. Safe, S. (1969). *Chem. Commun.*, 534
134. Bradshaw, T. K., Bowie, J. H. and White, P. Y. (1970). *Chem. Commun.*, 537
135. Johnstone, R. A. W. and Millard, B. J. (1966). *Z. Naturforsch.*, **21a,** 604
136. Yeo, A. N. H. and Williams, D. H. (1970). *Chem. Commun.*, 737
137. Saunders, M. and Rosenfeld, J. (1969). *J. Amer. Chem. Soc.*, **91,** 7756
138, Holmes, J. L. and McGillivray, D. (1970). *Can. J. Chem.*, **48,** 2791
139. Meisels, G. G., Park, J. Y. and Giessner, B. G. (1970). *J. Amer. Chem. Soc.*, **92,** 254
140. Bryce, W. A. and Kebarle, P. (1956). *Can. J. Chem.*, **34,** 1249
141. Meisels, G. G., Park, J. Y. and Giessner, B. G. (1970). *J. Amer. Chem. Soc.*, **92,** 1556
142. Williams, D. H. and Ronayne, J. (1967). *Chem. Commun.*, 1129
143. Yeo, A. N. H., Cooks, R. G. and Williams, D. H. (1968). *Org. Mass Spectrom.*, **1,** 910
144. Cole, W. G., Yeo, A. N. H. and Williams, D. H. (1968). *J. Chem. Soc. B,* 1284
145. Yeo, A. N. H., Cooks, R. G. and Williams, D. H. (1968). *Chem. Commun.*, 1269
146. Williams, D. H., Tam, S. W. and Cooks, R. G. (1968). *J. Amer. Chem. Soc.*, **90,** 2150
147. Williams, D. H., Ward, R. S., and Cooks, R. G. (1968). *J. Chem. Soc. B,* 522
148. McLafferty, F. W. and Schiff, L. J. (1969). *Org. Mass Spectrom.*, **2,** 757
149. Shapiro, R. H. and Tomer, K. B. (1969). *Org. Mass Spectrom.*, **2,** 579
150. Bowie, J. H., Simons, B. K., Donaghue, P. F. and Kallury, R. K. M. R. (1969). *Tetrahedron,* **25,** 3969
151. Yeo, A. N. H. (1970). *Chem. Commun.*, 1154
152. Cooks, R. G., Howe, I., Tam, S. W. and Williams, D. H. (1968). *J. Amer. Chem. Soc.*, **90,** 4064
153. Harrison, A. G. and Keyes, B. G. (1968). *J. Amer. Chem. Soc.*, **90,** 5046
154. Bursey, M. M. and Elwood, T. A. (1969). *J. Amer. Chem. Soc.*, **91,** 3812
155. Beynon, J. H., Job, B. E. and Williams, A. E. (1965). *Z. Naturforsch.*, **20a,** 883
156. Meyerson, S. and Corbin, J. L. (1965). *J. Amer. Chem. Soc.*, **87,** 3045
157. Shapiro, R. H., Tomer, K. B., Caprioli, R. M. and Beynon, J. H. (1970). *Org. Mass Spectrom.*, **3,** 1333
158. Shapiro, R. H., Tomer, K. B., Beynon, J. H. and Caprioli, R. M. (1970). *Org. Mass Spectrom.*, **3,** 1593
159. Shapiro, R. H. and Tomer, K. B. (1969). *Org. Mass Spectrom.*, **2,** 1175
160. Holmes, J. L. and Benoit, F. (1970). *Org. Mass Spectrom.*, **4,** 97
161. Howe, I. and McLafferty, F. W. (1970). *J. Amer. Chem. Soc.*, **92,** 3797
162. Macdonald, C. G., Shannon, J. S. and Sugowdz, G. (1963). *Tetrahedron Lett.*, 807
163. Budzikiewicz, H., Pelah, Z. and Djerassi, C. (1964). *Monatsh.*, **95,** 158
164. Green, M. M. and Schwab, J. (1968). *Tetrahedron Lett.*, 2955
165. Ward, R. S. and Williams, D. H. (1969). *J. Org. Chem.*, **34,** 3373
166. Benoit, F. and Holmes, J. L. (1971). *Can. J. Chem.*, **49,** 1161
167. Jennings, K. R. (1968). *Int. J. Mass Spectrom. Ion Phys.*, **1,** 227
168. Haddon, W. F. and McLafferty, F. W. (1968). *J. Amer. Chem. Soc.*, **90,** 4745
169. McLafferty, F. W. and Schuddemage, H. D. R. (1969). *J. Amer. Chem. Soc.*, **91,** 1866
170. Chupka, W. A. and Berkowitz, J. (1967). *J. Chem. Phys.*, **47,** 2921
171. Lifshitz, C. A., Shapiro, M. and Sternberg, R. (1969). *Israel J. Chem.*, **7,** 391
172. Dickinson, R. J. and Williams, D. H. (1971). *J. Chem. Soc. B,* 249
173. Holmes, J. L. and Benoit, F. (1970). *Chem. Commun.*, 1031